MATHEMATICS IN BIOLOGY

MATHEMATICS IN BIOLOGY

MATHEMATICS IN BIOLOGY

Markus Meister, Kyu Hyun Lee, and Ruben Portugues

The MIT Press
Cambridge, Massachusetts
London, England

The MIT Press would like to thank the anonymous peer reviewers who provided comments on drafts of this book. The generous work of academic experts is essential for establishing the authority and quality of our publications. We acknowledge with gratitude the contributions of these otherwise uncredited readers.

This book was set in Stone Serif and Stone Sans by Westchester Publishing Services. Printed and bound in the United States of America.

Library of Congress Cataloging-in-Publication Data

Names: Meister, Markus, author. | Lee, Kyu Hyun, 1946- author. | Portugues, Ruben, author.
Title: Mathematics in biology / Markus Meister, Kyu Hyun Lee, and Ruben Portugues.
Description: Cambridge, Massachusetts : The MIT Press, [2025] | Includes bibliographical references and index.
Identifiers: LCCN 2024017280 (print) | LCCN 2024017281 (ebook) | ISBN 9780262049405 (hardcover) | ISBN 9780262380829 (pdf) | ISBN 9780262380836 (epub)
Subjects: LCSH: Biomathematics. | Biology—Mathematical models.
Classification: LCC QH323.5 .M45 2025 (print) | LCC QH323.5 (ebook) | DDC 570.1/51—dc23/eng/20240723
LC record available at https://lccn.loc.gov/2024017280
LC ebook record available at https://lccn.loc.gov/2024017281

10   9   8   7   6   5   4   3   2   1

Contents

Introduction

This is a textbook on mathematical methods for advanced students in the life sciences. It emerged from a course that we started teaching at Harvard University in 2002. After a few years of this, the course was made compulsory for PhD students in molecular and cellular biology. What drove the biology faculty at Harvard to this step?

Biology has turned into a quantitative science. The core problems in the life sciences today involve large systems with many interacting components. As is true for all complex systems, a verbal description of phenomena and mechanisms can take you only so far. Eventually, the concepts must be expressed in the language of mathematics. There is a growing need to extract patterns from a vast sea of data, to formalize the relationships in a quantitative model, and to make testable and quantitative predictions for new experiments.

Unfortunately, most biologists trained today are unprepared for this change in the character of the discipline. At US universities, undergraduate programs generally require mathematical training only up to calculus. In many places, the biology major still serves as a refuge for students who want to be scientists while avoiding mathematics. Graduate programs rarely require additional mathematical training. But a cursory look at life science journals reveals articles from all research areas, which rely on mathematical methods beyond calculus. Many journals have been founded recently that specialize in quantitative biology. Thus the average PhD candidate in biology remains unprepared to appreciate much of the current research in the discipline.

The solution is straightforward: as in other areas of science and engineering, we need to teach mathematical methods to our biology students. In 2003, a widely cited report from the National Research Council (2003) made just that recommendation, along with specific suggestions for implementing it. This book emerged from nearly two decades of practical experience teaching mathematical methods to biology students.

The book covers three broad subjects: linear algebra, probability, and dynamical systems. We found that these areas provide a solid foundation for understanding the methods in today's quantitative biology, and for applying them creatively. Each area, in turn, is treated at three levels: basic principles, advanced topics, and applications. Motivations and examples are drawn from many areas of biology. Each chapter concludes with a series of exercises. Some of these involve simple scientific programming. Our overall goal is to present, in a volume of accessible size, a set of tools from different areas of mathematics, to demonstrate how they interact to support a scientific method, and to encourage the reader to start applying them creatively.

The style of the text is more practical than theoretical: We generally omit proofs, and gloss over the precise mathematical conditions for validity of a result. Our emphasis is on broad utility to the practicing scientist, not on the edge cases that occupy a mathematician. For the demanding reader, we offer links to further reading. At the same time, we assume that the reader is well versed with the Google search bar: An informal query like "Why is the delta function not a real function?" delivers a wealth of competent material and links for further study on the internet. Of course, the same is true for the main subject areas of the book. In this era, a "textbook" is perhaps most useful if it offers a coherent and well-organized guide to the vast ecosystem of online information.

We made a conscious choice to separate the mathematics from the applications. The basic and advanced chapters contain only minimal examples that serve to illustrate the mathematical principles without requiring any domain knowledge. In turn, the "applications" chapters cover problems from a wide range of fields, and in each case, they take some time to introduce the scientific background. A third component are the online materials (https://mitpress.mit.edu/mathinbio), which offer code to produce various graphs in the book, worked solutions to exercises, and expansions on additional topics as they emerge.

The book can be consumed in different modes: First, as a textbook in support of a formal course on mathematical methods. We taught this subject in a single twelve-week course, but this book spans considerably more material and could reasonably support two semesters. Second, readers may start by diving into an application chapter in their field of interest. The links there will lead to the relevant sections about the underlying mathematics. Finally, the book can serve as a reference work. Whatever the method, we hope that this project will help students to tackle problems that they might otherwise have avoided and to bring their research to a new level of quantitative richness.

Notation

- Lowercase letter with arrow (e.g., $\vec{x}$) denotes a vector object, independent of any basis.
- Bold lowercase letter (e.g., $\mathbf{x}$) denotes the components of a vector in a particular basis.
- x_i denotes the i-th component of the vector $\mathbf{x}$.
- Uppercase letter in calligraphic font (e.g., $\mathcal{L}$) denotes a linear operator, independent of any basis.
- Bold uppercase letter (e.g., $\mathbf{X}$) denotes a matrix, for example the components of a linear operator in a particular basis.
- $\dot{x}$ denotes the derivative of x with respect to time (i.e., $\frac{dx}{dt}$).
- $\hat{f}(\omega)$ denotes the Fourier transform of $f(t)$.
- $\delta(x)$ denotes the delta function.
- δ_{ij} denotes the Kronecker delta.
- $f * g$ denotes the convolution of functions f and g.
- $\equiv$ denotes a definition.
- $\square$ denotes the end of a proof, exercise, or example.

1 Elements of Calculus

Here, we revisit a number of elementary concepts and techniques that should be part of your repertoire before you start using advanced mathematical methods. We make no attempt to be exhaustive in this discussion, but focus on the ideas that are essential for the rest of the book. This chapter will serve as a reference for later chapters. This part of the book is not didactic, but enumerates subjects that you should know about. If they are unfamiliar, consult the sources from which you learned calculus. If you feel well prepared, skip this section or simply skim the headings.

1.1 Elementary Functions

Elementary functions are all real functions $f(x)$ of a real variable x. You should have a mental picture of how each of them behaves.

1.1.1 Powers

$$y = x^n, \text{ for } n > 0. \tag{1.1}$$

These all go to zero at $x = 0$ and diverge to $+\infty$ or $-\infty$ at $x = \pm\infty$.

$$y = x^{-n}, \text{ for } n > 0. \tag{1.2}$$

These all go to zero at $x = \pm\infty$ and diverge at $x = 0$.

1.1.2 Sinusoids

The classic oscillating functions. They repeat with a period of 2π. The cos is shifted along the x-axis relative to the sin by $\frac{\pi}{2}$. Sometimes the argument is measured in degrees, with $90° \equiv \pi/2$.

$$y = \sin x. \tag{1.3}$$

$$y = \cos x = \sin\left(x + \frac{\pi}{2}\right). \tag{1.4}$$

1.1.3 Exponentials

$$y = a^x = e^{\ln a \cdot x}. \tag{1.5}$$

The exponential $f(x) = e^x$ may be the most frequently used function in this book.

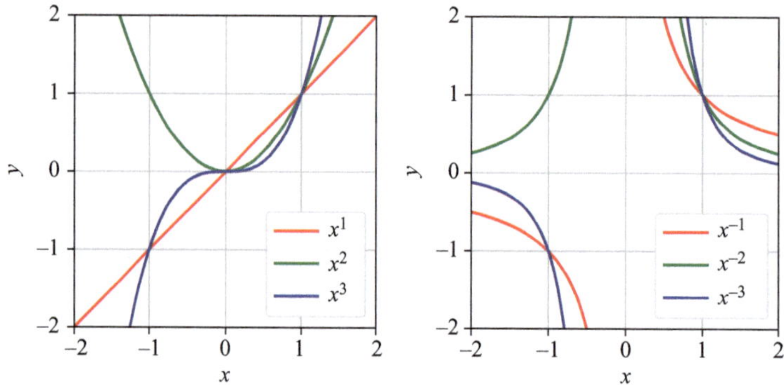

Figure 1.1
Power functions.

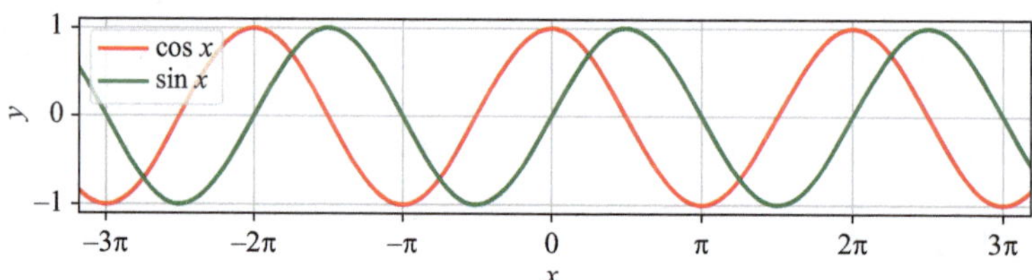

Figure 1.2
Sinusoids.

1.1.4 Logarithms

The inverse of the exponential, defined by $e^y = x$:

$$y = \ln x. \tag{1.6}$$

1.1.5 Symmetries

Some of these elementary functions are symmetric around the origin $x = 0$. In particular, one defines

$$\begin{aligned} \text{Even function:} \quad & f(-x) = f(x) \\ \text{Odd function:} \quad & f(-x) = -f(x). \end{aligned} \tag{1.7}$$

Exercise 1.1 Which of the functions encountered so far are odd or even? □

1.2 Differentiation

Often we are interested in how quickly the value of $y = f(x)$ changes as one moves along the x-axis. If a small move from x to $x + \Delta x$ causes a small change from y to $y + \Delta y$, then the slope $f(x)$ is approximately $\Delta y / \Delta x$. The **derivative** of $f(x)$ is what one gets by making Δx infinitesimally small:

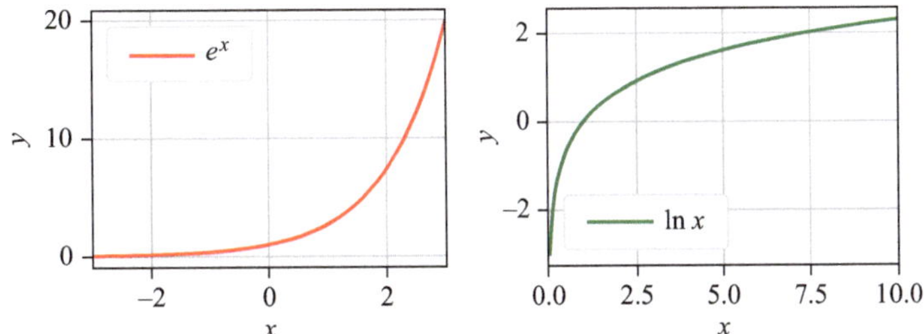

Figure 1.3
Exponential (left) and logarithm (right).

$$\frac{\mathrm{d}f}{\mathrm{d}x} = \lim_{\Delta x \to 0} \frac{f(x + \Delta x) - f(x)}{\Delta x}. \tag{1.8}$$

Depending on context, it is often called the "gradient," the "slope," or the "rate of change" of a function. Also, you may encounter a variety of shorthands, again depending on context:

$$\frac{\mathrm{d}f}{\mathrm{d}x} \equiv f' \equiv \dot{f} \equiv f_x. \tag{1.9}$$

1.2.1 Derivative from First Principles

If you were stranded on a desert island and forgot the slope of an elementary function, you might want to derive it from elementary things you know. For example, what is the slope of $f(x) = x^2$? You might write the following into the sand:

$$\frac{\mathrm{d}}{\mathrm{d}x} x^2 = \lim_{\Delta x \to 0} \frac{(x + \Delta x)^2 - x^2}{\Delta x} = \lim_{\Delta x \to 0} \frac{x^2 + 2x\Delta x + (\Delta x)^2 - x^2}{\Delta x} \tag{1.10}$$

$$= 2x + \lim_{\Delta x \to 0} \Delta x = 2x.$$

We don't recommend doing this routinely. You should have all the derivatives of elementary functions memorized as they are laid out next.

1.2.2 Derivatives of Elementary Functions

Function $f(x)$	Derivative $\frac{\mathrm{d}f}{\mathrm{d}x}$
x^n	nx^{n-1}
$\sin x$	$\cos x$
$\cos x$	$-\sin x$
e^x	e^x
$\ln x$	$\frac{1}{x}$

For functions that combine the elementary functions, a few rules help us find the derivative.

1.2.3 Sum Rule

For the sum of two functions, this is easy:

$$\frac{\mathrm{d}}{\mathrm{d}x}\left(f\left(x\right)+g\left(x\right)\right)=\frac{\mathrm{d}}{\mathrm{d}x}f\left(x\right)+\frac{\mathrm{d}}{\mathrm{d}x}g\left(x\right). \tag{1.11}$$

Let's omit the argument of the function, (x), to remove clutter and make the essence of the equation more visible:

$$\frac{\mathrm{d}}{\mathrm{d}x}\left(f+g\right)=\frac{\mathrm{d}f}{\mathrm{d}x}+\frac{\mathrm{d}g}{\mathrm{d}x}. \tag{1.12}$$

1.2.4 Product Rule

For the product of two functions,

$$\frac{\mathrm{d}}{\mathrm{d}x}\left(f\cdot g\right)=\frac{\mathrm{d}f}{\mathrm{d}x}g+\frac{\mathrm{d}g}{\mathrm{d}x}f. \tag{1.13}$$

1.2.5 Chain Rule

When two functions are chained together, like $y=h(x)=f(g(x))$,

$$\frac{\mathrm{d}}{\mathrm{d}x}\left(f\left(g\right)\right)=\frac{\mathrm{d}f}{\mathrm{d}g}\cdot\frac{\mathrm{d}g}{\mathrm{d}x}. \tag{1.14}$$

For example, what is the derivative of $h(x)=\sin(x^5)$?

$$\frac{\mathrm{d}h}{\mathrm{d}x}=\frac{\mathrm{d}}{\mathrm{d}g}\sin\left(g\right)\cdot\frac{\mathrm{d}}{\mathrm{d}x}x^5,\quad\text{where }g=x^5,$$
$$=\cos\left(x^5\right)\cdot 5x^4. \tag{1.15}$$

1.2.6 The Quotient Rule

The quotient rule follows from the chain and product rules, and it is worth memorizing on its own:

$$\frac{\mathrm{d}}{\mathrm{d}x}\left(\frac{f}{g}\right)=\frac{1}{g^2}\left(\frac{\mathrm{d}f}{\mathrm{d}x}g-f\frac{\mathrm{d}g}{\mathrm{d}x}\right). \tag{1.16}$$

1.2.7 Higher Derivatives

How fast does the slope of a function change? That will be the derivative of the derivative, which is called the "second derivative," and it is written as

$$\frac{\mathrm{d}^2 f}{\mathrm{d}x^2}\equiv\frac{\mathrm{d}}{\mathrm{d}x}\left(\frac{\mathrm{d}f}{\mathrm{d}x}\right). \tag{1.17}$$

Again, you will find other shorthands, like

$$\frac{\mathrm{d}^2 f}{\mathrm{d}x^2}\equiv f''\equiv\ddot{f}\equiv f_{xx}\equiv f^{(2)}. \tag{1.18}$$

The second derivative is sometimes called "curvature," although that term also has a slightly different mathematical meaning.

Higher derivatives are defined recursively: The nth derivative is the slope of the $(n-1)^{\text{st}}$, as follows:

$$\frac{\mathrm{d}^n f}{\mathrm{d}x^n} \equiv \frac{\mathrm{d}}{\mathrm{d}x}\left(\frac{\mathrm{d}^{n-1} f}{\mathrm{d}x^{n-1}}\right). \tag{1.19}$$

1.2.8 Taylor Series

Here's a truly profound insight: if you know all the derivatives of a function $f(x)$ in one particular point $x=a$, then you can compute the values of the function in an entire region surrounding a. The Taylor series is a power series designed to match all the derivatives of $f(x)$ in $x=a$:

$$\begin{aligned} f(x) &= f(a) + \frac{\mathrm{d}}{\mathrm{d}x}f(a)\cdot(x-a) + \frac{\mathrm{d}^2}{\mathrm{d}x^2}f(a)\cdot\frac{1}{2}(x-a)^2 + \dots \\ &= \sum_{n=0}^{\infty}\frac{1}{n!}f^{(n)}(a)\cdot(x-a)^n. \end{aligned} \tag{1.20}$$

For example, for the function $\sin x$ in $x=0$, the derivatives of successive order are $1, 0, -1, 0, 1, 0, \dots$. So the Taylor series is

$$\sin x = x - \frac{x^3}{3!} + \frac{x^5}{5!} - \dots \tag{1.21}$$

Of course, some restrictions apply: Generally, the Taylor series equals the function $f(x)$ within some **region of convergence** around a. If your function misbehaves (e.g., if it goes to infinity somewhere or it is not differentiable), then that will limit the region of convergence. Check with your local mathematician or the Google search bar for details.

A common use of the Taylor series is to approximate a function by a polynomial. For example, the tangent line to $f(x)$ in $x=a$ is often a pretty good approximation of $f(x)$ in the neighborhod of a. That tangent is equal to the first-order part of the Taylor series:

$$f(x) \approx f(a) + \frac{\mathrm{d}}{\mathrm{d}x}f(a)\cdot(x-a). \tag{1.22}$$

This is sometimes called "linearization of $f(x)$." For a better approximation, one can include the second-order term of the Taylor series, and so on with higher terms. We will see frequent uses of this.

1.2.9 Minima and Maxima

We are often interested in finding the maximum or minimum of a function. Collectively, these are also called **extrema**. If $f(x)$ has a local maximum at x, then obviously the slope $f'(x)$ must be zero, or else f would increase further either on the left or the right of x. Whether x is a maximum, minimum, or saddle point depends on the second derivative $f''(x)$:

$$x \text{ is a } \begin{cases} \text{maximum if } f'(x)=0 \text{ and } f''(x)<0. \\ \text{minimum if } f'(x)=0 \text{ and } f''(x)>0. \\ \text{saddle if } f'(x)=0 \text{ and } f''(x)=0. \end{cases} \tag{1.23}$$

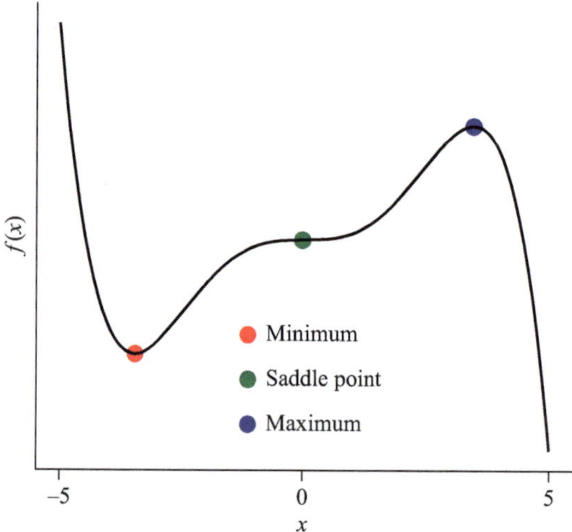

Figure 1.4
Stationary points of a function $f(x)$ are x-values with zero slope. They can be local maximum, minimum, or saddle points.

These conditions serve to find **local** extrema, but other kinds of arguments are needed to identify a function's **global** maximum and minimum.

1.2.10 Small Changes and Error Propagation

Every experimental measurement is associated with some uncertainty, commonly called the "error bar." Often we are in a situation where we have to estimate the effect of this experimental uncertainty on some derived quantity. For a quick evaluation, it helps to just use the first-order Taylor series of everything. If the errors are small relative to the absolute measurement, then that is usually sufficient.

For example, suppose that I measured the diameter of a spherical cell to $a = 10 \pm 0.5$ μm. What is my uncertainty about the volume of the cell? I remember that the volume is proportional to diameter cubed, $V \sim a^3$, so my uncertainty about the diameter $\Delta a = 0.5$ μm will propagate to an uncertainty in the volume of

$$\Delta V \sim \frac{\mathrm{d}}{\mathrm{d}a} a^3 \Delta a. \tag{1.24}$$

Dividing one by the other gives

$$\frac{\Delta V}{V} \approx \frac{3a^2 \Delta a}{a^3} = 3 \frac{\Delta a}{a} = 15 \text{ percent.} \tag{1.25}$$

Stated in words: the relative error in the volume $\Delta V/V$ is three times the relative error in the diameter $\Delta a/a$, and that amounts to 15 percent. Note that I didn't need to remember what constant relates the volume to the cube of the diameter.[1] To propagate

---

1. Maybe you forgot that while getting washed ashore on the desert island.

the uncertainty from the input quantity to the output quantity, we just need to know the general functional form of the relationship.

In general, if x is known to within some uncertainty Δx, then a derived quantity $f(x)$ is accurate to within

$$\Delta f \approx |f'(x)|\,\Delta x. \tag{1.26}$$

For some common derived quantities, this is summarized in the following table:

Function $f(x)$	Uncertainty in f	Rule
ax	$\frac{\Delta f}{f} = \frac{\Delta x}{x}$	This is the same relative error.
x^p	$\frac{\Delta f}{f} = p\frac{\Delta x}{x}$	Multiply the relative error by exponent p.
e^x	$\frac{\Delta f}{f} = \Delta x$	The relative error is the absolute error of x.
$\ln x$	$\Delta f = \frac{\Delta x}{x}$	The absolute error is the relative error of x.
$x+y$	$\Delta f = \Delta x + \Delta y$	Add the absolute errors.
xy	$\frac{\Delta f}{f} = \frac{\Delta x}{x} + \frac{\Delta y}{y}$	Add the relative errors.

1.3 Integration

Integration is the opposite process from differentiation. Formally, we say that

$$f(x) \text{ is an indefinite integral of } g(x) \text{ if } \frac{\mathrm{d}}{\mathrm{d}x}f(x) = g(x). \tag{1.27}$$

Two functions that differ by a constant C have the same derivative. So if $f(x)$ is an integral of $g(x)$, then $f(x) + C$ is as well, for all values of C.

1.3.1 Definite Integral
If $f(x)$ is an indefinite integral of $g(x)$, then the definite integral is as follows:

$$\int_a^b g(x)\,\mathrm{d}x = f(b) - f(a) \equiv [f(x)]_a^b. \tag{1.28}$$

A common interpretation of the definite integral $\int_a^b g(x)\,\mathrm{d}x$ is as "the area under the curve of $g(x)$ between $x = a$ and $x = b$." In that calculation, areas below the x-axis count as negative, as shown in figure 1.5.

1.3.2 Integrals of Elementary Functions
The integrals of elementary functions follow from inverting the table of derivatives from section 1.2. Learn them by heart:

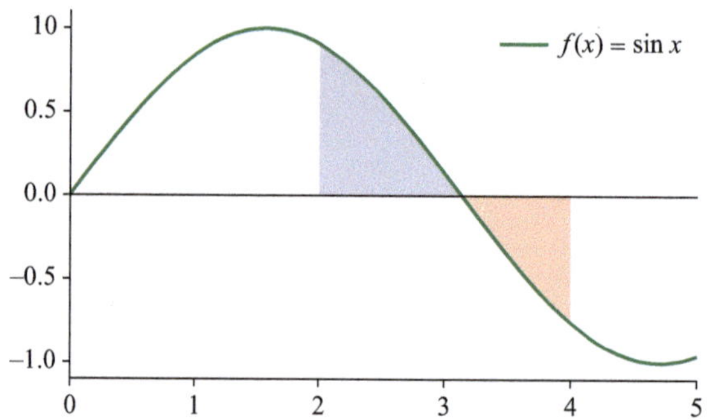

Figure 1.5
Area under the curve: The integral from $x=2$ to $x=4$ equals the blue area minus the red area.

Function $f(x)$	Integral $\int f(x)\,\mathrm{d}x$
x^n	$\frac{x^{n+1}}{n+1} + C$
$\sin x$	$-\cos x + C$
$\cos x$	$\sin x + C$
e^x	$e^x + C$
$\frac{1}{x}$	$\ln x + C$

1.3.3 Rules for Integration

- Rule for multiple and sum:

$$\int \left(\alpha f(x) + \beta g(x)\right)\,\mathrm{d}x = \alpha \int f(x)\,\mathrm{d}x + \beta \int g(x)\,\mathrm{d}x. \tag{1.29}$$

- Product rule (integration by parts):

$$\int \frac{\mathrm{d}f(x)}{\mathrm{d}x} g(x)\,\mathrm{d}x = f(x) \cdot g(x) - \int f(x) \cdot \frac{\mathrm{d}g(x)}{\mathrm{d}x}\,\mathrm{d}x. \tag{1.30}$$

- Chain rule (integration by substitution):

$$\int f(u(x)) \frac{\mathrm{d}u(x)}{\mathrm{d}x}\,\mathrm{d}x = \int f(u)\,\mathrm{d}u. \tag{1.31}$$

1.3.4 Tough Integrals

Unlike differentiation, which can always be accomplished by following the set of rules in section 1.2, integration has no deterministic algorithm. Some integrals that don't appear in section 1.3.2 can be solved with a clever substitution, using the Chain rule. For example, what is

$$\int \frac{1}{1+x^2}\,dx?$$

An inspired guess might lead you to define

$$x = \tan u = \frac{\sin u}{\cos u}, \tag{1.32}$$

which leads to

$$\frac{1}{1+x^2} = \cos^2 u$$

$$dx = \frac{1}{\cos^2 u}\,du \tag{1.33}$$

$$\int \frac{1}{1+x^2}\,dx = \int \cos^2 u \frac{1}{\cos^2 u}\,du = \int du = u = \arctan x.$$

In retrospect, the solution is deceptively simple, of course, but if you are like most people, then this kind of inspiration doesn't come naturally. Some academics make a living by finding clever substitutions for integrals. For the rest of us, though, the rule when faced with a tough integral is "Look it up." In the twentieth century, people owned handbooks like Abramowitz and Stegun (1964), but today, you can type the difficult integral into the Google search bar. If that doesn't help, you can always resort to numerical methods using the integration routines in your favorite scientific programming language.[2]

1.4 Differential Equations

Many problems are formulated as differential equations. The laws governing the process under study somehow relate a function to its derivatives. For example, we might postulate that the growth rate of a rabbit population is proportional to the number of rabbits that are already there:

$$\frac{d}{dt}N(t) = \alpha N(t). \tag{1.34}$$

Given this relationship, what are possible solutions for the function $N(t)$?

The analytical approach to solving differential equations is a rich and complex field of mathematics, and many books have been written on the topic–but this is not one of them. Realistic differential equations derived from concrete problems in the life sciences will often involve higher derivatives in nasty nonlinear functional forms. The section on nonlinear dynamics in chapter 11 will explain an approach to dealing with realistic differential equations that does not require an analytical solution.

---

2. If you're really trapped on a desert island, that's a tough break. Maybe move on to a different problem. Almost certainly, there is an amazing species of ant that you can observe while waiting to be rescued.

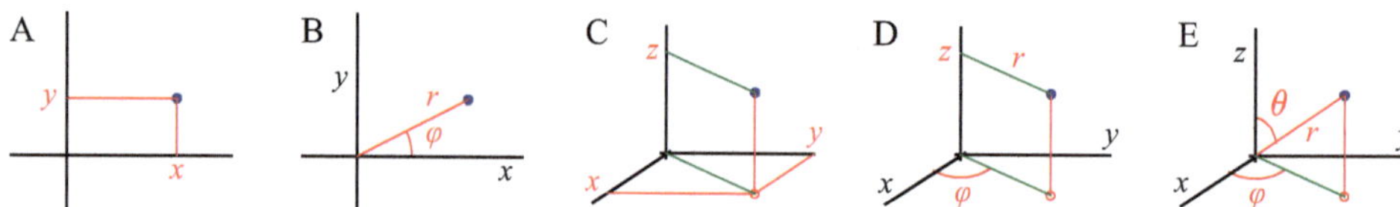

Figure 1.6
Coordinate systems: A: 2D Cartesian; B: 2D polar; C: 3D Cartesian; D: 3D cylindrical; E: 3D spherical.

However, we do ask the student to **memorize one differential equation**–namely, the one about the rabbits. Its solution is

$$N(t) = N(0)e^{\alpha t}, \tag{1.35}$$

where $N(0)$ is the number of rabbits at time $t=0$. We will see that this differential equation comes up over and over again.[3]

A slightly more general version is

$$\frac{\mathrm{d}}{\mathrm{d}t}f(t) - \alpha f(t) = g(t), \tag{1.36}$$

where $g(t)$ is a known function and $f(t)$ is unknown. The solution is

$$f(t) = f(0)e^{\alpha t} + \int_{t'=0}^{t} g(t')\, e^{\alpha(t-t')} \mathrm{d}t'. \tag{1.37}$$

1.5 Multiple Variables

So far, we have considered functions of a single variable, but many practical problems involve multiple variables. Examples include diffusion in three-dimensional (3D) space, chemical reactions with multiple species, and genomic data sets with thousands of variables per measurement.

1.5.1 Coordinate Systems

For problems involving two-dimensional (2D) or 3D space, an early decision involves the choice of coordinate system–namely, how points in the space should be specified. These coordinate systems are used frequently (figure 1.6):

Two dimensions:

- Cartesian (x, y); namely, the projections onto two orthogonal axes.
- Polar (r, φ), where r is the distance from the origin, and φ the angle around the origin relative to some reference.

---

3. Students in our course learn quickly that in response to the prompt "and what's the solution to this differential equation?" they simply need to shout "the exponential!" This is not guaranteed to work outside our lectures.

Three dimensions:

- Cartesian (x, y, z); namely the projections onto three orthogonal axes.
- Cylindrical (r, z, φ), where r is the distance from the z-axis, z is the projection onto the z-axis, and φ is the angle around the z-axis relative to some reference.
- Spherical (r, ϑ, φ), where r is the distance from the origin, ϑ the angle between the z-axis and the line to the point, and φ the angle around the z-axis relative to some reference.

The choice is usually dictated by the symmetries of the problem. For example, if the problem concerns diffusion in a square box, the Cartesian system will do fine. But if you are working on swarming bacteria in a circular petri dish, you might consider polar coordinates.

1.5.2   Partial Derivatives

Now one can ask separately how fast a function of multiple variables, $f(x, y)$, changes with one or the other of its arguments. This is expressed by the **partial derivative**:

$$\frac{\partial}{\partial x} f(x, y) = \lim_{\Delta x \to 0} \frac{f(x + \Delta x, y) - f(x, y)}{\Delta x}. \tag{1.38}$$

The **gradient** is a vector whose components are all the partial derivatives:

$$\nabla f = \left(\frac{\partial f}{\partial x}, \frac{\partial f}{\partial y} \right). \tag{1.39}$$

This vector points in the direction of the steepest ascent of the function f.

For an infinitesimal displacement of all the variables by the vector $\mathbf{d} = (dx, dy)$, the change in the function is

$$df = \frac{\partial f}{\partial x} \cdot dx + \frac{\partial f}{\partial y} \cdot dy = \nabla f \cdot \mathbf{d}, \tag{1.40}$$

where the last multiplication is known as the "dot product" between the gradient vector and the displacement vector. All the relationships in this section extend in obvious ways to more than two dimensions.

1.5.2.1   Chain rule for partial derivatives   Partial derivatives follow rules similar to those for normal derivatives. Suppose that a function $f(u, v)$ is expressed in terms of arguments $u(x, y)$ and $v(x, y)$, which themselves depend on the coordinates (x, y). What are the partial derivatives of f with respect to x and y?

$$\frac{\partial}{\partial x} f(u(x, y), v(x, y)) = \frac{\partial f}{\partial u} \frac{\partial u}{\partial x} + \frac{\partial f}{\partial v} \frac{\partial v}{\partial x}. \tag{1.41}$$

1.5.2.2   Higher partial derivatives   As expected, these are defined recursively. For example,

$$\frac{\partial^2 f}{\partial x^2} = \frac{\partial}{\partial x}\frac{\partial f}{\partial x}, \quad \frac{\partial^2 f}{\partial x \partial y} = \frac{\partial}{\partial x}\frac{\partial f}{\partial y}. \tag{1.42}$$

A common shorthand for partial derivatives places the respective variables in the subscript. For example,

$$f_{xx} \equiv \frac{\partial^2 f}{\partial x^2}, \quad f_{xy} \equiv \frac{\partial^2 f}{\partial x \partial y}. \tag{1.43}$$

1.5.2.3 Mixed derivatives commute When you take partial derivatives relative to different arguments, the order does not matter. For example,

$$\frac{\partial}{\partial x}\left(\frac{\partial}{\partial y} f(x,y)\right) = \frac{\partial}{\partial y}\left(\frac{\partial}{\partial x} f(x,y)\right). \tag{1.44}$$

1.5.2.4 Taylor series in 2D This polynomial approximation to a function extends naturally to higher dimensions. The Taylor series of $f(x,y)$ around a point (a,b) is the power series whose partial derivatives in (a,b) match all those of the function $f(x,y)$:

$$\begin{aligned}
f(x,y) = {}& f(a,b) \\
& + \frac{\partial f(a,b)}{\partial x}(x-a) + \frac{\partial f(a,b)}{\partial y}(y-b) \\
& + \frac{1}{2}\frac{\partial^2 f(a,b)}{\partial x^2}(x-a)^2 + \frac{\partial^2 f(a,b)}{\partial x \partial y}(x-a)(y-b) + \frac{1}{2}\frac{\partial^2 f(a,b)}{\partial y^2}(y-b)^2 \\
& + \dots
\end{aligned} \tag{1.45}$$

Again, the concept extends naturally to more than two dimensions.

1.5.2.5 Extrema in 2D At an extremum, the function must have zero gradient. As in the one-dimensional (1D) case, one needs an additional condition on the curvature to verify the presence of a local maximum or minimum. In 2D, the rules are

$$\begin{aligned}
\text{Maximum:} \quad & \tfrac{\partial f}{\partial x} = \tfrac{\partial f}{\partial y} = 0 \text{ and } f_{xx} < 0 \text{ and } f_{xx}f_{yy} - f_{xy}^2 > 0. \\
\text{Minimum:} \quad & \tfrac{\partial f}{\partial x} = \tfrac{\partial f}{\partial y} = 0 \text{ and } f_{xx} > 0 \text{ and } f_{xx}f_{yy} - f_{xy}^2 > 0. \\
\text{Saddle point:} \quad & \tfrac{\partial f}{\partial x} = \tfrac{\partial f}{\partial y} = 0 \text{ and } f_{xx}f_{yy} - f_{xy}^2 < 0.
\end{aligned} \tag{1.46}$$

1.5.3 Multivariate Integrals

The integral over multiple dimensions is performed by taking several integrals in sequence. For example, for a function of two variables $f(x,y)$, the double integral is

$$\int_y \int_x f(x,y)\,\mathrm{d}x\mathrm{d}y = \int_y \left(\underbrace{\int_x f(x,y)\,\mathrm{d}x}_{\text{Function of }y\text{ only}}\right)\mathrm{d}y. \tag{1.47}$$

When performing the inner integral $\int_x f(x,y)\,\mathrm{d}x$, y is taken to be constant. After performing that inner integral, the resulting expression no longer depends on x.

Geometrically, one can visualize $f(x, y)$ as a curved surface floating over the (x, y) plane. Then the integral in equation (1.47) is the volume under that surface. As in the earlier 1D case (figure 1.5), if the function dips below the plane in to negative values, then those volumes are negative.

In general, one may switch the order of integration in equation (1.47). However, if the integration limits of one variable depend on the other variable, then care is required. For example, the following integral covers the lower triangle of the unit square where $0 < y < x < 1$:

$$\int_{x=0}^{1} \int_{y=0}^{x} f(x, y) dy dx = \int_{y=0}^{1} \int_{x=y}^{1} f(x, y) dx dy. \tag{1.48}$$

Note the change in the integration limits.

1.5.3.1 Integrals of separable functions If $f(x, y)$ separates into a product of a function of x and a function of y, i.e.,

$$f(x, y) = u(x) v(y) \tag{1.49}$$

then the double integral becomes a simple product of single integrals

$$\int_y \int_x f(x, y) \, dx dy = \left(\int_x u(x) \, dx \right) \left(\int_y v(y) \, dy \right). \tag{1.50}$$

1.5.3.2 Change of coordinates for integration When moving from one set of coordinates (x, y) to new coordinates (u, v), the volume integrals become

$$\int_y \int_x f(x, y) \, dx dy = \int_v \int_u f(u, v) \frac{\partial (x, y)}{\partial (u, v)} \, du dv, \tag{1.51}$$

where

$$\frac{\partial (x, y)}{\partial (u, v)} \equiv \det \begin{bmatrix} \frac{\partial x}{\partial u} & \frac{\partial x}{\partial v} \\ \frac{\partial y}{\partial u} & \frac{\partial y}{\partial v} \end{bmatrix}. \tag{1.52}$$

This latter term is known as the **determinant of the Jacobian matrix**, and it takes into account how a small surface element $dx \cdot dy$ gets transformed into the corresponding surface element $du \cdot dv$.

1.6 Complex Numbers

All the numbers coming out of measuring devices are real numbers; in fact, any instrument connected to a computer just produces integers. So it is a legitimate question why research scientists should have to deal with complex numbers at all. The main reason is that some of the mathematical methods that we use are formulated much more elegantly and conveniently if they are extended to the complex plane. Of course, the ultimate predictions for experimental outcomes have to return to the realm of real numbers.

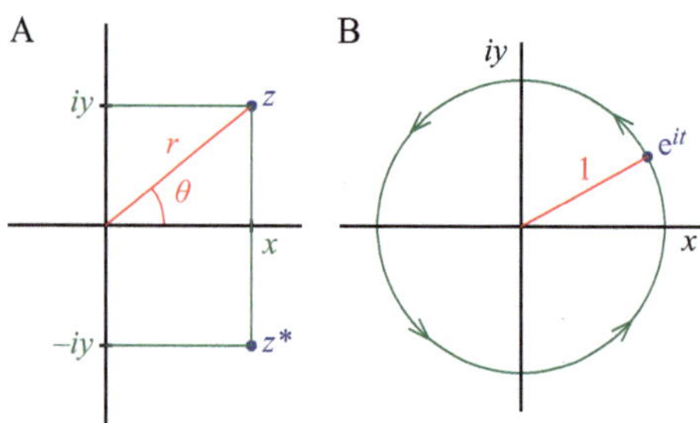

Figure 1.7
Complex numbers. A: standard form $z = x + iy$, exponential form $z = re^{i\vartheta}$, and complex conjugate $z = x - iy$. B: the complex phasor e^{it}.

1.6.1 Definitions

A complex number $z = x + iy$ has a real part x and an imaginary part y. The imaginary number i is defined as $\sqrt{-1}$. Complex numbers are often represented as points on a plane, with the real part on the horizontal axis and the imaginary part on the vertical axis (figure 1.7A).

$$
\begin{aligned}
&\text{Standard form:} && z = x + iy, \text{ where } i^2 = -1 \\
&\text{Real part of } z\text{:} && \mathrm{Re}\,(z) \equiv x \\
&\text{Imaginary part of } z\text{:} && \mathrm{Im}\,(z) \equiv y \\
&\text{Complex conjugate of } z\text{:} && z^* = x - iy
\end{aligned}
\tag{1.53}
$$

1.6.2 Euler's Formula

$$
e^{i\vartheta} = \cos\vartheta + i\sin\vartheta.
\tag{1.54}
$$

Euler's formula is a deep result and a good part of the reason why complex numbers are useful. You can verify it by writing out the Taylor series in ϑ for both sides of the equation, see (1.21).

1.6.3 Exponential Form

Given Euler's formula (equation 1.54), one can also express a complex number as $z = re^{i\vartheta}$, where r is the length of the line from the origin to z and ϑ is the angle between that line and the real axis (figure 1.7A):

$$
\begin{aligned}
&\text{Definition:} && z = re^{i\vartheta} \\
&\text{Modulus:} && r = |z| = \sqrt{z \cdot z^*} \\
&\text{Phase or argument:} && \vartheta = \arg z \\
&\text{Relation to standard form:} && x = r\cos\vartheta, \quad y = r\sin\vartheta \\
&\text{Multiplication:} && z_1 \cdot z_2 = \left(r_1 \cdot e^{i\vartheta_1}\right)\left(r_2 \cdot e^{i\vartheta_2}\right) = r_1 \cdot r_2 \cdot e^{i(\vartheta_1 + \vartheta_2)}
\end{aligned}
\tag{1.55}
$$

1.6.4 Phasors

The complex function of time

$$
f(t) = e^{it}
\tag{1.56}
$$

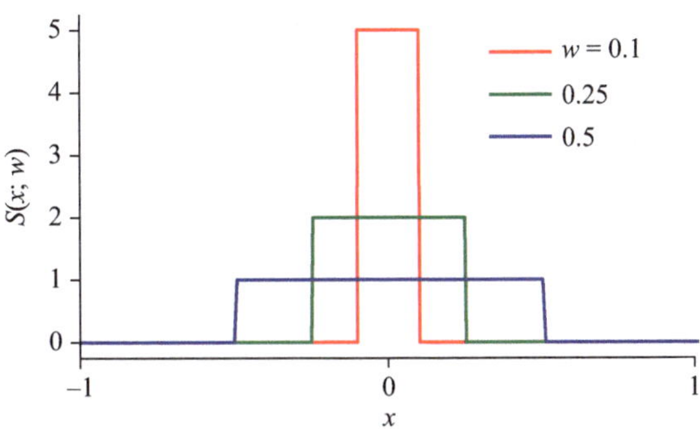

Figure 1.8
Increasingly narrow pulse with area = 1.

has modulus 1 and phase t. From Euler's formula, one sees that as time goes on, $f(t)$ traces the unit circle in the complex plane (figure 1.7B). It orbits counter-clockwise around the origin and completes an orbit whenever t increases by 2π. Such periodic functions are called **phasors** and will play an important role in Fourier analysis.

1.7 The Delta Function

In many applications, we deal with very short, compact pulses of something. For example, an auditory click is a very short pulse in time. A bolus of particles injected at one location is a compact pulse in space. The **delta function** $\delta(x)$ is a mathematical idealization of a narrow pulse.

For example, you can start with a square pulse around $x=0$ with width w and area 1:

$$S(x; w) = \begin{cases} \frac{1}{w} & \text{if } |x| < w/2 \\ 0 & \text{otherwise.} \end{cases} \tag{1.57}$$

Then gradually make the pulse narrower but at the same time taller, so the area under the curve always remains 1. In the limit of zero width, that pulse turns into the delta function:[4]

$$\delta(x) = \lim_{w \to 0} S(x; w). \tag{1.58}$$

Loosely, $\delta(x)$ is zero everywhere except at the origin, and it has an integral of 1:

$$\delta(x) = 0 \text{ for } x \neq 0$$

$$\int_{-\infty}^{\infty} \delta(x)\, dx = 1. \tag{1.59}$$

---

4. Warning: Mathematicians don't consider this a proper function. In practice, although the treatment here is sloppy, we have never gotten in trouble with it.

The delta function can be seen as the derivative of the step function, which is also called the **Heaviside function**:[5]

$$\delta(x) = \frac{d}{dx}H(x), \quad H(x) = \begin{cases} 0, x < 0 \\ 1, x \geqslant 0 \end{cases}. \tag{1.60}$$

When integrated against another function, the delta function acts as a "selector":

$$\int_{-\infty}^{\infty} f(x)\,\delta(x-a)\,dx = f(a). \tag{1.61}$$

Of particular value are representations of the delta function as sums or integrals of complex phasors, such as

$$\delta(x) = \frac{1}{2\pi} \int_{k=-\infty}^{\infty} e^{ikx}\,dk \tag{1.62}$$

and

$$\delta(x) = \frac{1}{2\pi} \sum_{n=-\infty}^{\infty} e^{inx}, \quad \text{for } x \in [-\pi, \pi]. \tag{1.63}$$

---

5. Again, you might get fired from a math department for saying this. But remember that the delta function was invented by physicists for their own purposes.

LINEAR SYSTEMS

2   Basics of Linear Algebra

Mathematics is the art of reducing any problem to linear algebra.
  —William Stein

For the whole is nothing but the sum of its parts.
  —Anonymous linear algebrist

2.1 Motivation

The mention of "linear algebra" immediately conjures up columns and arrays of numbers. But the relevance of this field goes much beyond recipes for shuffling and combining lists of numbers. Fundamentally, linear algebra comes into play whenever we deal with a collection of objects for which addition makes some sense. For example, two sounds can superpose to make another, two chemical solutions can be mixed together, and two images may be overlaid to make another. Linear algebra studies the representation of such objects and the transformations between them. Only in that context do the columns and arrays of numbers become important. The next section illustrates how linear systems analysis can connect phenomena from radically different parts of biology.

2.1.1 How Do Living Things Respond to Their Environment?

To pursue the question of how living things respond to their environment, scientists often perform some measurement of a system's response $r(t)$ as a function of time, while varying some kind of environmental stimulus $s(t)$. Then one would like to summarize how the response depends on the stimulus. Almost always, the simplest model for that relationship is "linearity." In fact, this is the default model, and one needs to explore it fully before considering anything fancier. The linearity assumption says that the system's response is just a weighted sum of the stimuli it experienced in the past:

$$r(t) = \sum_{u=-\infty}^{t} s(u)k(t-u). \tag{2.1}$$

Here, $s(u)$ is the stimulus at some earlier time u, and $k(t-u)$ is the weight attached to that stimulus, which happened a time $t-u$ ago.

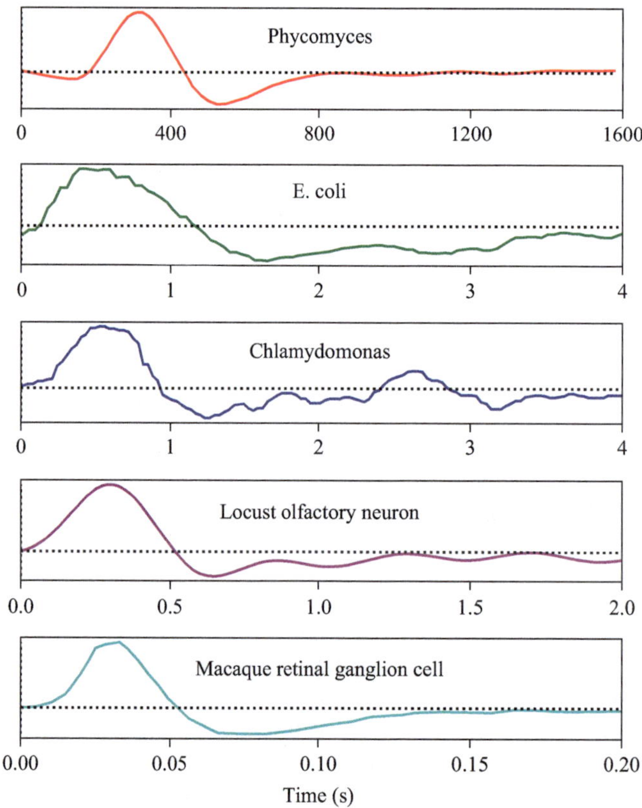

Figure 2.1
Impulse responses of five biological systems.

If time is taken to be continuous, rather than discrete, then the sum turns into an integral:

$$r(t) = \int_{u=-\infty}^{t} s(u)k(t-u)\,\mathrm{d}u. \tag{2.2}$$

This integral is called a "convolution," and we will encounter it frequently in linear systems analysis. The function $k(t)$ is called the system's "impulse response" because that's how the system reacts to a short stimulus pulse at time $t = 0$.

Figure 2.1 shows the measured impulse responses $k(t)$ of five very different biological systems:

- *Phycomyces* is a fungus that grows a fine, hairlike structure several centimeters above the colony to disperse its spores. Somehow the process that grows this single-cell filament can sense both gravity and light. Here, we are looking at the response of the growth rate to an impulse of light intensity (Lipson, 1975). After a delay of 150 s, the growth accelerates, peaks at 300 s, and then decelerates again, falling below the starting value from 400 to 800 s.
- *Escherichia coli* is a bacterium, one of many residents of the human gut. It propels itself using helical flagella driven by a rotary motor. Using a variety of receptor proteins on its cell surface, the bacterium senses the concentrations of important

chemicals in the environment and steers itself toward or away from them. It does so by regulating the direction of rotation of the flagellar motor. When the motor turns counterclockwise (CCW), the bacterium swims straight; when it turns clockwise, the bacterium executes a tumble that reorients it randomly. Here, we are looking at the bacterial response to a short pulse of the amino acid aspartate at time 0 (Block et al., 1982). The motor bias (i.e., the probability of turning CCW) increases for about a second, but then returns and dips below the starting value for a few seconds. Alternatively, one can interpret this curve $k(t)$ as the weight applied to stimuli at different times in the past. The bacterium measures the aspartate concentration in the past 1 s and subtracts from that the concentration over the preceding 2 s. In other words, the bacterium measures the rate of increase of the aspartate concentration and continues swimming straight for as long as that is positive.

- *Chlamydomonas reinhardtii* is a single-celled alga that swims using two cilia. It relies on photosynthesis and can steer toward or away from light, which it senses via an eye spot. The plot shows the response of the flagellar beat pattern to an impulse of light at time 0 (Josef et al., 2005). Again, one sees a positive response for about 1 s, followed by a dip below the baseline for the following few seconds.

- The locust *Schistocerca americana* senses odors in the air with a pair of large antennae. The primary receptor neurons in the antenna send their processes to second-stage neurons in the animal's brain. Here, we are recording the electrical response of such a cell to a whiff of the odor octanol at time 0 (Geffen et al., 2009). The neuron gets excited for about 0.5 s, but then its activity dips below baseline for a few seconds.

- The retina of the monkey *Macaca fascicularis* is very similar to our own. It senses light with photoreceptor cells and processes the signal through several layers of neurons before sending it through the optic nerve to the brain. The plot for this in the figure shows the response of one such optic nerve fiber to a brief pulse of light (Chichilnisky and Kalmar, 2002). After a short delay, the neuron's firing rate increases briefly, peaks at about 40 ms, and then drops below the baseline level for about 100 ms.

Although they are drawn from dramatically different organisms, all these kernel functions have something in common: the biphasic shape. The system places a strong positive weight on the most recent stimuli, but negative weights on stimuli that happened at earlier times. In some cases, the positive and negative weights cancel, such that the integral of $k(t)$ is close to zero. It appears that all these systems respond primarily to *change* in the stimulus—namely, the difference between the immediate past and the more distant past. There are several interpretations for this, depending on the function in question. For one, if an external variable doesn't change, then it is not threatening, and the system can continue operating as it is. But a sudden change is interesting, as it may require adjustment, such as initiating pursuit or escape. An alternative (not exclusive) argument is that in a slowly changing world, the rate of change predicts the future. In that sense, the biphasic impulse responses enable a predictive adjustment.

The kernels in figure 2.1 also differ in an important respect: the time scale of the response kernel (e.g. the distance between the negative and positive peaks). In a sense, this reflects the characteristic speed of the organism's life. For example, one can say that *E. coli* smells about 20 times slower than we see. As it happens, different senses in our nervous system also operate on different time scales. How our brain integrates those data into a single coherent stream of processing remains a deep puzzle.

Finally, on a technical level, it is interesting that only one of these curves was measured using an actual impulse stimulus. The others were derived from step stimuli,

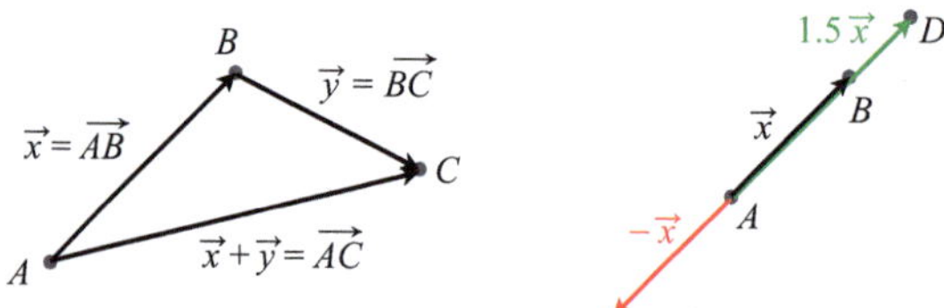

Figure 2.2
Left: Addition of displacement vectors. Right: Scalar multiplication and inverse of a vector.

or oddly enough, from completely random flicker stimuli. We will soon find out how this is possible.

Linear systems analysis pervades not only the study of living things, but also the design and understanding of instruments that we build for biological observations. Moreover, many of the mathematical methods in parts II and III of this book rely on linear algebra. Consequently, we start this book by laying the groundwork with important concepts and results from this field.

2.2 Vector Space

A vector space $\mathbb{V}$ is a set of objects $\vec{x}$, called **vectors**, on which two operations can be performed:

- **Scalar multiplication:** A vector multiplied by a number λ is also a vector in the space: $\lambda\vec{x} \in \mathbb{V}$.
- **Addition:** The sum of two vectors is also a vector, $\vec{x} + \vec{y} \in \mathbb{V}$.

Example 2.1 Displacements are vectors. If a particle moves from point A to point B, we call the directed line $\overrightarrow{AB}$ the displacement vector. One can represent that with an arrow pointing from A to B (figure 2.2).

Suppose this is followed by a second displacement from B to C; then the total displacement is $\overrightarrow{AC}$. This defines formally what we mean by the sum of two arrows $\overrightarrow{AC} = \overrightarrow{AB} + \overrightarrow{BC}$: place the tail of one arrow at the tip of the other arrow and evaluate the combined tail-to-tip arrow.

Similarly, suppose that another particle is displaced λ times as far as the first. That defines the scalar multiplication $\overrightarrow{AD} = \lambda\,\overrightarrow{AB}$: make an arrow that points in the same direction but is λ times as long.

We will often use arrows in a plane for examples of vector spaces and their operations. $\square$

Example 2.2 The forces acting on a particle are vectors. Scalar multiplication is well defined: we know what it means to exert three times that force. And if you apply two forces simultaneously, we call the result the sum of the two forces. $\square$

Example 2.3 Functions of time $f(t)$ are vectors, such as the time series of voltage values that comes out of a scientific instrument. You can obviously multiply a function by a scalar, and $\lambda f(t)$ is another function. Similarly, the sum of two functions $f(t) + g(t)$ is another function. $\square$

The operation of addition has to satisfy two more conditions:

- **Null vector:** There exists a null vector $\vec{0}$ such that $\vec{x} + \vec{0} = \vec{x}$ for every vector $\vec{x}$.
- **Inverse vector:** For every vector $\vec{x}$, there exists an inverse $\overrightarrow{-x}$, which when added to $\vec{x}$ gives the null vector, $\vec{x} + \overrightarrow{-x} = \vec{0}$.

In all practical cases where there is an intuitive notion of scalar multiplication and addition, these two conditions are met.

Exercise 2.1 What are null and inverse vectors in the various examples given here? $\square$

2.3 Basis Sets

2.3.1 Linear Independence

Given a set of vectors $\{\vec{x}_1, \vec{x}_2, \ldots, \vec{x}_n\}$ and a set of numbers $\{\lambda_1, \lambda_2, \ldots, \lambda_n\}$, the vector

$$\vec{y} = \sum_{i=1}^{n} \lambda_i \vec{x}_i \tag{2.3}$$

is called a **linear combination** of $\overrightarrow{x_i}$ (figure 2.3).

A set of vectors $\{\vec{x}_1, \vec{x}_2, \ldots, \vec{x}_n\}$ is called **linearly independent** if none of the vectors can be written as a linear combination of the others. In that case, there exists no set of numbers $\{\lambda_1, \lambda_2, \ldots, \lambda_n\}$ such that

$$\sum_{i=1}^{n} \lambda_i \vec{x}_i = \vec{0}. \tag{2.4}$$

A set of vectors $\{\vec{e}_1, \vec{e}_2, \ldots, \vec{e}_n\}$ is called a **basis set** of the vector space if two conditions are met:

1. Every vector in the space can be expressed as a linear combination of the $\vec{e}_i$, and
2. The set of vectors $\{\vec{e}_1, \vec{e}_2, \ldots, \vec{e}_n\}$ is linearly independent.

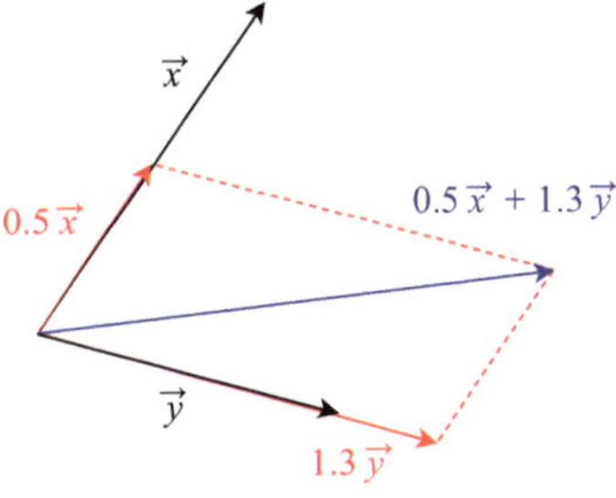

Figure 2.3
A linear combination in the space of arrows on the plane.

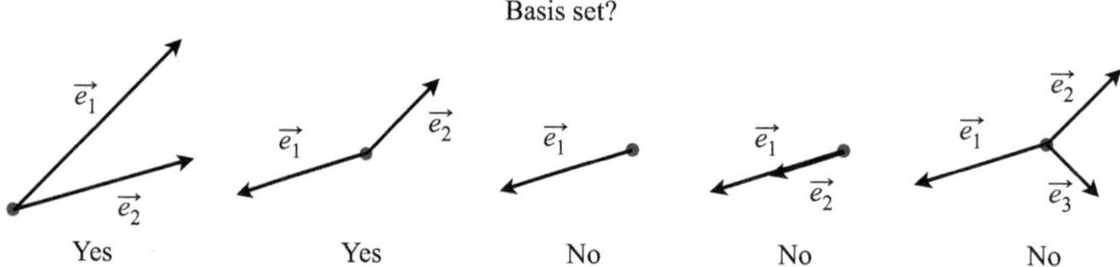

Figure 2.4
What makes a basis set for the space of arrows on the plane?

Example 2.4 For the space of displacement vectors in the two-dimensional (2D) plane, any two vectors that are not colinear will form a basis set (figure 2.4). A single vector cannot be a basis set because it does not serve to express all vectors in the space. Any set of three vectors is not a basis set because they are linearly dependent: one can be expressed as a linear combination of the other two. $\square$

2.3.2 Dimensionality
The **dimension** of a vector space $\mathbb{V}$ is defined as the number of vectors in any basis set of $\mathbb{V}$. In the example 2.4 of displacement vectors in a plane, that number is 2. So this is a case of a two-dimensional vector space.[1]

2.3.3 Component Representation of a Vector
Suppose that we are given a particular basis set $\{\vec{e}_1, \vec{e}_2, \ldots, \vec{e}_n\}$. Then any vector $\vec{x}$ in the space can be expressed uniquely as a linear combination of the basis vectors:

$$\vec{x} = \sum_{i=1}^{n} x_i \vec{e}_i. \tag{2.5}$$

So in this basis, the vector $\vec{x}$ is identified by a column of numbers:

$$\mathbf{x} = \begin{bmatrix} x_1 \\ x_2 \\ \vdots \\ x_n \end{bmatrix}. \tag{2.6}$$

These are called the **components** of $\vec{x}$ in the basis $\{\vec{e}_i\}$.[2]

In any basis $\{\vec{e}_i\}$, the addition of vectors and scalar multiplication correspond to simple operations on the list of components: componentwise sum and componentwise multiplication with a number:

---

1. You probably suspected this, but now we have a formal definition of dimensionality that also will work in cases that aren't completely obvious.
2. It is useful to distinguish the vector as an object, $\vec{x}$, from its list of components, $\mathbf{x}$, in any given basis. Many of the results in linear algebra can be expressed and remembered without resorting to any specific basis.

Vector Object	Component Representation
$\vec{x}$	$\mathbf{x} = \begin{bmatrix} x_1 \\ \vdots \\ x_n \end{bmatrix}$
$\vec{x} + \vec{y}$	$\mathbf{x} + \mathbf{y} = \begin{bmatrix} x_1 + y_1 \\ \vdots \\ x_n + y_n \end{bmatrix}$
$\lambda\vec{x}$	$\lambda\mathbf{x} = \begin{bmatrix} \lambda x_1 \\ \vdots \\ \lambda x_n \end{bmatrix}$

2.4 Linear Operators

A linear operator $\mathcal{M}$ on vector space $\mathbb{V}$ is a function that turns vector $\vec{x} \in \mathbb{V}$ into another vector in the same space: $\mathcal{M}(\vec{x}) \in \mathbb{V}$.[3]

It also has to satisfy the following **linearity condition**:

$$\mathcal{M}(\lambda\vec{x} + \mu\vec{y}) = \lambda\mathcal{M}(\vec{x}) + \mu\mathcal{M}(\vec{y}). \tag{2.7}$$

Example 2.5 In the space of displacement arrows on the plane, the instruction "Rotate the arrow by 30 degrees counterclockwise" is a linear operator (figure 2.5). Note that the result is another displacement arrow. □

Example 2.6 In the space of functions of time $f(t)$, "take the time-derivative" is a linear operator: $\mathcal{M}(f(t)) = \frac{\mathrm{d}}{\mathrm{d}t}f(t)$. □

Exercise 2.2 Verify that both these operators meet the linearity condition. □

Exercise 2.3 Explain why this is not a linear operator: $\mathcal{M}(f(t)) = f(t) + 3$. □

Some notes on nomenclature: In what follows, we will often use the term "operator" to mean "linear operator." Also, the output of an operation $\mathcal{M}(\vec{x})$ is often called the **image** of $\vec{x}$.

---

3. A broader definition of linear operators would allow mappings from one space into a space of a different dimension. This leads to the algebra of nonsquare matrices. However, we find that the most pertinent applications are covered by the present definition and operations on square matrices.

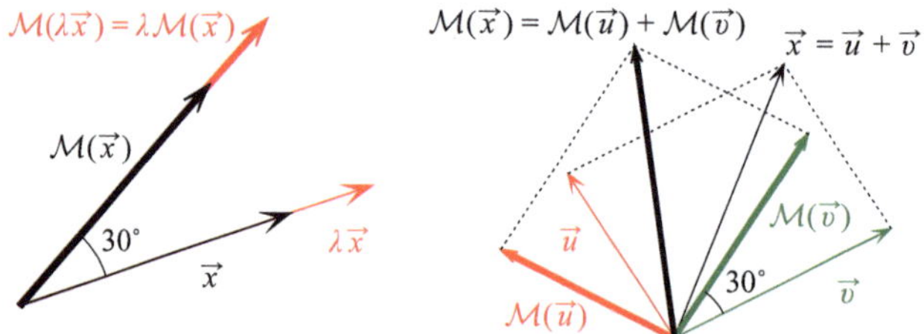

Figure 2.5
A linear operator in the space of arrows on the plane: "Rotate the arrow 30 degrees counterclockwise." The constructions illustrate why this operator satisfies the two linearity condition.

2.4.1 Component Representation of an Operator

Suppose again that we are using a particular basis set $\{\vec{e}_1, \vec{e}_2, \ldots, \vec{e}_n\}$. If a vector $\vec{x}$ has components $\{x_i\}$ in that basis, what are the components of the image $\mathcal{M}(\vec{x})$? Here, we can exploit the linearity property:

$$\mathcal{M}(\vec{x}) = \mathcal{M}\left(\sum_{i=1}^{n} x_i \vec{e}_i\right)$$
$$= \sum_{i=1}^{n} x_i \mathcal{M}(\vec{e}_i). \tag{2.8}$$

If we know the images $\mathcal{M}(\vec{e}_i)$ of all the basis vectors, we can compute the image of any arbitrary vector from its components. So suppose that the image of basis vector $\vec{e}_j$ has components M_{ij}; namely,

$$\mathcal{M}(\vec{e}_j) = \sum_{i=1}^{n} M_{ij} \vec{e}_i. \tag{2.9}$$

The numbers M_{ij} are the **components of operator** $\mathcal{M}$ in the basis $\{\vec{e}_i\}$. They are usually listed as a **matrix**; namely, a 2D array of numbers:

$$\mathbf{M} = \begin{bmatrix} M_{11} & M_{12} & \ldots & M_{1n} \\ M_{21} & M_{22} & \ldots & M_{2n} \\ \vdots & \vdots & \vdots & \vdots \\ M_{n1} & M_{n2} & \ldots & M_{nn} \end{bmatrix}. \tag{2.10}$$

Note that the columns of $\mathbf{M}$ are simply the components of the images of the basis vectors: if $\vec{f}_j = \mathcal{M}(\vec{e}_j)$, then

$$\mathbf{f}_j = \begin{bmatrix} M_{1j} \\ M_{2j} \\ \vdots \\ M_{nj} \end{bmatrix}. \tag{2.11}$$

Given those numbers, any given vector $\vec{x}$ gets imaged into

$$\vec{y} = \mathcal{M}(\vec{x}) = \sum_{j=1}^{n} x_j \mathcal{M}(\vec{e}_j)$$

$$= \sum_{j=1}^{n} x_j \sum_{i=1}^{n} M_{ij} \vec{e}_i$$

$$= \sum_{i=1}^{n} \left(\sum_{j=1}^{n} M_{ij} x_j \right) \vec{e}_i. \tag{2.12}$$

So the components of the image vector are

$$y_i = \sum_{j=1}^{n} M_{ij} x_j, \tag{2.13}$$

and one writes this shorthand

$$\mathbf{y} = \mathbf{M} \cdot \mathbf{x}. \tag{2.14}$$

This defines the **multiplication of a matrix and a vector.**[4]

Figure 2.6 shows a simple graphic mnemonic to remember the rules of matrix multiplication $\mathbf{y} = \mathbf{M}\mathbf{x}$:

- Write the components of $\mathbf{M}$ in a matrix; remember that the first index in M_{ij} is for the row, and the second is for the column.
- Write the vector $\mathbf{x}$ in a column to the top right of $\mathbf{M}$.
- Match each element of $\mathbf{x}$ with an element in the first row of $\mathbf{M}$, as shown.
- Multiply them and sum the results. That makes the first component of $\mathbf{y}$. Then do the same with the second row of $\mathbf{M}$ to get the second component of $\mathbf{y}$. And the process continues to the last row.

Example 2.7 Again, using the arrows on the plane, consider the linear operator

$$\mathcal{M} = \text{flip the arrow across the vertical axis.}$$

We get to choose a basis set for this problem, and following a hunch, we pick one basis vector $\vec{e}_1$ to be an arrow that lies on the vertical axis, and then a second vector $\vec{e}_2$ perpendicular to it (figure 2.7). With this choice, it is easy to see what the images of the two basis vectors are:

$$\mathcal{M}(\vec{e}_1) = 1 \times \vec{e}_1 + 0 \times \vec{e}_2$$

$$\mathcal{M}(\vec{e}_2) = 0 \times \vec{e}_1 - 1 \times \vec{e}_2. \tag{2.15}$$

---

4. As in the case of vectors, we use a different notation for the operator $\mathcal{M}$ and for the matrix $\mathbf{M}$ of its components in any given basis.

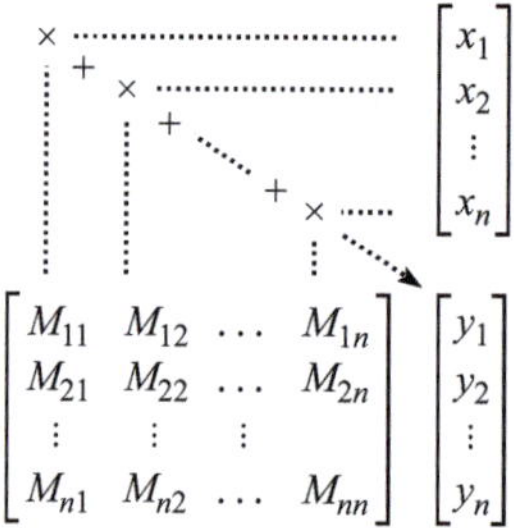

Figure 2.6
A mnemonic for multiplying a matrix by a vector.

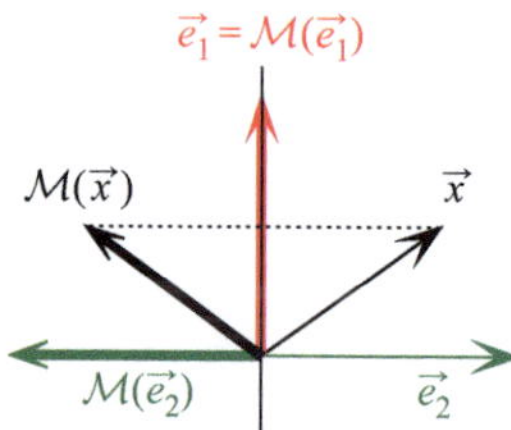

Figure 2.7
A reflection operator for arrows on the plane.

So the component representation of $\mathcal{M}$ in this basis is simply

$$\mathbf{M} = \begin{bmatrix} 1 & 0 \\ 0 & -1 \end{bmatrix}.$$
(2.16)

$\square$

2.4.2 Operator Algebra

As for vectors, one can multiply an operator by a number or add two operators together.
The definitions are straightforward:

- **Scalar multiplication:** The operator $(\lambda\mathcal{M})$ is defined by

$$(\lambda\mathcal{M})(\vec{x}) = \lambda \cdot \mathcal{M}(\vec{x}).$$
(2.17)

- **Addition:** Similarly, $(\mathcal{L} + \mathcal{M})$ is defined by

$$(\mathcal{L} + \mathcal{M})(\vec{x}) = \mathcal{L}(\vec{x}) + \mathcal{M}(\vec{x}).$$
(2.18)

Exercise 2.4 Confirm that these are in fact operators that respect the linearity condition
(equation (2.7)). $\square$.

One can also multiply two operators.

Multiplication: The product operator $\mathcal{L}\mathcal{M}$ means: first perform $\mathcal{M}$ and then perform $\mathcal{L}$ on the result:

$$(\mathcal{L}\mathcal{M})(\vec{x}) = \mathcal{L}(\mathcal{M}(\vec{x})). \qquad (2.19)$$

In general, the order of the two operations matters. In special cases, the order doesn't matter, and then one says that the two operators **commute:**

$$\mathcal{L}\mathcal{M} = \mathcal{M}\mathcal{L}, \text{ if } \mathcal{L} \text{ and } \mathcal{M} \text{ commute.} \qquad (2.20)$$

Exercise 2.5 For arrows in a plane, consider the rotation operator

$$\mathcal{M}_\alpha = \text{rotate the error by an angle } \alpha.$$

Show that $\mathcal{M}_\alpha$ and $\mathcal{M}_\beta$ commute for all angles α, β. $\square$

Exercise 2.6 Now consider arrows in a three-dimensional (3D) space. Say that

$$\mathcal{M} = \text{``Rotate the arrow by 30 degrees about the } x\text{-axis''}, \text{ and}$$
$$\mathcal{L} = \text{``Rotate the arrow by 70 degrees about the } y\text{-axis''}$$

Do $\mathcal{L}$ and $\mathcal{M}$ commute? $\square$

2.4.3 Identity and Inverse Operators

The trivial operator that images every vector into itself is called the **identity operator:**

$$\mathcal{I}(\vec{x}) = \vec{x}, \text{ for all vectors } \vec{x}. \qquad (2.21)$$

More interesting is the notion of an **inverse operator**, which undoes whatever another operator did. The inverse of operator $\mathcal{M}$ is denoted as $\mathcal{M}^{-1}$. Concatenating an operator with its inverse produces the identity operator

$$\mathcal{M}\mathcal{M}^{-1} = \mathcal{I}. \qquad (2.22)$$

However, depending on $\mathcal{M}$, the inverse $\mathcal{M}^{-1}$ may not exist.

Example 2.8 For arrows on the plane, consider the reflection operator of figure 2.7. This operator is its own inverse because reflecting the image of an arrow returns the original arrow. So $\mathcal{M}^{-1} = \mathcal{M}$. Obviously, this is a special case. $\square$

Example 2.9 Again, for arrows on the plane, consider the following operator:

$$\mathcal{M} = \text{project the arrow onto the horizontal line.}$$

This is a perfectly reasonable operation (figure 2.8), but it cannot be inverted. Knowing the image $\mathcal{M}(\vec{x})$, it is impossible to recover $\vec{x}$ because there are infinitely many other vectors that produce the same image. $\square$

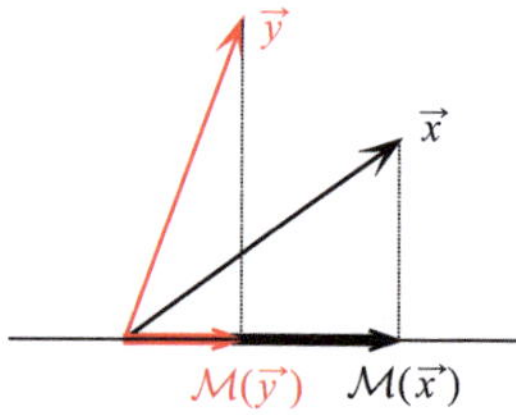

Figure 2.8
A projection operator for arrows on the plane.

If an operator $\mathcal{M}$ has an inverse, then $\mathcal{M}$ is called **regular**. If it does not, it is called **singular**.

If $\mathcal{M}$ is singular, then there exists at least one nonzero vector whose image is the null vector, so $\mathcal{M}(\vec{x}) = \vec{0}$.

Exercise 2.7 For the projection operator (figure 2.8) find an arrow that gets imaged into $\vec{0}$. □

2.5 Matrix Algebra

Suppose again that we have chosen a specific set of basis vectors $\{\vec{e}_i\}$ for the space. Now every operator on this space is represented by a matrix of components. How are these matrix representations affected by the algebraic operations in section 2.4.2? For scalar multiplication and addition, the results are straightforward:

- **Scalar multiplication:** If $\mathcal{P} = \lambda \mathcal{M}$, then $P_{ij} = \lambda M_{ij}$.
- **Addition:** If $\mathcal{P} = \mathcal{L} + \mathcal{M}$, then $P_{ij} = L_{ij} + M_{ij}$.

What about the product of two operators $\mathcal{P} = \mathcal{L}\mathcal{M}$? We can analyze this question by allowing $\mathcal{P}$ to act on an arbitrary vector and shuffling indices around:

$$
\begin{aligned}
\mathcal{P}(\vec{x}) &= \sum_{i=1}^{n} \sum_{j=1}^{n} P_{ij} x_j \vec{e}_i \\[1mm]
&= \mathcal{L}\left(\mathcal{M}(\vec{x})\right) \\[1mm]
&= \mathcal{L}\left(\sum_{k=1}^{n}\left(\sum_{j=1}^{n} M_{kj} x_j\right)\vec{e}_k\right) \\[1mm]
&= \sum_{i=1}^{n}\left(\sum_{k=1}^{n} L_{ik}\left(\sum_{j=1}^{n} M_{kj} x_j\right)\right)\vec{e}_i \\[1mm]
&= \sum_{i=1}^{n}\sum_{j=1}^{n}\left(\sum_{k=1}^{n} L_{ik} M_{kj}\right) x_j \vec{e}_i.
\end{aligned}
\tag{2.23}
$$

By comparing the first and last lines, one gets the components of $\mathcal{P}$.

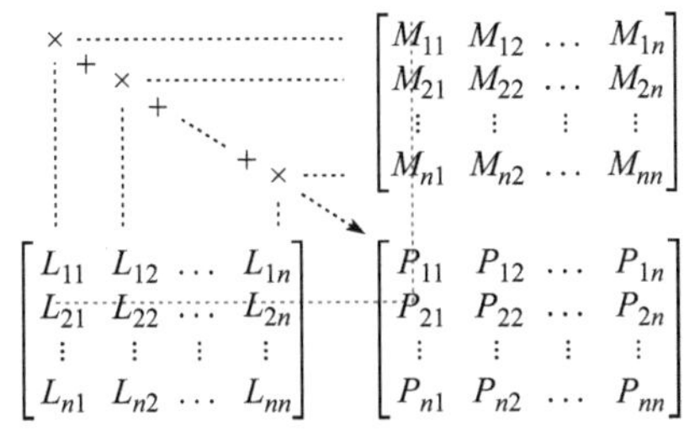

Figure 2.9
A mnemonic for multiplying a matrix by a matrix, $\mathbf{P} = \mathbf{L} \cdot \mathbf{M}$. For a given target element, like P_{21}, choose the corresponding row of $\mathbf{L}$ and column of $\mathbf{M}$ (dotted lines), then multiply and add as indicated.

Product: If $\mathcal{P} = \mathcal{L}\mathcal{M}$, then

$$P_{ij} = \sum_{k=1}^{n} L_{ik} M_{kj}. \tag{2.24}$$

This defines the **multiplication of two matrices.** We write that matrix product as simply

$$\mathbf{P} = \mathbf{L} \cdot \mathbf{M}. \tag{2.25}$$

As in the case of matrix $\times$ vector, one can recall the index operations for matrix multiplication with a mnemonic, see figure 2.9.

2.5.1 Identity Matrix
Recall the identity operator $\mathcal{I}$, which maps every vector onto itself. This applies in particular to each of the basis vectors, $\mathcal{I}(\vec{e}_i) = \vec{e}_i$. So the component representation of the identity is the **identity matrix**:

$$\mathbf{I} = \begin{bmatrix} 1 & 0 & \cdots & 0 \\ 0 & 1 & \cdots & 0 \\ \vdots & \vdots & \ddots & \vdots \\ 0 & 0 & \cdots & 1 \end{bmatrix}. \tag{2.26}$$

The matrix components here are

$$I_{jk} = \delta_{jk}, \tag{2.27}$$

using the definition of the **Kronecker delta:**

$$\delta_{jk} = \begin{cases} 1 & \text{if } j = k \\ 0 & \text{otherwise.} \end{cases} \tag{2.28}$$

This component representation of the identity is the same regardless of the choice of basis set.

2.5.2 Inverse Matrix

The inverse of a matrix $\mathbf{A}$ is another matrix $\mathbf{A}^{-1}$, whose product with $\mathbf{A}$ makes the identity matrix:

$$\mathbf{A} \cdot \mathbf{A}^{-1} = \mathbf{I}. \tag{2.29}$$

As discussed already, certain operators cannot be inverted. The matrix corresponding to such a **singular** operator has no inverse.

Example 2.10 Let's try to find an inverse matrix for

$$\mathbf{S} = \begin{bmatrix} 1 & 0 \\ 0 & 0 \end{bmatrix}$$

$$\begin{bmatrix} 1 & 0 \\ 0 & 0 \end{bmatrix} \cdot \begin{bmatrix} a & b \\ c & d \end{bmatrix} = \begin{bmatrix} 1 & 0 \\ 0 & 1 \end{bmatrix}$$

Clearly, there is no way to satisfy the equality for the bottom-right element of the identity matrix: $0 \cdot b + 0 \cdot d = 1$. $\square$

Is there a way to tell whether a matrix has an inverse from the list of its components? Yes, you can calculate its determinant.

2.5.3 Determinant

The **determinant** of a matrix is a single number that determines (surprise!) several of its properties. Its general definition is somewhat laborious (as discussed next), but it is easy and useful to remember the expression for one-dimensional (1D), 2D, and 3D matrices. In particular,

$$\det \begin{bmatrix} a \end{bmatrix} = a \tag{2.30}$$

$$\det \begin{bmatrix} a & b \\ c & d \end{bmatrix} = ad - bc \tag{2.31}$$

$$\det \begin{bmatrix} a & b & c \\ d & e & f \\ g & h & i \end{bmatrix} = aei + bfg + cdh - afh - bdi - ceg. \tag{2.32}$$

Figure 2.10 offers is a graphic mnemonic for determinants of 2D and 3D matrices. The general expression for the determinant is

$$\det \mathbf{A} = \sum_{\mathbf{P}} (-1)^{T(\mathbf{P})} \prod_{i=1}^{n} a_{i,P_i}, \tag{2.33}$$

where

$$\mathbf{P} = (P_1, \ldots, P_n) \tag{2.34}$$

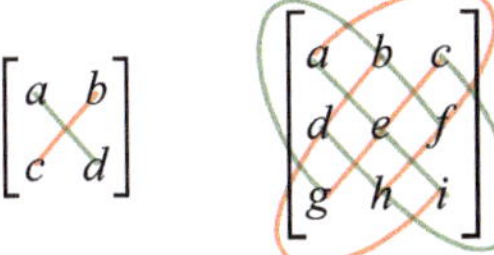

Figure 2.10
To calculate the determinant, multiply the terms along each of the diagonals, and then add
the green diagonals and subtract the red diagonals.

is a permutation (or reordering) of the numbers $(1,\ldots,n)$. Any such permutation
can be constructed by starting with the order $(1,\ldots,n)$ and applying several pairwise
interchanges (or "transpositions") of two numbers. $T(\mathbf{P})$ is the number of transpositions
required to get the particular permutation $\mathbf{P}$. The term

$$\prod_{i=1}^{n} a_{i,P_i} \tag{2.35}$$

is the product of n matrix elements, of which exactly one is chosen from each row and
each column. The permutation element P_i spells out which column goes with row i.
These terms are then summed over all possible permutations $\mathbf{P}$.[5]

Some interesting properties of the determinant are as follows:

- The determinant is **independent of the chosen basis**. This means that the determinant is a property of the operator, regardless of how you choose to represent that
  operator. For example, the determinant of the reflection operator for arrows on the
  plane (see example 2.7) is -1. Properties of an operator that are independent of the
  basis are also called **invariant**.
- The **determinant of the identity operator** is 1:

$$\det \mathbf{I} = 1. \tag{2.36}$$

From the form of the identity matrix in equation (2.26), you can see that the only
nonzero product term in equation (2.35) in the formula for the determinant (2.33)
is the one containing all the diagonal elements.

- The **determinant of a diagonal matrix**

$$\mathbf{D} = \begin{bmatrix} d_1 & 0 & \cdots & 0 \\ 0 & d_2 & \cdots & 0 \\ \vdots & \vdots & \ddots & \vdots \\ 0 & 0 & \cdots & d_n \end{bmatrix} \tag{2.37}$$

---

5. Computers do not use this formula to evaluate a determinant because it would require of
order $n!$ operations. The standard algorithm can do it in order n^3 operations.

is simply the product of the elements on the diagonal

$$\det \mathbf{D} = \prod_{i=1}^{n} d_i. \tag{2.38}$$

- The **determinant of a matrix product** is the product of the individual determinants:

$$\det(\mathbf{AB}) = \det \mathbf{A} \det \mathbf{B}. \tag{2.39}$$

Note this does not apply to addition:

$$\det(\mathbf{A} + \mathbf{B}) \neq \det \mathbf{A} + \det \mathbf{B}. \tag{2.40}$$

- The **determinant of the inverse matrix** (if it exists) is

$$\det \mathbf{A}^{-1} = \frac{1}{\det \mathbf{A}}. \tag{2.41}$$

This follows directly from equations (2.29), (2.36), and (2.39).

2.5.4 Singular Matrices

Let us return now to the question whether a matrix has an inverse. The last relation (2.41) shows that a matrix with a zero determinant cannot have an inverse. Such matrices are called **singular**, as with the associated operators.

Here are a few equivalent **ways to tell when a matrix A is singular**:

- $\det \mathbf{A} = 0$.
- There exists a vector $\mathbf{x}$ such that $\mathbf{A} \cdot \mathbf{x} = \mathbf{0}$.
- The column vectors of $\mathbf{A}$ are linearly dependent.
- The row vectors of $\mathbf{A}$ are linearly dependent.

2.5.5 Inverse Formula

From the theory of determinants, one can derive a general formula for the inverse of a matrix. In practice, it really helps to remember the formula for two dimensions:

$$\begin{bmatrix} a & b \\ c & d \end{bmatrix}^{-1} = \frac{1}{ad - bc} \begin{bmatrix} d & -b \\ -c & a \end{bmatrix}. \tag{2.42}$$

Here is a simple mnemonic: Exchange the elements along the diagonal (a and d); flip the sign of the elements on the antidiagonal (b and c); and divide by the determinant.

2.6 Change of Basis

Often, it is convenient to choose a new set of basis vectors for the space. Obviously, that will change all the components of vectors and operators. The relationship between the old and new components is called a **basis transform**.

Suppose that the old basis is the set of vectors $\{\vec{e}_1, \ldots, \vec{e}_n\}$ and the new basis is $\{\vec{e}_1', \ldots, \vec{e}_n'\}$, which are expressed in terms of the old basis vectors as

$$\vec{e}_j' = \sum_{i=1}^{n} S_{ij}\, \vec{e}_i. \tag{2.43}$$

This defines the **transformation matrix:**

$$\mathbf{S} = [S_{ij}]. \tag{2.44}$$

Note the columns of the transformation matrix are simply the components of the new basis vectors as expressed in the old basis.

2.6.1 Transform of a Vector

Now we can express an arbitrary vector $\vec{x}$ in both the old and the new bases:

$$
\begin{aligned}
\vec{x} &= \sum_{i=1}^{n} x_i \vec{e}_i \\
&= \sum_{j=1}^{n} x_j' \vec{e}_j' \\
&= \sum_{j=1}^{n} x_j' \sum_{i=1}^{n} S_{ij} \vec{e}_i \\
&= \sum_{i=1}^{n} \left(\sum_{j=1}^{n} S_{ij} x_j' \right) \vec{e}_i.
\end{aligned}
\tag{2.45}
$$

Comparing the first and last lines, one finds that

$$x_i = \sum_{j=1}^{n} S_{ij} x_j' \tag{2.46}$$

or, in matrix form,

$$\mathbf{x} = \mathbf{S} \cdot \mathbf{x}'. \tag{2.47}$$

Therefore, the components $\mathbf{x}'$ in the new basis are transformed from those in the old basis by

$$\mathbf{x}' = \mathbf{S}^{-1} \cdot \mathbf{x}. \tag{2.48}$$

Note that the transformation matrix $\mathbf{S}$ is regular because its columns are the new basis vectors, and those must be linearly independent; see section 2.5.4. Therefore, the inverse $\mathbf{S}^{-1}$ is guaranteed to exist.

2.6.2 Transform of an Operator

Similarly, one can express the output of an operator in either the old or the new basis, and perform the same type of index acrobatics. One finds that the matrix for an operator transforms from $\mathbf{M}$ in the old basis to $\mathbf{M}'$ in the new basis as follows:

$$\mathbf{M}' = \mathbf{S}^{-1} \cdot \mathbf{M} \cdot \mathbf{S}. \tag{2.49}$$

Example 2.11 Let us revisit the reflection operator for arrows on a plane: "Reflect about the vertical axis." In example 2.7, we chose a convenient basis, in which $\vec{e}_1$ lies on the vertical line and $\vec{e}_2$ perpendicular to it. What if, instead, we choose $\vec{e}_1'$ and $\vec{e}_2'$ rotated 45 degrees from the previous set (see figure 2.11)?

The new basis vectors expressed in the old basis have the components

$$\vec{e}_1' : \begin{bmatrix} \frac{1}{\sqrt{2}} \\ \frac{1}{\sqrt{2}} \end{bmatrix} \text{ and } \vec{e}_2' : \begin{bmatrix} -\frac{1}{\sqrt{2}} \\ \frac{1}{\sqrt{2}} \end{bmatrix}. \tag{2.50}$$

So the transformation matrix is

$$\mathbf{S} = \begin{bmatrix} \frac{1}{\sqrt{2}} & -\frac{1}{\sqrt{2}} \\ \frac{1}{\sqrt{2}} & \frac{1}{\sqrt{2}} \end{bmatrix}. \tag{2.51}$$

The inverse is

$$\mathbf{S}^{-1} = \begin{bmatrix} \frac{1}{\sqrt{2}} & \frac{1}{\sqrt{2}} \\ -\frac{1}{\sqrt{2}} & \frac{1}{\sqrt{2}} \end{bmatrix}, \tag{2.52}$$

which can be verified easily.

The operator expressed in the old basis is

$$M = \begin{bmatrix} 1 & 0 \\ 0 & -1 \end{bmatrix}. \tag{2.53}$$

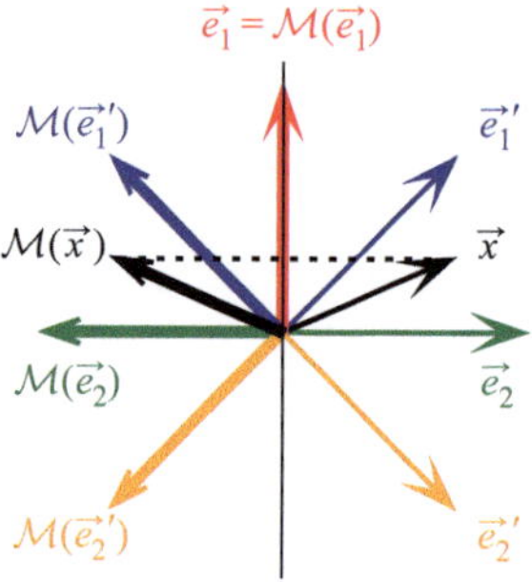

Figure 2.11
The reflection operator for arrows on the plane with two alternative basis sets.

Therefore, the components in the new basis are

$$\mathbf{M}' = \mathbf{S}^{-1} \cdot \mathbf{M} \cdot \mathbf{S}$$

$$= \begin{bmatrix} \frac{1}{\sqrt{2}} & \frac{1}{\sqrt{2}} \\ -\frac{1}{\sqrt{2}} & \frac{1}{\sqrt{2}} \end{bmatrix} \cdot \begin{bmatrix} 1 & 0 \\ 0 & -1 \end{bmatrix} \cdot \begin{bmatrix} \frac{1}{\sqrt{2}} & -\frac{1}{\sqrt{2}} \\ \frac{1}{\sqrt{2}} & \frac{1}{\sqrt{2}} \end{bmatrix} \tag{2.54}$$

$$= \begin{bmatrix} 0 & -1 \\ -1 & 0 \end{bmatrix}.$$

We can verify this result by checking how the operator acts on the new basis vectors:

$$\mathcal{M}(\vec{e}_1') = 0 \times \vec{e}_1' - 1 \times \vec{e}_2' \text{ and } \mathcal{M}(\vec{e}_2') = -1 \times \vec{e}_1' + 0 \times \vec{e}_2', \tag{2.55}$$

and remembering that the matrix for an operator has the images of the basis vectors as its columns, so that

$$\mathbf{M}' = \begin{bmatrix} 0 & -1 \\ -1 & 0 \end{bmatrix}. \tag{2.56}$$

$\square$

2.7   Scalar Product

For many applications, we need a way to define the length of a vector and the angle between two vectors. The scalar product[6] is a function that turns a pair of vectors $(\vec{x}, \vec{y})$ into a generally complex number, usually denoted as $\mathbf{x} \cdot \mathbf{y}$ or, more sloppily, $\mathbf{xy}$.[7]

To qualify as a **scalar product**, this function needs to fulfill certain formal conditions—namely,

$$\vec{y} \cdot \vec{x} = (\vec{x} \cdot \vec{y})^* \text{ (the complex conjugate).}$$

$$\vec{x} \cdot (\lambda \vec{y} + \mu \vec{z}) = \lambda \vec{x} \cdot \vec{y} + \mu \vec{x} \cdot \vec{z}. \tag{2.57}$$

$$\vec{x} \cdot \vec{x} \geq 0, \text{ with identity only if } \vec{x} = 0.$$

Note that the first two conditions imply that

$$(\lambda \vec{x}) \cdot \vec{y} = \lambda^* (\vec{x} \cdot \vec{y}), \tag{2.58}$$

whereas

$$\vec{x} \cdot (\lambda \vec{y}) = \lambda (\vec{x} \cdot \vec{y}). \tag{2.59}$$

---

6. Sometimes this is called an **inner product**.

7. Note that this rounds out a panoply of multiplication types that we have encountered already: number × number, number × vector, number × matrix, matrix × vector, matrix × matrix, and finally vector × vector.

This is an important detail because we will often encounter vectors with complex coefficients.

In any given vector space, there are infinitely many forms of a scalar product that meet these conditions. One chooses a particular form based on the needs of the analysis.

Example 2.12 Consider the vector space of arrows on a plane, together with an arbitrary basis. If the component representations of vectors $\vec{x}, \vec{y}$ are $\begin{bmatrix} x_1 \\ x_2 \end{bmatrix}, \begin{bmatrix} y_1 \\ y_1 \end{bmatrix}$, then the following is a valid scalar product:

$$\vec{x} \cdot \vec{y} = x_1 y_1 + x_2 y_2. \tag{2.60}$$

$\square$

Example 2.13 In the vector space of certain special functions, the following is a useful scalar product:

$$f \cdot g = \int_{-\infty}^{\infty} f^*(x) e^{-x^2} g(x) \, dx. \tag{2.61}$$

$\square$

Exercise 2.8 Show that this satisfies the conditions in equation (2.57). $\square$

2.7.1 Metric

Given a scalar product, we can introduce a **metric**, referring to a way of measuring distances and angles in the space:

- The **length of a vector** $\vec{x}$ is defined as

$$|\vec{x}| = \sqrt{\vec{x} \cdot \vec{x}}. \tag{2.62}$$

- The **angle** α between two vectors $\vec{x}$ and $\vec{y}$ is defined by

$$\cos \alpha = \frac{\vec{x} \cdot \vec{y}}{|\vec{x}||\vec{y}|}. \tag{2.63}$$

On the other hand, if you already have a way of measuring lengths and angles, that defines a scalar product. For example, for arrows on a plane, we are already comfortable measuring length with a ruler and angles with a protractor. That determines the corresponding scalar product between any two arrows because

$$\vec{x} \cdot \vec{y} = |\vec{x}||\vec{y}| \cos \alpha, \tag{2.64}$$

where $\alpha = \angle(\vec{x}, \vec{y})$ is the angle between the two arrows.

2.7.2 Orthonormal Basis Set

A vector $\vec{x}$ is called a **unit vector** or **normalized** if its length equals 1: $\sqrt{\vec{x} \cdot \vec{x}} = 1$. Two vectors $\vec{x}$ and $\vec{y}$ are called **orthogonal** if their scalar product is zero: $\vec{x} \cdot \vec{y} = 0$.

For any given choice of scalar product, there exists a set of basis vectors $\{\vec{e}_i\}$ that are **orthonormal**. This means that the vectors are all mutually orthogonal and each has a length of 1:

$$\vec{e}_i \cdot \vec{e}_j = \delta_{ij}.$$ (2.65)

In such an orthonormal basis, the scalar product is a very simple function of the vector components:

$$\begin{aligned}
\vec{x} \cdot \vec{y} &= \left(\sum_i x_i \vec{e}_i \right) \cdot \left(\sum_j y_j \vec{e}_j \right) \\
&= \sum_i \sum_j x_i^* y_j \delta_{ij} \\
&= \sum_i x_i^* y_i.
\end{aligned}$$ (2.66)

This algebraic form of the scalar product, given in equation (2.66) is often called the **dot product**.

Example 2.14 (Cartesian coordinates) In regular 3D space, where lengths are measured with a ruler and angles with a protractor, a very common way to represent vectors is the "Cartesian coordinate system," where the basis vectors are orthogonal vectors of length 1 that lie on the x-, y-, and z-axes. The coordinates (a_x, a_y, a_z) of a vector $\vec{a}$ are the coefficients in that basis. Why is that such a useful basis? Because in that basis, the conventional length of a vector is simply $\sqrt{a_x^2 + a_y^2 + a_z^2}$ and two vectors $\vec{a}$ and $\vec{b}$ are at a right angle if $a_x b_x + a_y b_y + a_z b_z = 0$.

Of course, one could choose a different basis, in which the basis vectors have different lengths or are nonorthogonal. The physical system will behave the same way; for example, the planets will continue to follow the same orbits around the Sun. But the laws that govern the motion of the planets, which rely on conventional lengths and angles, all take a more complicated form when expressed in that new basis. $\square$

There are, in fact, infinitely many of these orthonormal basis sets. Say that $\mathbf{S}$ is the transformation matrix from the old orthonormal basis set to a new basis set, such that

$$\vec{e}_j' = \sum_{i=0}^{n} S_{ij} \, \vec{e}_i.$$ (2.67)

Then the scalar product of the new basis vectors is

$$\begin{aligned}
\vec{e}_i' \cdot \vec{e}_j' &= \left(\sum_k S_{ki} \vec{e}_k \right) \cdot \left(\sum_l S_{lj} \vec{e}_l \right) \\
&= \sum_k \sum_l S_{ki}^* S_{lj} \delta_{kl} \\
&= \sum_k S_{ki}^* S_{kj}.
\end{aligned}$$ (2.68)

For the new basis to also be orthonormal, therefore, we need

$$\sum_k S_{ki}^{*} S_{kj} = \delta_{ij}, \tag{2.69}$$

or in matrix language,

$$\mathbf{S}^{\dagger} = \mathbf{S}^{-1}. \tag{2.70}$$

Here, $\mathbf{S}^{\dagger}$ is called the **adjoint** matrix of $\mathbf{S}$, defined as

$$S^{\dagger}_{ik} = S_{ki}^{*}. \tag{2.71}$$

A matrix that satisfies $\mathbf{S}^{\dagger} = \mathbf{S}^{-1}$ is called a **unitary matrix**. Any unitary basis transform preserves the orthonormality of basis vectors, so the scalar product in the new basis retains the simple form

$$\vec{x} \cdot \vec{y} = \sum_i x_i^* y_i = \sum_i x_i'^* y_i'. \tag{2.72}$$

Example 2.15 For the Cartesian coordinate system in example 2.14, any rigid **rotation of the axes is a unitary transform**. The basis vectors now point in different directions from before, but they remain orthonormal. So the scalar product follows the same simple dot product formula (2.66). $\square$

2.8 Special Matrix Properties

Sometimes we will encounter matrices with other special properties that arise from certain symmetries of the problem at hand. Here is a dictionary of related terms:

A matrix A is called ...	If ...
The **transpose** of $\mathbf{B}$, $\mathbf{A} = \mathbf{B}^{\top}$	$a_{ij} = b_{ji}$
The **complex conjugate** of $\mathbf{B}$, $\mathbf{A} = \mathbf{B}^{*}$	$a_{ij} = b_{ij}^{*}$
The **adjoint** of $\mathbf{B}$, $\mathbf{A} = \mathbf{B}^{\dagger} \equiv \mathbf{B}^{*\top}$	$a_{ij} = b_{ji}^{*}$
Real	$\mathbf{A}^{*} = \mathbf{A}$
Symmetric	$\mathbf{A}^{\top} = \mathbf{A}$
Hermitian or self-adjoint	$\mathbf{A}^{\dagger} = \mathbf{A}$
Unitary	$\mathbf{A}^{\dagger} = \mathbf{A}^{-1}$
Orthogonal	$\mathbf{A}^{\top} = \mathbf{A}^{-1}$
Normal	$\mathbf{A}^{\dagger}\mathbf{A} = \mathbf{A}\mathbf{A}^{\dagger}$
Idempotent	$\mathbf{A}\mathbf{A} = \mathbf{A}$

2.9 Eigenvalues and Eigenvectors

Most of the time, when an operator acts on a vector, the resulting image points in a different direction; for example, see figure 2.11. However, for every operator, there exists at least one special vector that gets imaged into a multiple of itself, such that

$$\mathcal{M}(\vec{x}) = \lambda \vec{x}. \tag{2.73}$$

In that case, $\vec{x}$ is called an **eigenvector** of $\mathcal{M}$, and λ is the corresponding **eigenvalue**.

Note that if $\vec{x}$ is an eigenvector of $\mathcal{M}$, then so is every multiple $c\vec{x}$ because by the linearity of operators,

$$\mathcal{M}(c\vec{x}) = c\mathcal{M}(\vec{x}) = c(\lambda\vec{x}) = \lambda(c\vec{x}). \tag{2.74}$$

Effectively, the eigenvector identifies a **direction** in the vector space.

Example 2.16 Take the reflection operator for arrows on the plane in figure 2.7. Here, the vector $\vec{e}_1$ gets imaged into itself, so it is an eigenvector with eigenvalue 1. The vector $\vec{e}_2$ gets imaged into its opposite, so it is again an eigenvector, but with eigenvalue -1. $\square$

Exercise 2.9 Take the projection operator for arrows on the plane in figure 2.8. What are its eigenvectors? $\square$

From these examples, it already appears that the eigenvectors and their eigenvalues serve to characterize an operator: the eigenvectors label certain special directions in space and the eigenvalues spell out what the operator does along those directions. Note that the set of eigenvalues is independent of whatever basis set is chosen to represent the operator. They are **invariant** properties of the operator, independent of the specific coordinate system you might be working in.

2.10 The Characteristic Equation

In any chosen basis set, the eigenvector condition expressed in equation (2.73) is

$$\mathbf{Mx} = \lambda\mathbf{x}, \tag{2.75}$$

which turns into

$$(\mathbf{M} - \lambda\mathbf{I})\,\mathbf{x} = \mathbf{0}. \tag{2.76}$$

As laid out in section 2.5.4, this has a nonzero solution $\mathbf{x}$ only if

$$\det(\mathbf{M} - \lambda\mathbf{I}) = 0. \tag{2.77}$$

This is called the **characteristic equation** for the matrix $\mathbf{M}$.[8] The left side is an nth order polynomial in λ, where n is the dimensionality of the space. If one knows the

---

8. Recall that the determinant is invariant under a basis transform; see section 2.5.3. So the characteristic equation is the same in any basis. It is another invariant aspect of an operator.

polynomial's roots λ_i, one can write the characteristic equation as

$$(\lambda - \lambda_1)^{k_1} \dots (\lambda - \lambda_m)^{k_m} = \prod_{i-1}^{m} (\lambda - \lambda_i)^{k_i} = 0. \tag{2.78}$$

Here, k_i is called the "multiplicity of the ith root λ_i," or equivalently, the "degeneracy of the eigenvalue λ_i." The multiplicities add up to the dimensionality of the space n:

$$\sum_{i=1}^{m} k_i = n. \tag{2.79}$$

Example 2.17 What are the eigenvalues of

$$\mathbf{M} = \begin{bmatrix} 0 & 1 \\ 1 & 0 \end{bmatrix}? \tag{2.80}$$

The characteristic equation is

$$\begin{aligned} \det(\mathbf{M} - \lambda \mathbf{I}) &= \det \begin{bmatrix} -\lambda & 1 \\ 1 & -\lambda \end{bmatrix} \\ &= (-\lambda) \times (-\lambda) - 1 \times 1 \\ &= (\lambda - 1)(\lambda + 1) = 0. \end{aligned} \tag{2.81}$$

So the roots are 1 and -1, each with a multiplicity of 1. $\square$

Exercise 2.10 Find the eigenvalues of the projection operator

$$\mathbf{M} = \begin{bmatrix} 1 & 0 \\ 0 & 0 \end{bmatrix}. \tag{2.82}$$

$\square$

Exercise 2.11 What are the eigenvalues and their multiplicities for the identity operator? $\square$

In two dimensions, $n = 2$, the characteristic equation is a quadratic. So for any matrix

$$\mathbf{M} = \begin{bmatrix} a & b \\ c & d \end{bmatrix}, \tag{2.83}$$

we can solve the eigenvalues using the quadratic formula as follows:

$$\begin{aligned} \det(\mathbf{M} - \lambda \mathbf{I}) &= \det \begin{bmatrix} a - \lambda & b \\ c & d - \lambda \end{bmatrix} \\ &= \lambda^2 - \lambda(a + d) + (ad - bc) = 0, \end{aligned} \tag{2.84}$$

which leads to

$$\lambda_\pm = -\frac{T}{2} \pm \sqrt{\left(\frac{T}{2}\right)^2 - D}, \tag{2.85}$$

where

$$T = a + d \tag{2.86}$$

is the **trace** of the matrix and

$$D = ad - bc \tag{2.87}$$

is its **determinant**.

In the special case where $\left(\frac{T}{2}\right)^2 = D$, the two solutions are identical, $\lambda_+ = \lambda_-$, so there is a single eigenvalue with a multiplicity of 2.

Throughout this book, we will use the analytical expression in (2.85) for eigenvalues in two dimensions because it helps us develop an analysis topic further. For higher dimensions $n > 2$, there is little use in trying an analytical solution, and you will mostly just compute the eigenvalues numerically using your favorite scientific programming package.

2.10.1 Eigenvectors

Once an eigenvalue λ is found, one can plug it into the equation

$$(\mathbf{M} - \lambda \mathbf{I})\,\mathbf{x} = 0 \tag{2.88}$$

and solve that for the eigenvector $\mathbf{x}$ using standard algebra.

The eigenvectors have the following important properties:

- Each eigenvalue has at least one associated eigenvector.
- If an eigenvalue has multiplicity k, then it will have at most k associated eigenvectors.
- Eigenvectors for different eigenvalues are linearly independent.

Example 2.18 Returning to example 2.17,

$$\mathbf{M} = \begin{bmatrix} 0 & 1 \\ 1 & 0 \end{bmatrix} \tag{2.89}$$

with eigenvalues $\lambda = 1$ and -1.

For $\lambda = +1$, the eigenvector must satisfy

$$(\mathbf{M} - \mathbf{I})\,\mathbf{x} = \begin{bmatrix} -1 & 1 \\ 1 & -1 \end{bmatrix} \mathbf{x} = 0. \tag{2.90}$$

This is solved by

$$\mathbf{x} = \begin{bmatrix} 1 \\ 1 \end{bmatrix} \tag{2.91}$$

or any multiple of that vector.

For $\lambda = -1$, we get

$$(\mathbf{M} + \mathbf{I})\,\mathbf{x} = \begin{bmatrix} 1 & 1 \\ 1 & 1 \end{bmatrix} \mathbf{x} = 0, \tag{2.92}$$

which leads to

$$\mathbf{x} = \begin{bmatrix} 1 \\ -1 \end{bmatrix} \tag{2.93}$$

or any multiple thereof. $\square$

Exercise 2.12 Following on example 2.18, note that we have come across an operator with eigenvalues 1 and -1 before, namely, the reflection operator of example 2.16. This makes us suspect that the present matrix $\mathbf{M}$ represents a reflection operator as well. Show that this is the case, and determine the line of reflection. $\square$

2.11 Diagonalizing a Matrix

If an operator has n linearly independent eigenvectors, then one can use them as a powerful new basis set $\{\vec{e}_1, \ldots, \vec{e}_n\}$. In that basis, the matrix representing the operator is **diagonal**:

$$\mathbf{A} = \begin{bmatrix} \lambda_1 & 0 & \ldots & 0 \\ 0 & \lambda_2 & \ldots & 0 \\ \vdots & \vdots & \ddots & \vdots \\ 0 & 0 & \ldots & \lambda_n \end{bmatrix}. \tag{2.94}$$

The reason for this is easy to see: Recall that the columns of the matrix are the images of the basis vectors. But the basis vectors were chosen such that $\mathcal{A}(\vec{e}_i) = \lambda_i \vec{e}_i$. Hence, the column vectors have coefficients only along the diagonal of the matrix.

This basis is called the **eigenbasis** of the operator. Working in this basis can offer huge advantages. Suppose, for example, that you have to apply this operator repeatedly to many vectors in the space. In the eigenbasis, computing $\mathbf{A}\mathbf{x}$ requires only n multiplications, whereas in any other basis, that takes n^2 multiplications and additions. Suppose now that your vectors have dimension 10^6, perhaps because they are megapixel images. Suddenly the difference between n and n^2 distinguishes an easy project from an impossible one.

Diagonalization is a powerful tool that we will encounter many times. As a rule, if your problem is defined by a square matrix of some kind, it almost always pays to transform your coordinates into a basis where the matrix in question is diagonal.

Extension: In general, the operator in question may not have n independent eigenvectors, but one can at least partially diagonalize its matrix to the following form:

$$
A = \begin{bmatrix}
\lambda_1 & \times & \times & 0 & \cdots & 0 & 0 \\
0 & \lambda_1 & \times & 0 & \cdots & 0 & 0 \\
0 & 0 & \lambda_1 & 0 & \cdots & 0 & 0 \\
\hline
0 & 0 & 0 & \lambda_2 & \cdots & 0 & 0 \\
\vdots & \vdots & \vdots & \vdots & \ddots & \vdots & \vdots \\
\hline
0 & 0 & 0 & 0 & \cdots & \lambda_m & \times \\
0 & 0 & 0 & 0 & \cdots & 0 & \lambda_m
\end{bmatrix}. \tag{2.95}
$$

That matrix is zero except for a number of blocks along the diagonal, one for each eigenvalue λ_i. Each block has the size of the eigenvalue's multiplicity k_i. Its values are λ_i along the diagonal and zero below the diagonal. $\square$

2.11.1 Hermitian Operators

A matrix is called **Hermitian** if

$$
A^\dagger = A. \tag{2.96}
$$

Note that **real symmetric** matrices are a special instance of this (see section 2.8). Some powerful relationships apply to Hermitian operators, such as the following:

- A Hermitian matrix has all real eigenvalues: $\lambda_i \in \mathbb{R}$.
- A Hermitian matrix has n mutually orthogonal eigenvectors: $e_i e_j = \delta_{ij}$.

This means that the eigenvectors of A are guaranteed to form a basis, and—even more useful—that basis can be made orthonormal, which preserves the simple form of the scalar product (see section 2.7.2).

2.11.2 Other Useful Results Using Eigensystems

- **Commuting matrices:** If two matrices A and B are diagonalizable (i.e. each has n independent eigenvectors), and they commute (i.e. $AB = BA$), then they share all the same eigenvectors. So they can be diagonalized in the same basis.
- **Positive matrix:** If a matrix has all positive elements, then its largest eigenvalue is real and positive, and so are all the components of the corresponding eigenvector.
- **Product of the eigenvalues:** The product of all the eigenvalues equals the determinant of the matrix:

$$
\prod_{i=1}^{n} \lambda_i = \det A. \tag{2.97}
$$

In this product, an eigenvalue λ_i with multiplicity k_i appears k_i times.

- **Sum of the eigenvalues:** The sum of all the eigenvalues equals the trace of the matrix:

$$
\sum_{i=1}^{n} \lambda_i = \operatorname{Tr} A, \tag{2.98}
$$

which is defined as the sum of the diagonal elements:

$$\mathrm{Tr}\,\mathbf{A} \equiv \sum_{i=1}^{n} a_{ii}. \tag{2.99}$$

Note that the eigenvalues of an operator are independent of the basis set. So the trace, like the determinant, is an invariant property of an operator.

- **Powers of a matrix:** If matrix A has eigenvalue λ with eigenvector $\mathbf{e}$, then matrix $\mathbf{A}^k$ has eigenvalue λ^k with that same eigenvector $\mathbf{e}$. Suppose that A can be diagonalized by a basis transform S:

$$\mathbf{S}^{-1}\mathbf{A}\mathbf{S} = \mathbf{D} = \begin{bmatrix} \lambda_1 & 0 & \cdots & 0 \\ 0 & \lambda_2 & \cdots & 0 \\ \vdots & \vdots & \ddots & \vdots \\ 0 & 0 & 0 & \lambda_n \end{bmatrix}, \tag{2.100}$$

where D is a diagonal matrix containing the eigenvalues of A along the diagonal. Because $\mathbf{A}^k$ has all the same eigenvectors, that same basis transform also diagonalizes $\mathbf{A}^k$:

$$\mathbf{S}^{-1}\mathbf{A}^k\mathbf{S} = \mathbf{D}^k = \begin{bmatrix} \lambda_1^k & 0 & \cdots & 0 \\ 0 & \lambda_2^k & \cdots & 0 \\ \vdots & \vdots & \ddots & \vdots \\ 0 & 0 & 0 & \lambda_n^k \end{bmatrix}. \tag{2.101}$$

This leads to a simple expression for the powers of matrix $\mathbf{A}^k$:

$$\mathbf{A}^k = \mathbf{S}\mathbf{D}^k\mathbf{S}^{-1}. \tag{2.102}$$

- **Function of a matrix:** By extension, one can define a function of a diagonizable matrix. Suppose the function $f(x)$ has a Taylor series

$$f(x) = \sum_{k=0}^{\infty} c_k x^k. \tag{2.103}$$

Then one defines the corresponding matrix function as follows:

$$f(\mathbf{A}) = \sum_k c_k \mathbf{A}^k. \tag{2.104}$$

Using the diagonalizing transform S and equation (2.102), one gets

$$f(\mathbf{A}) = \mathbf{S} \cdot f(\mathbf{D}) \cdot \mathbf{S}^{-1}. \tag{2.105}$$

The matrix D has the eigenvalues λ_i along the diagonal. Therefore, $f(\mathbf{D})$ has the values $f(\lambda_i)$ on the diagonal. Finally, this yields the function $f(\mathbf{A})$ as follows:

$$f(\mathbf{A}) = \mathbf{S} \cdot \begin{bmatrix} f(\lambda_1) & \cdots & 0 \\ \vdots & \ddots & \vdots \\ 0 & \cdots & f(\lambda_n) \end{bmatrix} \cdot \mathbf{S}^{-1}. \tag{2.106}$$

Example 2.19 Consider vector trajectory $\mathbf{x}(t)$, governed by the linear first-order differential equation

$$\frac{\mathrm{d}}{\mathrm{d}t}\mathbf{x} = \mathbf{A}\mathbf{x}. \tag{2.107}$$

This is the multivariable analog of the single-variable rabbit equation (1.34). If $\mathbf{A}$ is diagonalizable, then one can immediately write the solution as

$$\mathbf{x}(t) = \mathbf{x}(0)\mathrm{e}^{\mathbf{A}t}. \tag{2.108}$$

The dynamics of this solution are governed by the eigenvalues $\{\lambda_i\}$ of $\mathbf{A}$. It will involve some additive combination of exponential growth (if $\lambda > 0$), exponential decay ($\lambda < 0$), and rotation (λ complex). $\square$

3.1 Linear Systems Analysis

In the preceding chapters, we have occasionally considered examples from the vector space of functions, where the vectors are continuous functions of time or position. We will focus now on this situation because it encompasses many applications.

3.1.1 What Is a Linear System?

A linear system $\mathcal{L}$ is an operator that turns an input function into an output function, doing so in a linear way. Specifically, if f_1 and f_2 are two possible inputs with the respective outputs $\mathcal{L}(f_1)$ and $\mathcal{L}(f_2)$, the output of the system to the combined input $f_1 + f_2$ will be the sum of the individual outputs:

$$f_1(x) + f_2(x) \quad \xrightarrow{\text{linear system, } \mathcal{L}} \quad \mathcal{L}\left[f_1(x) + f_2(x)\right] = \mathcal{L}\left[f_1(x)\right] + \mathcal{L}\left[f_2(x)\right]. \tag{3.1}$$

An example of a linear system operating in space is a fluorescence microscope. Here, the input function is the distribution of fluorescence intensity within the sample. The output function is the distribution of intensity in the (x, y) plane of the camera. Much intricate microscope design goes into devising the transformation between input and output. However, the microscope is guaranteed to meet the condition of linearity: twice the amount of fluorescence in the sample will produce twice the intensity in the image. An example of a linear system operating in the time domain is an acoustical environment that transforms a sound waveform emitted by the source to another sound heard at the receiver. Electrical filters in an amplifier and image-sharpening tools in a graphics package are further examples of linear systems.

Note that linearity is a rather special condition that certainly is not met by all systems. However, as illustrated here, for many devices or natural processes, one can assert linear behavior from a basic consideration of the system's function, at least over some range of its operating conditions.

3.1.2 The Superposition Principle

The linearity condition expressed in equation (3.1) offers a powerful approach to linear systems analysis. Suppose that we need to predict a system's response to a novel input. If we can express the input function f as a superposition of the more elementary inputs f_1 and f_2, whose outputs are known individually, the desired solution follows immediately as the sum of those elementary outputs. More generally, we will encounter many instances where a problem can be decomposed into constituent subproblems, which are solved individually and then superposed to a solution of the larger problem. This approach is known as the "principle of superposition." One might say that the mantra of linear systems analysis is: "The whole is the sum of its parts."

3.1.3 Translational Invariance

A linear system is **translation-invariant** if its behavior is the same everywhere in the domain. For example, in the case of the microscope mentioned in section 3.1.1, this means that if we take an image of a sample and then shift it in the camera plane, we should get the same result as if we shifted the original sample on the stage and then took its image. Furthermore, this should be true for all possible shifts. In that case, the optical system would be called "invariant" under spatial translations. If the independent variable is time and the operation of the linear system does not depend on absolute time, this invariance under time translation is also called **stationarity**.

Formally, let $\mathcal{L}$ denote the linear system and let $\mathcal{S}$ be the operation that shifts the input or output functions. The statement that the image of a shifted input is the same as the shifted image of the original input can be written mathematically as

$$\mathcal{L}\left(\mathcal{S}\left(\text{input}\right)\right) = \mathcal{S}\left(\mathcal{L}\left(\text{input}\right)\right), \tag{3.2}$$

or schematically,

$$\mathcal{L}\mathcal{S} = \mathcal{S}\mathcal{L}. \tag{3.3}$$

So a linear operator is translation invariant if it commutes with the shift operator (see equation 2.20).

Example 3.1 Consider the simple scaling operator

$$\mathcal{L}\left[f(x)\right] = \lambda f(x) \tag{3.4}$$

and the shift operator

$$\mathcal{S}\left[f(x)\right] = f(x - a). \tag{3.5}$$

Clearly,

$$\begin{aligned}
\mathcal{L}\left[\mathcal{S}\left[f(x)\right]\right] &= \mathcal{L}\left[f(x - a)\right] \\
&= \lambda f(x - a) \\
&= \mathcal{S}\left[\mathcal{L}\left[f(x)\right]\right].
\end{aligned} \tag{3.6}$$

$\square$

Note that this is not always the case. For example, a microscope has a limited aperture, and shifting the sample beyond the edge obviously produces a different image. Even within the aperture, optical imaging follows slightly different rules for off-axis rays than right on the optical axis. Engineers do their best to minimize this, and for many applications, we can regard the microscope as fully translation-invariant in the image plane. Note also that a system may be translation-invariant but nonlinear:

For example, this is the case for certain electronic or software filters used in data processing.

3.1.4 Impulse Response

We can begin to understand how a linear system functions by probing it with an input consisting of an impulse. An **impulse** is a function that is very localized in space or time. In the case of a microscope, one might use a single, tiny light-emitting particle. In the case of an acoustic system, one might use a very short click. Mathematically, we will use the idealized impulse represented by the delta function: an infinitely brief and infinitely tall pulse with an integral of 1 (see section 1.7). This function is denoted by $\delta(x)$ for an impulse in space, or $\delta(t)$ for an impulse in time. Naturally, all the results that we will encounter apply equally to linear systems that use some other independent variable.

When presented with an impulse $\delta(t)$ as input, the linear system generates an output that is aptly named the **impulse response function**, usually denoted as $h(t)$:

$$\delta(t) \xrightarrow{\text{linear system } \mathcal{L}} h(t). \tag{3.7}$$

When considering translation-invariant linear systems, the same impulse response function specifies how the system acts everywhere along the independent variable. An impulse $\delta(t - t_1)$ delivered at time t_1 will yield the output $h(t - t_1)$, regardless of the offset t_1. In what follows, we will focus on linear systems that are translation-invariant without spelling this out every time.

It is a fundamental result that **any translation-invariant linear system is completely characterized by its impulse response function**. In the next section, we will see why this is so.

3.1.5 Convolutions

Suppose that we want to predict the response of a linear system to an arbitrary input $f(t)$, and we already know its impulse response $h(t)$. Using the superposition principle, we can proceed as follows:

- Decompose the input function $f(t)$ into a sum of impulses.
- Identify the response to each impulse.
- Sum up all these responses to get the total response.

For the first step, we begin by approximating the function $f(t)$ as a sum of square pulses, as illustrated in figure 3.1. As one makes these pulses narrower, this approximation eventually becomes exact. In the limit of zero width, each pulse is a delta function (see section 1.7) and $f(t)$ becomes an integral over delta functions:

$$f(t) = \int_{-\infty}^{\infty} f(t')\delta(t - t')\,dt'. \tag{3.8}$$

We know that a translation-invariant linear system transforms each of the input pulses into an impulse response according to

$$\delta(t - t') \xrightarrow{\text{linear system } \mathcal{L}} h(t - t'). \tag{3.9}$$

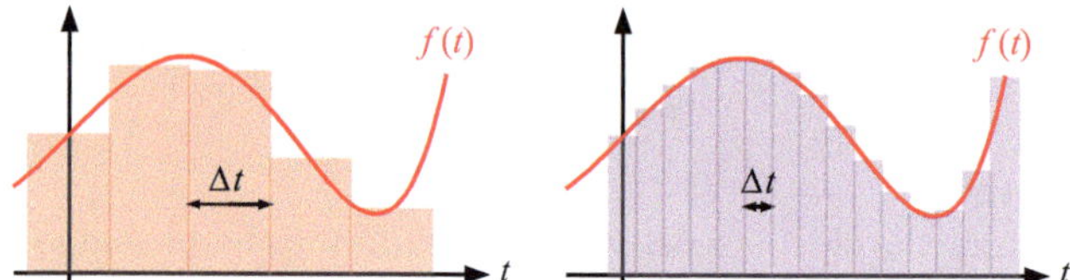

Figure 3.1
A function approximated by square pulses. The function $f(t)$ can be approximated by the rectangular bars of width Δt. As the width decreases, the approximation improves. In the limit of infinitesimally thin rectangles, the approximation becomes exact.

Furthermore, the linearity condition implies that a weighted sum of such impulses will transform into a similar weighted sum of impulse responses. This leads to the desired result:

$$f(t) = \int f(t')\delta(t-t')\,dt' \xrightarrow{\text{linear system }\mathcal{L}} g(t) = \int f(t')h(t-t')\,dt'. \tag{3.10}$$

Integrals of the type encountered in equation (3.10) arise so frequently that they are given their own name: **convolutions**. The convolution of two functions f and h is often written as $f * h$ and defined as

$$(f*h)(t) = \int_{t'=-\infty}^{\infty} f(t')h(t-t')\,dt'. \tag{3.11}$$

To summarize, **the output of a translation-invariant linear system is given by the convolution of the input with the impulse response**. The steps that were followed to reach this result are illustrated again in figure 3.2.

3.1.6 Eigenfunctions of a Linear System

We have seen in previous chapters that a linear operator is fully characterized by its eigenvalues and eigenvectors, and problems can often be solved easily by transforming into the eigenbasis of the relevant operator. Now that we are considering linear systems, it is therefore natural to ask: What are the eigenvectors of a linear system? Because a linear system acts on functions, we are looking for special functions that satisfy the eigenvector condition

$$\mathcal{L}[f(x)] = \lambda f(x). \tag{3.12}$$

These are often called **eigenfunctions**.

Here, we encounter one of the most far-reaching ideas in applied mathematics: **all translation-invariant linear systems share a common set of eigenfunctions, and they are the complex exponentials**,

$$f(t) = e^{i\omega t}, \tag{3.13}$$

also known as **complex phasors** (section 1.6.4). To verify this, let us consider a general linear system $\mathcal{L}$ with the impulse response function $h(t)$. Its action on the complex

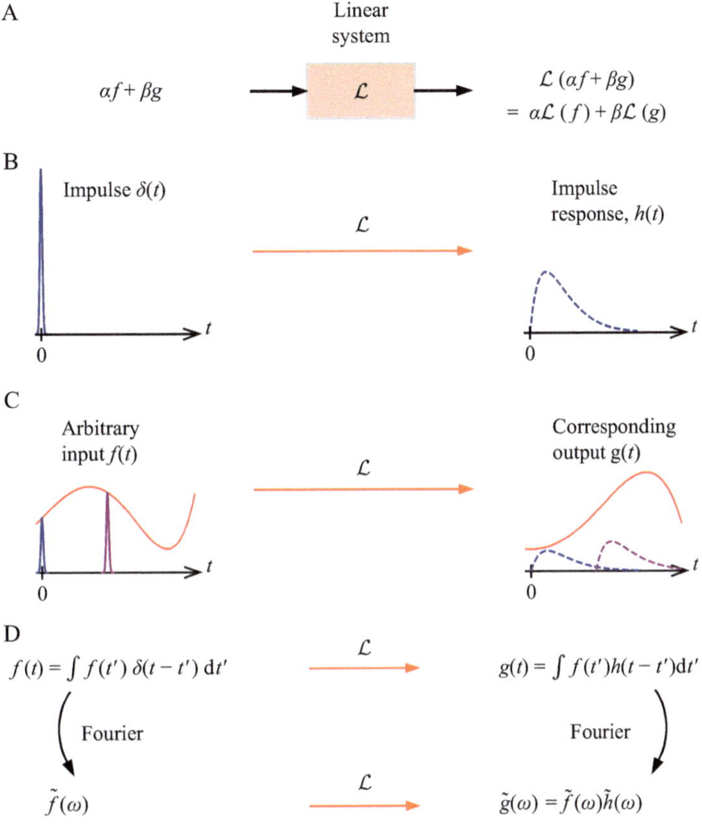

Figure 3.2
Concepts in linear systems analysis. (A) A linear system mapping input (left) to output (right). (B) An impulse input is mapped to the impulse response. (C) The superposition principle guarantees that one can decompose the input into pulses, process the pulses separately, and then superpose the resulting impulse responses in the output. (D) This yields the convolution of the input with the impulse response. In the Fourier domain, this amounts to a multiplication by the transfer function.

exponential with frequency ω is given by the convolution

$$\mathcal{L}[e^{i\omega t}] = \int_{-\infty}^{\infty} e^{i\omega t'} h(t - t')\, dt'$$

$$= \int_{-\infty}^{\infty} e^{i\omega(t-u)} h(u)\, du \tag{3.14}$$

$$= e^{i\omega t} \int_{-\infty}^{\infty} e^{-i\omega u} h(u)\, du.$$

The last line of equation (3.14) shows that $\mathcal{L}$ transforms the function $e^{i\omega t}$ into a multiple of itself, and this is true for any frequency ω. The associated eigenvalue depends on ω and is given by

$$\hat{h}(\omega) = \int_{-\infty}^{\infty} e^{-i\omega u} h(u)\, du. \tag{3.15}$$

This result has far-reaching implications. It means that a sinusoid wave, when sent through a linear system, will turn into a sinusoid of the same frequency. The system may only alter the amplitude or the phase of the sinusoid. Measuring the gain and the phase shift experienced by a sinusoid at frequency ω will yield the modulus and phase of the complex number $\hat{h}(\omega)$. This explains why sine waves are so beloved as test signals: for probing electric circuits, for testing the behavior of an acoustical system, or even in psychophysical experiments on the human visual system.

With this in mind, and considering our prior experience with the use of eigenbases, one expects to gain a powerful advantage by transforming any given function into this basis of eigenfunctions. And because that basis is the same for every possible linear system, this basis transform is used ubiquitously. It is called the **Fourier transform**, and it is the subject of the next section.

3.2 Fourier Transforms

3.2.1 Complex Exponentials and the Fourier Transform

We now know that the complex exponentials:

$$f(x) = e^{ikx} \tag{3.16}$$

are eigenfunctions of any linear system that operates on a function of x. Furthermore, one can show that the complex exponentials form a complete basis for the space of functions. This means that any reasonable function $f(x)$ can be written as a linear superposition of complex exponentials:

$$f(x) = \frac{1}{2\pi} \int_{-\infty}^{\infty} \hat{f}(k) e^{ikx} \, dk. \tag{3.17}$$

Here, $\hat{f}(k)$ denotes the weight assigned to the complex exponential with frequency k, the integral sums over all contributions from different frequencies, and the factor $\frac{1}{2\pi}$ is added merely as a matter of convention. The function $\hat{f}(k)$ is known as the **Fourier transform** of $f(x)$ and is given by

$$\hat{f}(k) = \int_{-\infty}^{\infty} f(x) e^{-ikx} \, dx. \tag{3.18}$$

The reconstruction of the function $f(x)$ from its Fourier transform $\hat{f}(k)$ by equation (3.17) is called the **inverse Fourier transform**.[1] You can verify these relations by inserting equation (3.18) back into equation (3.17); see exercise 5.15.

---

1. Note that some sources choose an alternative convention for the factor $1/2\pi$, such as by splitting it into two factors $1/\sqrt{2\pi}$, which appear in alternative forms of equations (3.17) and (3.18). There is even some variation in the sign of the exponent: for example, the book *Numerical Recipes* and the scientific programming package Igor use e^{-ikx} in equation (3.17) and e^{ikx} in equation (3.18). As a result, the command FFT produces different answers in MATLAB and Igor. User beware!

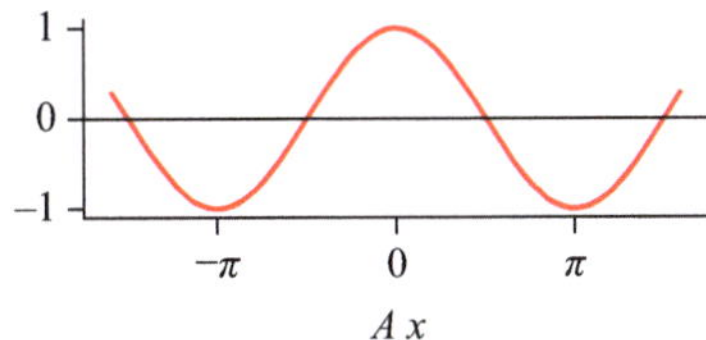

Figure 3.3
A cosine function.

3.2.1.1 Conjugate variables The Fourier transform as defined by equation (3.18) maps a function $f(x)$ into another function $\hat{f}(k)$. The variable k is called the **conjugate variable** to x. If x refers to a spatial distance, then k is called a "spatial frequency." Similarly, if $f(t)$ is a function of time t, then the Fourier transform $\hat{f}(\omega)$ is a function of the temporal frequency ω. Again, t and ω are conjugate variables.

3.2.2 Examples of Fourier Transforms
We will start to familiarize ourselves with the Fourier transform by computing some simple examples.

Example 3.1 Cosine

What is the Fourier transform of

$$f(x) = \cos(Ax)? \tag{3.19}$$

Recall the relation

$$\cos(Ax) = \frac{1}{2}\left(e^{iAx} + e^{-iAx}\right). \tag{3.20}$$

So the Fourier transform is

$$\begin{aligned}
\hat{f}(k) &= \int_{-\infty}^{\infty} f(x)e^{-ikx}\mathrm{d}x \\
&= \int_{-\infty}^{\infty} \frac{1}{2}\left(e^{iAx} + e^{-iAx}\right)e^{-ikx}\mathrm{d}x \\
&= \pi\left(\delta(A-k) + \delta(A+k)\right),
\end{aligned} \tag{3.21}$$

where in the last line, we again used equation (1.62) for the delta function. The two delta functions show that the cosine contains complex exponentials at only two frequencies, A and $-A$.

Example 3.2 Exponential pulse

Consider an exponentially decaying pulse (figure 3.4)

$$f(t) = \begin{cases} 0 & t < 0 \\ \frac{1}{\tau}e^{-\frac{t}{\tau}} & t \geq 0. \end{cases} \tag{3.22}$$

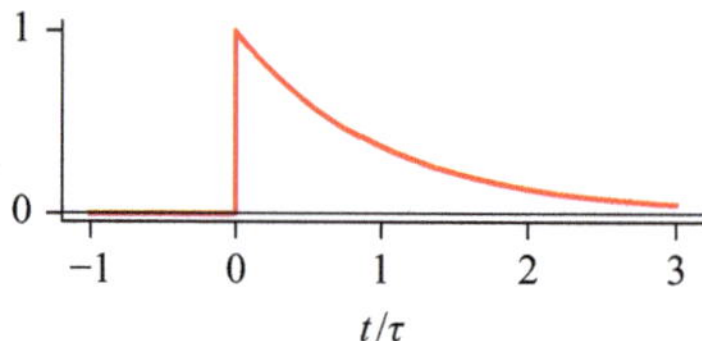

Figure 3.4
An exponential function.

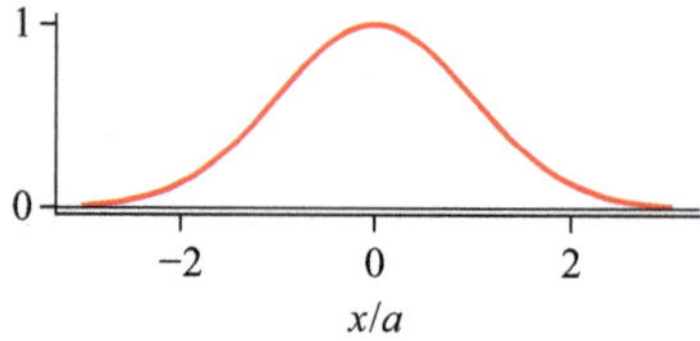

Figure 3.5
A Gaussian function.

What is its Fourier transform?

Again, one simply inserts the function in question into the definition of the Fourier transform given in equation (3.18) to obtain the integral

$$
\begin{aligned}
\hat{f}(\omega) &= \int_{-\infty}^{\infty} f(t)\mathrm{e}^{-i\omega t}\,\mathrm{d}t \\
&= \int_{0}^{\infty} \frac{1}{\tau}\mathrm{e}^{-\frac{t}{\tau}}\mathrm{e}^{-i\omega t}\,\mathrm{d}t \\
&= \left[-\frac{\frac{1}{\tau}\,\mathrm{e}^{-(\frac{1}{\tau}+i\omega)t}}{\frac{1}{\tau}+i\omega} \right]_{t=0}^{\infty} \\
&= \frac{1}{1+i\omega\tau}.
\end{aligned}
\tag{3.23}
$$

Example 3.3 Gaussian

What is the Fourier transform of a Gaussian function (figure 3.5)

$$
f(x) = \mathrm{e}^{-x^2/2a^2}\ ?
\tag{3.24}
$$

Again, starting with the definition of the Fourier transform, we obtain

$$
\begin{aligned}
\hat{f}(k) &= \int_{-\infty}^{\infty} f(x)\mathrm{e}^{-ikx}\,\mathrm{d}x \\
&= \int_{-\infty}^{\infty} \mathrm{e}^{-x^2/2a^2}\mathrm{e}^{-ikx}\,\mathrm{d}x \\
&= \int_{-\infty}^{\infty} \mathrm{e}^{-\frac{\left(x+ika^2\right)^2}{2a^2}}\mathrm{e}^{\frac{-k^2a^2}{2}}\,\mathrm{d}x \\
&= \sqrt{2\pi}\,a\,\mathrm{e}^{-k^2a^2/2}.
\end{aligned}
\tag{3.25}
$$

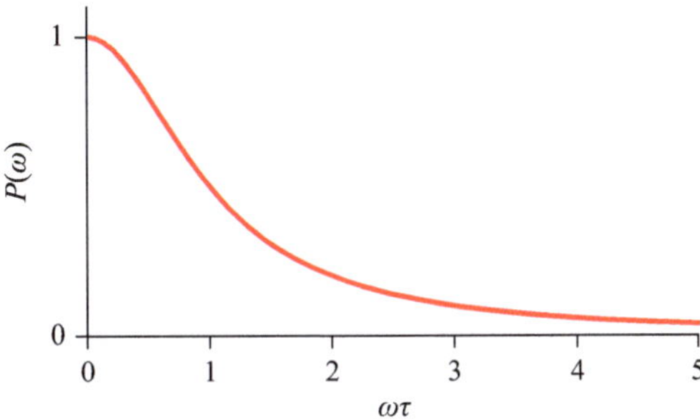

Figure 3.6
Power spectrum of an exponential pulse.

Here, the last step uses a common contour integral in the complex plane. In fact, contour integration is a useful tool for solving Fourier transform integrals. In many cases, however, one can also proceed with the regular tools of real analysis. See, for example, exercise 5.20.

Note that **the Fourier transform of a Gaussian is also a Gaussian function**, but in the domain of frequency. If the width of the original function $f(x)$ is a, the width of the Fourier transform $\hat{f}(k)$ is $1/a$. So a broad-peaked function will have a narrow-peaked Fourier transform and vice versa. This is a general result that extends beyond the Gaussian shape. It serves to explain many phenomena where precision in one domain is inversely related to the precision in the conjugate domain. Examples are the uncertainty principle in quantum mechanics and the limits of resolution in optical instruments.

3.2.3   The Power Spectrum of a Signal

In many instances, the square of a function is proportional to the power or energy of the signal represented by that function. For example, if $v(t)$ denotes the velocity of a particle of mass m, then $\frac{1}{2}mv^2$ is its kinetic energy. If $I(t)$ is the current flowing through resistor R, then RI^2 is the electric power dissipated. On that basis, one defines the total power P_{total} in a signal $f(t)$ as the integral of its squared amplitude:

$$P_{\text{total}} = \int_{-\infty}^{\infty} |f(t)|^2 \, dt. \tag{3.26}$$

Parseval's theorem states that this integral can also be performed in frequency space, with the same result:

$$P_{\text{total}} = \frac{1}{2\pi} \int_{-\infty}^{\infty} |\hat{f}(\omega)|^2 \, d\omega. \tag{3.27}$$

The function

$$P(\omega) = |\hat{f}(\omega)|^2 \tag{3.28}$$

is called the **power spectrum** of the signal. It tells us how the power is distributed across different frequencies ω.

Example 3.4 Continuing example of section 3.2, what is the power spectrum of an exponentially decaying pulse

$$f(t) = \frac{1}{\tau}\,\mathrm{e}^{-\frac{t}{\tau}}, t \geq 0? \tag{3.29}$$

Using the result in equation (3.23) for the Fourier transform of this function, we find

$$P(\omega) = |\hat{f}(\omega)|^2 = \frac{1}{1 + \omega^2\tau^2}. \tag{3.30}$$

Notice that at low frequencies, as $\omega \to 0$, $P(\omega) \to 1$. On the other hand, at high frequencies $\omega \to \infty$, $P(\omega)$ declines to zero with a dependence $\sim 1/\omega^2$. The function in equation (3.29) and its power spectrum in equation (3.30) appear in many practical applications.

3.2.4 The Fourier Transform Simplifies Many Operations

We will now consider various operations that can be performed on a function $f(t)$: differentiation, integration, and translation. We will see that in all these cases, the Fourier transform $\hat{f}(\omega)$ changes in a very simple way.

3.2.4.1 Differentiation

When one takes the derivative of a function, the effect on its Fourier transform is a simple multiplication by the conjugate variable:

$$\frac{\mathrm{d}f}{\mathrm{d}t} \xrightarrow{F.T.} i\omega\hat{f}(\omega). \tag{3.31}$$

To check this claim we start again with the definition of the Fourier transform,

$$\begin{aligned}
\widehat{\frac{\mathrm{d}f}{\mathrm{d}t}} &= \int_{-\infty}^{\infty} \left(\frac{\mathrm{d}f}{\mathrm{d}t}\right) \mathrm{e}^{-i\omega t}\,\mathrm{d}t \\
&= \left[f(t)\mathrm{e}^{-i\omega t}\right]_{-\infty}^{\infty} + i\omega \int_{-\infty}^{\infty} f(t)\mathrm{e}^{-i\omega t}\,\mathrm{d}t \\
&= i\omega\hat{f}(\omega),
\end{aligned} \tag{3.32}$$

where in the last step, we have assumed that $f(t)$ vanishes at $t = \pm\infty$.[2]

Higher derivatives in the time domain simply result in more factors of $i\omega$ in the conjugate frequency domain. Differential equations in the time domain become algebraic equations in the frequency domain (see exercise 5.18).

3.2.4.2 Integration

When one integrates a function, the effect on its Fourier transform is simple division by the conjugate variable:

---

2. This is a hint that a mathematically correct treatment of the Fourier transform involves certain restrictions on the space of functions. In practical scientific computing, of course, most functions are known over only a finite time interval, and therefore they can be set to zero at $t = \pm\infty$.

$$\int^{t} f(t')\,dt' \xrightarrow{F.T.} \frac{1}{i\omega}\hat{f}(\omega) + C\delta(\omega), \tag{3.33}$$

where C is an arbitrary integration constant.

We can check this quickly by applying the differentiation rule in equation (3.31) to both sides of equation (3.33):

$$f(t) \xrightarrow{F.T.} \hat{f}(\omega) + i\omega C\delta(\omega) = \hat{f}(\omega), \tag{3.34}$$

because $\omega\delta(\omega) = 0$ everywhere.

3.2.4.3 Time translation Consider shifting a function $f(t)$ to the left by a time interval s. Again, the effect on the Fourier transform is rather simple:

$$f(t+s) \xrightarrow{F.T.} e^{i\omega s}\hat{f}(\omega). \tag{3.35}$$

So translation of a function by a time interval s results in a change in phase of the Fourier transform by the exponential factor $e^{i\omega s}$. The proof is left for exercise 5.12.

3.2.4.4 Convolution Recall that the convolution of two functions $f(t)$ and $h(t)$ is defined by

$$g(t) = (f * h)(t) = \int_{-\infty}^{+\infty} f(t')h(t-t')\,dt'. \tag{3.36}$$

We learned in section 3.1.6 that the complex exponentials are eigenfunctions of the convolution operator. As a consequence, convolutions simplify dramatically in the frequency domain. **The Fourier transform of a convolution is the product of the two Fourier transforms:**

$$\int_{-\infty}^{+\infty} f(t')h(t-t')\,dt' \xrightarrow{F.T.} \hat{f}(\omega)\hat{h}(\omega). \tag{3.37}$$

A quick check that this is indeed the case gives the following result:

$$\begin{aligned}
\widehat{(f * h)}(t) &= \int_{-\infty}^{+\infty}\int_{-\infty}^{+\infty} f(t-s)h(s)e^{-i\omega t}\,ds\,dt \\
&= \int_{-\infty}^{+\infty}\int_{-\infty}^{+\infty} f(t-s)h(s)e^{-i\omega(t-s)}e^{-i\omega s}\,ds\,dt \\
&= \int_{-\infty}^{+\infty}\int_{-\infty}^{+\infty} f(u)h(s)e^{-i\omega u}e^{-i\omega s}\,ds\,du \\
&= \hat{f}(\omega)\,\hat{h}(\omega).
\end{aligned} \tag{3.38}$$

This result, known as the **convolution theorem**, is of central importance and underlies many practical uses of the Fourier transform. Convolutions are ubiquitous in scientific computing, and because it is numerically much easier to multiply two functions than to integrate them, the convolution is almost always performed in the frequency domain. We will comment on this further in section 3.2.7, when discussing the discrete Fourier transform (DFT).

3.2.5 Linear Systems in the Frequency Domain

In section 3.1.3, we discussed translationally invariant linear systems that turn an input function into an output function:

$$f(t) \xrightarrow{\mathcal{L}} g(t). \tag{3.39}$$

We saw that the output $g(t)$ can be found as the convolution of the input $f(t)$ with the system's impulse response $h(t)$:

$$g(t) = (f * h)(t) = \int_{-\infty}^{+\infty} f(t')h(t-t')\,\mathrm{d}t'. \tag{3.40}$$

Now we also know that the convolution of two functions becomes a simple multiplication in the frequency domain. We can therefore take the Fourier transform of both sides of equation (3.40) and obtain

$$\hat{g}(\omega) = \hat{f}(\omega)\hat{h}(\omega). \tag{3.41}$$

The function $\hat{h}(\omega)$ is the Fourier transform of the impulse response function, and it is called the **transfer function** of the linear system. **In the frequency domain, the output of a linear system is simply the input multiplied by the transfer function.**

How does a linear system modify the power spectrum of the input? Recall that the power spectrum $P(\omega)$ of a function f is defined as

$$P(\omega) = |\hat{f}(\omega)|^2. \tag{3.42}$$

Using equation (3.41), one sees that the power spectra of the input and output functions are related by

$$P_{\text{out}}(\omega) = |\hat{g}(\omega)|^2 = |\hat{h}(\omega)|^2|\hat{f}(\omega)|^2 = |\hat{h}(\omega)|^2 P_{\text{in}}(\omega). \tag{3.43}$$

So the power spectrum of the output is simply that of the input multiplied by the square of the transfer function. For that reason,

$$P_{\mathcal{L}} = |\hat{h}(\omega)|^2 \tag{3.44}$$

is sometimes called the **power spectrum of the linear system** $\mathcal{L}$.

3.2.6 The Fourier Series

Consider a function that repeats periodically, with period L:

$$f(x) = f(x+L). \tag{3.45}$$

Like any other function, this can be expressed as a superposition of complex exponentials. However, the special constraint of periodicity implies that this expansion will need only those exponentials that are themselves periodic with period L. These are the functions

$$e^{2\pi i \frac{n}{L} x} \tag{3.46}$$

with integer n. They form a basis set for all functions $f(x)$ defined over the interval $0 \le x < L$. In particular, one can decompose $f(x)$ as the **Fourier series**:

$$f(x) = \sum_{k=-\infty}^{\infty} \hat{f}_k e^{2\pi i \frac{k}{L} x}. \tag{3.47}$$

In equation (3.47), the **Fourier coefficients** $\hat{f}_k$ are given by

$$\hat{f}_k = \frac{1}{L} \int_0^L f(x)\, e^{-2\pi i \frac{k}{L} x}\, dx. \tag{3.48}$$

Again, we can verify these relations by inserting equation (3.47) back into equation (3.48):

$$\begin{aligned}
f(x) &= \sum_{k=-\infty}^{\infty} e^{2\pi i \frac{k}{L} x} \frac{1}{L} \int_0^L f(x')\, e^{-2\pi i \frac{k}{L} x'}\, dx', \\
&= \int_0^L f(x')\, \frac{1}{L} \sum_{k=-\infty}^{\infty} e^{2\pi i \frac{k}{L}(x-x')}\, dx', \\
&= \int_0^L f(x')\, \delta(x-x')\, dx', \\
&= f(x).
\end{aligned} \tag{3.49}$$

Here, we have used equation (1.63) for the delta function:

$$\frac{1}{L} \sum_{k=-\infty}^{\infty} e^{2\pi i \frac{k}{L}(x-y)} = \delta(x-y). \tag{3.50}$$

Example 3.5 Periodic square pulses

Let us evaluate the Fourier series of a function that delivers a square pulse of width $2c$ with a period of 2π, as in figure 3.7.

Over the interval $-\pi < x \le \pi$, the function is given by

$$f(x) = \begin{cases} 1 & |x| < c \\ 0 & c \le |x| \le \pi. \end{cases} \tag{3.51}$$

We wish to write the function as a series in equation (3.47), where the coefficients are given by equation (3.48). By inserting equation (3.51) into equation (3.48), we

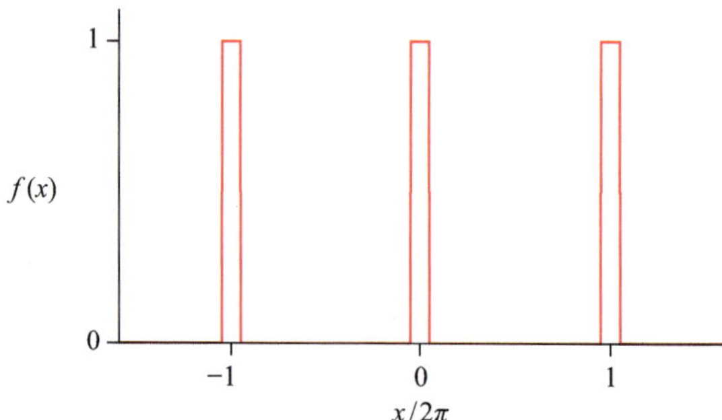

Figure 3.7
A periodic pulse function.

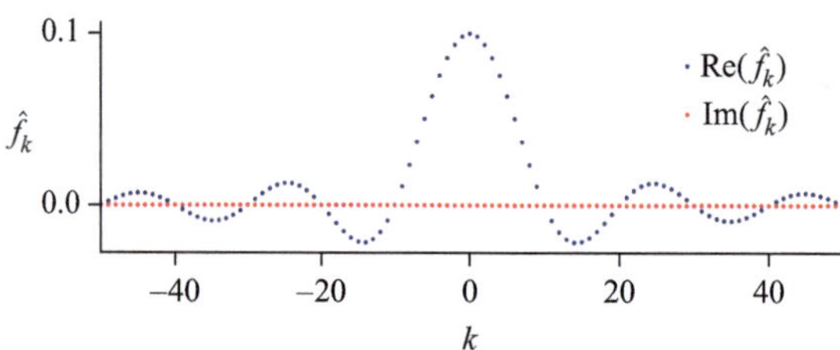

Figure 3.8
The Fourier series of a periodic pulse function.

obtain

$$\hat{f}_k = \frac{1}{2\pi} \int_{-c}^{c} e^{-ikx} dx$$

$$= \frac{1}{2\pi} \left[\frac{-1}{ik} e^{-ikx} \right]_{-c}^{c}$$

$$= \frac{i}{2\pi k} \left[e^{-ikc} - e^{ikc} \right]$$

$$= \frac{1}{\pi} \frac{\sin(kc)}{k}. \tag{3.52}$$

Therefore,

$$f(x) = \frac{1}{\pi} \sum_{k=-\infty}^{\infty} \frac{\sin(kc)}{k} e^{ikx}. \tag{3.53}$$

These Fourier coefficients $\hat{f}_k$ in figure 3.8 are all real. This occurs in the special case where the function $f(x)$ is even relative to $x = 0$. The proof of this is left for exercise 5.21.

Also, note that the Fourier coefficients have a hump-shaped dependence on k, with the largest values at small k. And the width of that hump is inversely related to c, the width of the pulse in the time domain. This is a general result (see also example 3.2): **the Fourier transform of a broad-peaked function is a narrow-peaked function and vice versa.**

3.2.6.1 Real Fourier series   The Fourier series can also be expressed in terms of sines and cosines instead of complex exponentials. The two formats are entirely equivalent, and one can choose among them depending on convenience.

By inserting

$$e^{2\pi i \frac{k}{L} x} = \cos\left(2\pi \frac{k}{L} x\right) + i \sin\left(2\pi \frac{k}{L}\right) \tag{3.54}$$

into equations (3.47) and (3.48), one obtains the alternative formulation

$$f(x) = \frac{A_0}{2} + \sum_{k=1}^{\infty} \left(A_k \cos\left(2\pi \frac{k}{L} x\right) + B_k \sin\left(2\pi \frac{k}{L} x\right)\right), \tag{3.55}$$

where the coefficients are given by

$$A_k = \frac{2}{L} \int_0^L f(x) \, \cos\left(2\pi \frac{k}{L} x\right) \, dx$$
$$B_k = \frac{2}{L} \int_0^L f(x) \, \sin\left(2\pi \frac{k}{L} x\right) \, dx. \tag{3.56}$$

Note that when $f(x)$ is a real function, as is often the case for practical problems, the coefficients A_k and B_k are also real. This is one attraction of the expansion into sines and cosines in equation (3.55) over that using complex exponentials in equation (3.47).

Continuing the preceding example 3.5, one can expand the same pulse function in terms of sines and cosines as

$$f(x) = \frac{c}{\pi} + \frac{2}{\pi} \sum_{k=1}^{\infty} \frac{\sin(kc)}{k} \cos(kx). \tag{3.57}$$

Note that there is no contribution from the sines in this expansion. This is because the function $f(x)$ is even about the origin, whereas the sines are all odd.

3.2.7 The Discrete Fourier Transform

In real-world applications, although quantities vary continuously, they are sampled at discrete points in time or space. For example, a voltage may be sampled 10,000 times a second, or an image may be taken with a camera sensor that has regularly spaced pixels a few microns apart. Analysis software for these kinds of data generally acts on arrays of numbers that represent values sampled at discrete and regularly spaced intervals. For these applications, one uses a version of the Fourier transform known as the discrete Fourier transform (DFT).

Consider sampling a function $f(t)$ at discrete intervals of width Δt. The resulting array of numbers is

$$f_j = f(j \, \Delta t), \qquad 0 \le j < n. \tag{3.58}$$

Then one can expand the data f_j into complex exponentials as

$$f_j = \frac{1}{n} \sum_{k=0}^{n-1} \widehat{f_k} \cdot e^{2\pi i \frac{kj}{n}}, \tag{3.59}$$

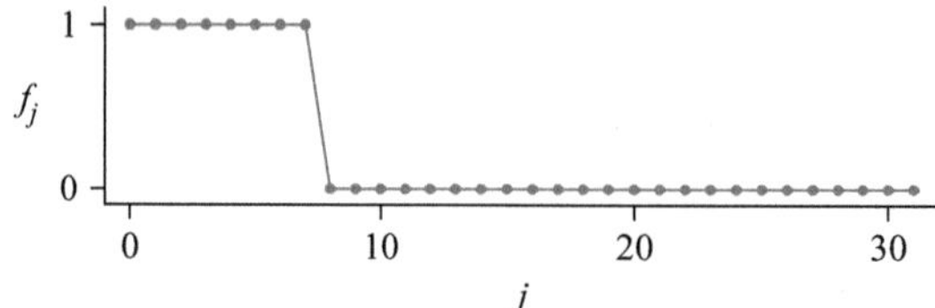

Figure 3.9
A square pulse function sampled at discrete times.

where the Fourier coefficients $\hat{f}_k$ are given by

$$\hat{f}_k = \sum_{j=0}^{n-1} f_j \cdot e^{-2\pi i \frac{kj}{n}}. \tag{3.60}$$

So the DFT turns a function with n values at discrete time points into n amplitudes at different frequencies. If the time points range over

$$t = 0, \Delta t, \ldots, (n-1)\Delta t, \tag{3.61}$$

then the frequencies range over

$$\omega = 0, 2\pi/n\Delta t, \ldots, (n-1)2\pi/n\Delta t. \tag{3.62}$$

Example 3.6 DFT of a square pulse

The signal in figure 3.9 consists of $n = 32$ samples, of which the first $m = 8$ are high and the remainder are low.

$$f_j = \begin{cases} 1, & j = 0, \ldots, m-1 \\ 0, & j = m, \ldots, n-1 \end{cases} \tag{3.63}$$

The DFT is

$$\begin{aligned}
\hat{f}_k &= \sum_{j=0}^{n-1} f_j \exp\left(-2\pi i \frac{k}{n} j\right) \\
&= \sum_{j=0}^{m-1} \exp\left(-2\pi i \frac{k}{n} j\right) \\
&= \frac{1 - \exp\left(-2\pi i \frac{k}{n} m\right)}{1 - \exp\left(-2\pi i \frac{k}{n}\right)},
\end{aligned} \tag{3.64}$$

where we used the geometric series

$$\sum_{j=0}^{m-1} x^j = \frac{1 - x^m}{1 - x}. \tag{3.65}$$

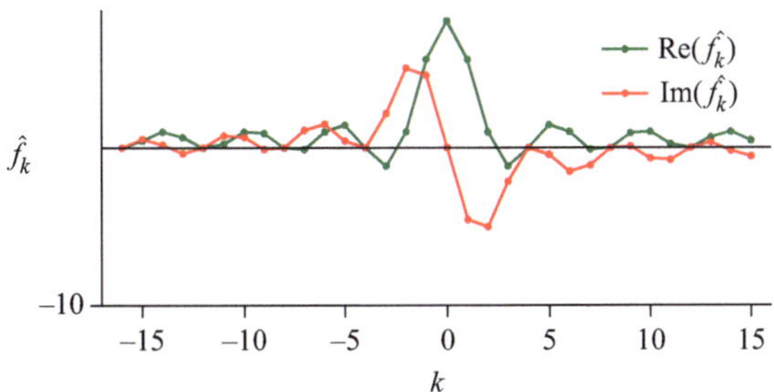

Figure 3.10
The discrete Fourier transform of a square pulse.

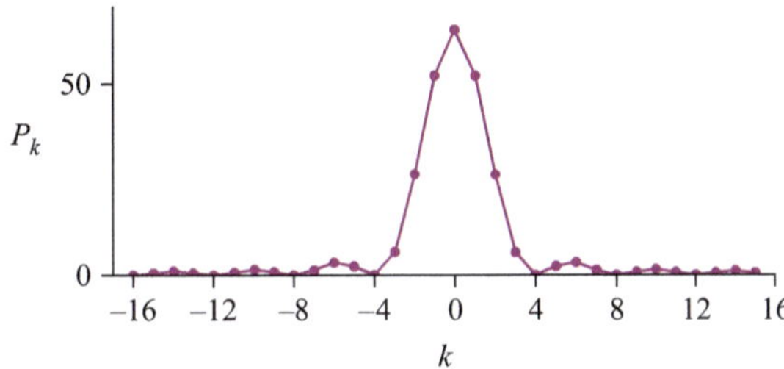

Figure 3.11
The power spectrum of a square pulse.

In figure 3.10, we have adopted the common convention of labeling frequencies with $k > n/2$ by negative values, namely, $k - n$. This is equivalent because $\exp\left(-2\pi i \frac{k}{n} j\right) = \exp\left(-2\pi i \frac{k-n}{n} j\right)$.

These amplitudes are generally complex numbers. If the original function has real values (as in this example), then the Fourier amplitudes obey certain symmetries that are apparent in figure 3.10: The real part is an even function of frequency, and the imaginary part is an odd function.

3.2.7.1 Power spectrum from DFT All the concepts developed earlier in this chapter for continuous functions have direct analogs for discrete functions. For example, the **power spectrum of a discretely sampled function** $[f_j]$ is given by

$$P_k = |\hat{f}_k|^2. \tag{3.66}$$

Continuing the example 3.5, the power spectrum of a square pulse becomes

$$P_k = |\hat{f}_k|^2 = \left| \frac{1 - \exp\left(-2\pi i \frac{k}{n} m\right)}{1 - \exp\left(-2\pi i \frac{k}{n}\right)} \right|^2. \tag{3.67}$$

As seen in figure 3.11, this power spectrum is an even function: The power at positive and negative frequencies is identical. This symmetry is a consequence of the fact that the original function f_j was real-valued.

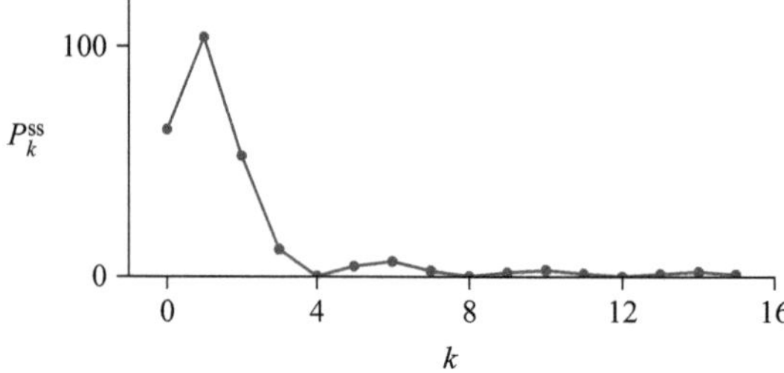

Figure 3.12
The single-sided power spectrum of a square pulse.

For real-valued functions f_j, one therefore often reports the **single-sided power spectrum**:

$$P_k^{ss} = P_k + P_{n-k}. \tag{3.68}$$

As shown in figure 3.12, this is plotted for positive frequency only, and it represents the sum of power at a given frequency and its negative.

3.2.7.2 Convolution and the DFT Convolutions require some special treatment for discretely sampled functions, if one wants to use the convolution theorem.

Suppose that the functions f_j and h_j extend over n samples. Then one defines the **circular convolution** of f and h as

$$g_j = (f * h)_j = \sum_{k=0}^{n-1} f_k h_{(j-k)\bmod n}. \tag{3.69}$$

Here, the mod n means that the functions "wrap around," or repeat periodically. One can visualize this by imagining that the functions f, g, and h all exist on a closed circle, such that sample f_{n-1} is followed again by f_0, and so on.

With the circular convolution defined in this way, the **convolution theorem** applies: If f, g, and h satisfy equation (3.69), then

$$\hat{g}_j = \hat{f}_j \, \hat{h}_j. \tag{3.70}$$

In numerical work, one must take care to remember the circular nature of the discrete convolution. Generally, the signals or images that we measure do not repeat periodically. Then it may be inappropriate to pretend that beyond the last measurement point, the signal simply wraps around to start at the beginning. Instead, we may wish to set the signal to zero outside the measurement interval. This is accomplished by making the array longer and "padding" it with zeroes. One then adjusts the length of zero padding so the circular convolution can never produce awkward results. If both functions have nonzero values over the entire range $0, \ldots, n-1$, then a safe procedure is to double the length of the array and pad values $n, \ldots, 2n-1$ with zeroes. See exercise 5.25.

3.2.7.3 The Fast Fourier Transform As has been pointed out already, convolutions happen ubiquitously in scientific computing, for example in filtering signals or processing images. At any given moment, a good fraction of all the world's computers are engaged in convolutions. We can now consider the computational cost of this processing. If we compute a convolution according to equation (3.69), that requires $\sim n^2$ multiplications and additions. If, instead, we compute it in the frequency domain according to equation (3.70), that requires only n multiplications, potentially promising huge savings. However, the signals f and h must be brought into the frequency domain first, and the result g then returned to the time domain, a total of three Fourier transforms. According to equation (3.59), the Fourier transform itself requires $\sim n^2$ multiplications and additions. That would seem a lot more effort than simply computing the convolution in the time domain!

Here, we encounter one of the game-changing inventions in computer science: The fast Fourier transform (FFT) algorithm was developed by James Cooley and John Tukey in 1965, though several previous inventions heralded its arrival. It turns out that by clever decomposition of the exponential terms, the Fourier transform in equation (3.60) and its inverse in equation (3.59) can be accomplished using only $\sim n \log n$ operations. Thus, the total cost of performing the convolution in the frequency domain comes to only a few times $n \log n$. When n is several million, as for a typical digital photograph, the difference between n^2 and $n \log n$ amounts to many orders of magnitude. This enormous amount of savings makes possible digital filters and convolutions that we could otherwise only dream about. All scientific programming packages include FFT routines.

3.2.8 The Fourier Transform as a Change of Basis

We have alluded repeatedly to the idea that the Fourier transform of a function is simply its representation in an alternative basis. Here, we explore this connection to concepts from linear algebra in more depth.

Consider the space of all functions consisting of n samples in time. Any such function f can be represented in the time domain by a vector listing the values of the samples:

$$\mathbf{f} = [f_j], \qquad j = 0, \ldots, n-1. \tag{3.71}$$

Clearly, these function vectors live in n-dimensional space. Let us consider an alternative basis in this space, defined by the n basis vectors

$$\mathbf{e}'_k = \left[\frac{1}{\sqrt{n}} e^{2\pi i \frac{k}{n} j} \right], \qquad k = 0, \ldots, n-1, \tag{3.72}$$

where j is the index of the vector component. Also, we will use the conventional scalar product from equation (2.66) in this space:

$$\mathbf{f} \cdot \mathbf{g} = \sum_{j=0}^{n-1} f_j^* g_j. \tag{3.73}$$

Under this scalar product, the new basis vectors are orthonormal:

$$\begin{aligned} \mathbf{e}'_k \cdot \mathbf{e}'_l &= \frac{1}{n} \sum_{j=0}^{n-1} e^{-2\pi i \frac{k}{n} j} e^{2\pi i \frac{l}{n} j} \\ &= \frac{1}{n} \sum_{j=0}^{n-1} e^{-2\pi i \frac{l-k}{n} j} \\ &= \delta_{kl}. \end{aligned} \tag{3.74}$$

The transformation matrix in equation (2.44) for the change to the new basis is composed of the coefficients of the new basis vectors:

$$\mathbf{S} = \left[\mathbf{e}'_0, \mathbf{e}'_1, \ldots, \mathbf{e}'_{n-1} \right]$$

$$S_{jk} = \frac{1}{\sqrt{n}} e^{2\pi i \frac{k}{n} j}. \tag{3.75}$$

Because its column vectors are orthonormal, $\mathbf{S}$ is unitary:

$$\mathbf{S}^{-1} = \mathbf{S}^{\dagger}. \tag{3.76}$$

Therefore, the coefficients of $\mathbf{f}$ in the new basis are

$$\mathbf{f}' = \mathbf{S}^{\dagger} \mathbf{f} \tag{3.77}$$

and, vice versa,

$$\mathbf{f} = \mathbf{S} \mathbf{f}'. \tag{3.78}$$

Now we recognize that equation (3.60) for the DFT of f is simply a scaled version of this basis transform:

$$\widehat{\mathbf{f}} = \sqrt{n}\, \mathbf{f}' = \sqrt{n}\, \mathbf{S}^{-1} \mathbf{f}. \tag{3.79}$$

Using the same methods, one can derive the Fourier transform and the Fourier series of periodic functions; in both cases, the space of functions has infinite dimensions.

3.2.9 Summary of Fourier Methods

Table 3.1 summarizes the transforms introduced here, and it serves to highlight the close relationships among the three versions.

We emphasize again that there exist different conventions regarding the allocations of the factors $\frac{1}{2\pi}$, $\frac{1}{L}$, and $\frac{1}{n}$ between the forward and the inverse Fourier transforms.

Table 3.1
Fourier methods

	Fourier Transform	Fourier Series	DFT
Function	$f(x),\ -\infty < x < \infty$	$f(x),\ 0 \le x < L$	$f_j,\ 0 \le j < n$
Transform	$\hat{f}(k),\ -\infty < k < \infty$	$\hat{f}_k,\ -\infty < k < \infty$	$\hat{f}_k,\ 0 \le k < n$
Forward	$\hat{f}(k) = \int\limits_{-\infty}^{\infty} f(x)e^{-ikx}\,dx$	$\hat{f}_k = \int\limits_{-\infty}^{\infty} f(x)e^{-2\pi ikx}\,dx$	$\hat{f}_k = \sum\limits_{k=0}^{n} f_j e^{-2\pi i \frac{k}{n} j}$
Inverse	$f(x) = \frac{1}{2\pi}\int\limits_{-\infty}^{\infty} \hat{f}(k)e^{ikx}\,dk$	$f(x) = \frac{1}{L}\sum\limits_{k=-\infty}^{\infty} \hat{f}_k e^{2\pi ikx}$	$f_j = \frac{1}{n}\sum\limits_{k=0}^{n} \hat{f}_k e^{2\pi i \frac{k}{n} j}$
Purpose	Analyze continuous functions.	Analyze periodic functions.	Numerical work on discrete functions.

4　Applications of Linear Systems

In this chapter, we will consider various applications of linear systems analysis in the biological sciences. Before embarking on specific cases, though, it is worth setting up a basic framework that will allow us to think about what all these cases have in common.

As discussed in chapter 3, a linear system converts an input signal into an output signal such that the overall transformation is linear (figure 4.1). In some applications, the linear system is the actual biological entity under study, whose properties are mysterious at the outset, and we seek to understand how it converts its inputs to outputs. In other cases, the linear system represents a measuring instrument whose transform is well known to us, and we need to reconstruct its input signal from recordings of the output signal. In yet other applications, the linear system may be a data analysis tool, and we need to optimize its properties for a certain task we want to perform. For example, the input or output signals may be corrupted by "noise," by which we mean some random process that is not under the control of the experimenter. In this case, we can design a linear filter to optimally suppress the noise in favor of the signal.

In this chapter, we present concrete examples of all these applications, and the reader will gain the experience to adapt similar logic to her own problems. We begin by considering some tools of biological imaging, which illustrate these concepts nicely. Then we broaden the discussion to various methods of signal processing, including noise suppression and signal reconstruction. Eventually, this brings us back to computational methods in imaging.

4.1　Microscopy

The use of microscopy is widespread in the life sciences for the purpose of observing and measuring phenomena at the scale of a cell or smaller. All microscopy techniques rely on a similar principle: irradiating a sample with waves or particles, observing the scattered radiation, and using it to make inferences about the sample. Here, we discuss light microscopy, although other probes such as electrons or sound are also used.

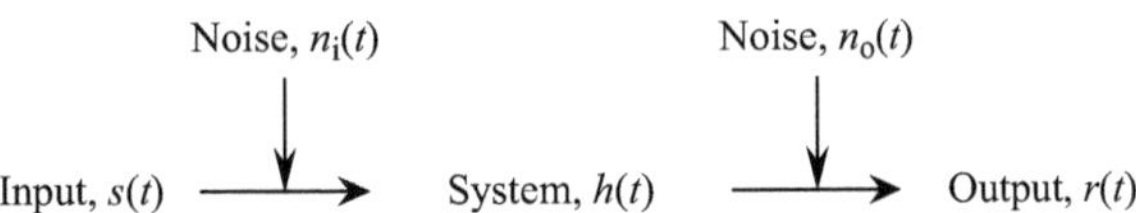

Figure 4.1
Input signals are linearly processed to output signals. Both input and output may be corrupted by noise.

If you have spent time around microscopists, you have likely heard them brag about the numerical aperture of their objectives. And the price of those objectives is closely related to their aperture. What makes a large aperture so desirable? Here, we show that a larger aperture creates a smaller blur in the resulting image. And the relationship between the blur and the aperture is given by ... a Fourier transform.

4.1.1 A Simple Microscope

Figure 4.2 shows the basic anatomy of a simple microscope, which consists of two convex lenses. Consider a point source of light in the sample, positioned in the focal plane of the **objective lens**. The source emits rays of light in all directions. The objective lens transforms that divergent bundle of rays into a parallel beam of rays above the objective. That parallel beam then enters the **tube lens**, which transforms it back into a convergent bundle of rays that intersect in a single point on the image plane. In this way, a point in the sample gets mapped into a point in the image plane. However, the size of the image is magnified relative to the sample. The geometric construction shows that the **magnification ratio** M is simply the ratio of the focal lengths:

$$M = \frac{f_{\text{tube}}}{f_{\text{obj}}}. \tag{4.1}$$

This simple picture (figure 4.2) suggests that by using tube lenses with longer focal distances, we could achieve arbitrarily large magnifications and observe smaller and smaller structures. Unfortunately, this is not the case. In practice, a point source

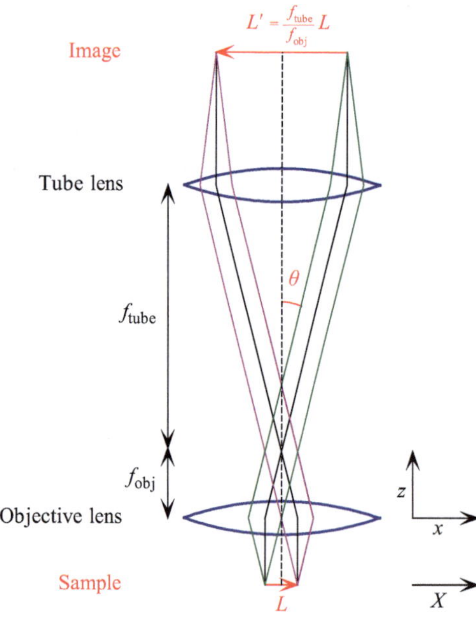

Figure 4.2
A simple microscope consisting of two lenses generates an enlarged image of the sample. $f_{\text{tube}}, f_{\text{obj}}$ = focal lengths of tube lens and objective lens, respectively. Dashed line = optical axis. X = distance from optical axis in the sample plane. x = distance from the optical axis in the plane of the objective.

in the sample plane produces a fuzzy **blur circle** in the image plane. And the size of the blur grows proportionally with the magnification, along with the image. This blur limits the resolution of small structure in the sample, and that resolution remains constant with magnification. However, it turns out that the blur gets smaller the bigger the diameter of the objective lens is. To understand what is happening, we need to move beyond the framework of geometric optics, where we draw light rays as straight lines, and consider the wave nature of light.

4.1.2 Diffraction at an Aperture

Perhaps the simplest optical element is an aperture: an opaque screen with a hole to let light through. When light hits such an aperture, rays outside the aperture are blocked, whereas those inside the aperture continue (figure 4.3). However, this light is diffracted, which means that some of it gets bent into slightly different directions. Now we will derive how much the light gets diffracted.

Consider a wavefront of parallel light rays impinging perpendicularly on an aperture of size a, as in figure 4.3. Below the aperture, the light propagates upward (in the z-direction) as a **planar electromagnetic wave**. The electric field has an amplitude of

$$
\begin{aligned}
E(z, t) &= \cos\left(kz - \omega t\right) \\
&= \mathrm{Re}\left(\exp\left(ikz\right)\exp(-i\omega t)\right).
\end{aligned}
$$
(4.2)

Here, $k = \frac{2\pi}{\lambda}$ is called the wavenumber, λ is the wavelength, and ω is the frequency of the light. It is often convenient to represent the wave by its complex **phasor** (section 1.6.4):

$$
\exp\left(ikz\right)\exp(-i\omega t),
$$
(4.3)

while keeping in mind that the electric field is the real part of that complex number. Furthermore, for what follows, we may simply consider a snapshot in time, and thus drop the factor $\exp(-i\omega t)$, which is identical everywhere in space.

How does the light behave on the other side of the aperture? Here, one can use Huygens' principle, which states that every point on a wavefront itself becomes a point source of a secondary wave (the blue dots in figure 4.3). Light propagation above the aperture is simply the linear superposition of these secondary waves (section 3.1.2). What then is the complex phasor for light diffracted at angle θ from the original direction of propagation? We must sum up the wavelets coming from all the points within the aperture, each of which has to traverse a slightly different distance d to travel in direction θ (figure 4.3). These path differences Δd lead to differences in phase $k\Delta d$. If that phase difference amounts to π, then the two wavelets cancel each other out because $e^{ikd} + e^{i(kd+\pi)} = 0$. This is called **destructive interference**. Because the phase differences depend on direction θ, we will find different amounts of light scattered in different directions.[1]

---

1. Technically, this treatment is correct in the **Fraunhofer approximation** of diffraction— namely, at distances much larger than the aperture size, and for small angle θ. This applies to the situation in a microscope.

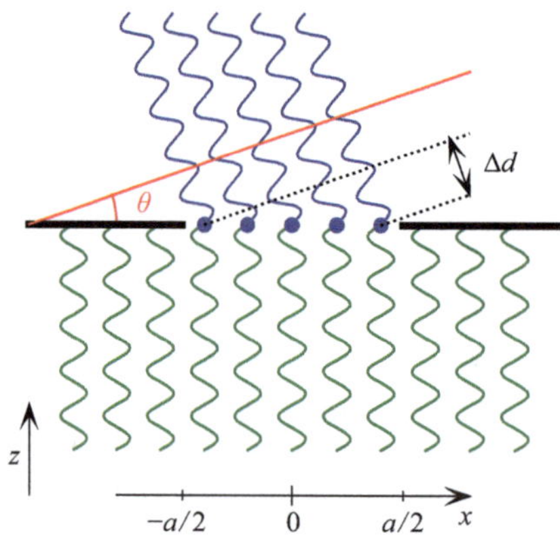

Figure 4.3
Diffraction at an aperture. A plane electromagnetic wave reaches the aperture from below
(green). The aperture is a slit of width a. The wavefront above the aperture is the superposition of the wavelets emitted by all the Huygens oscillators (blue dots) within the
aperture. For propagation in a particular diffraction angle θ, the wavelets travel different
distances; and Δd illustrates the difference in path lengths for two such wavelets.

For a wavelet starting at position x within the aperture, the phase difference relative
to the wavelet at $x = 0$ is

$$\Delta d(x) = x \sin \theta. \tag{4.4}$$

Integrating the contributions from the entire aperture $-a/2 < x < a/2$, we obtain the
complex phasor of the diffracted wave in direction θ as

$$
\begin{aligned}
E(\theta) &= \int_{-\frac{a}{2}}^{\frac{a}{2}} e^{ik \sin \theta \, x} \, dx \\
&= \int_{-\infty}^{\infty} A(x) e^{ik \sin \theta \, x} \, dx,
\end{aligned}
\tag{4.5}
$$

where we defined $A(x)$ as the profile of the aperture:

$$A(x) = \mathrm{rect}(x/a) = \begin{cases} 1, & |x| < a/2 \\ 0, & \text{otherwise.} \end{cases} \tag{4.6}$$

Note that amplitude $E(\theta)$ is simply **the Fourier transform of the aperture** $A(x)$
evaluated at the frequency $k \sin(\theta)$. From equation (4.5), one gets

$$
\begin{aligned}
E(\theta) &= \frac{1}{ik \sin \theta} \left(e^{ik \frac{a}{2} \sin \theta} - e^{-ik \frac{a}{2} \sin \theta} \right) \\
&= a \frac{\sin \left(\frac{ka}{2} \sin \theta \right)}{\frac{ka}{2} \sin \theta} \\
&= a \, \mathrm{sinc} \left(\frac{\pi a}{\lambda} \sin \theta \right),
\end{aligned}
\tag{4.7}
$$

where the last step used the definition of the sinc function

$$\mathrm{sinc}(x) = \frac{\sin x}{x}. \tag{4.8}$$

The light intensity is given by the square of the amplitude; therefore

$$I(\theta) = a^2 \mathrm{sinc}^2 \left(\frac{\pi a}{\lambda} \sin \theta \right). \tag{4.9}$$

How is all this related to the microscope of figure 4.2? Our idealized ray-optics model has the objective lens turning a point source into a parallel bundle of rays. Now we know that the bundle cannot be perfectly parallel because the objective lens has a finite diameter a. Instead, the bundle will contain a range of angles, and that distribution is given by equation (4.9). The tube lens in turn converts that small range of angles into a range of positions in the image plane, producing a small blur in the image.

How does that blur affect the spatial resolution of the microscope? For that purpose, it is best to project the effects back into the sample plane, where the spatial resolution really matters. Specifically, let us consider a point source of light at $X = 0$ in the sample plane. The objective lens turns that into rays with a distribution of angles $I(\theta)$. But a ray traveling at angle θ seems to come from a point of the sample at $X = f_{\mathrm{obj}} \sin \theta$ (see figure 4.2). Therefore, the objective lens effectively smears a pointlike sample into a blur profile:

$$B(X) = I \left(\arcsin \frac{X}{f_{\mathrm{obj}}} \right) \propto \mathrm{sinc}^2 \left(\frac{\pi a}{\lambda} \frac{X}{f_{\mathrm{obj}}} \right). \tag{4.10}$$

This blur function is plotted in figure 4.4A. It has a bell-shaped profile with small side lobes away from the center. The first zeros on either side of the maximum occur at $X = \pm f_{\mathrm{obj}} \lambda / a$. We can take this half-width of the blur spot as a measure of spatial resolution in the sample plane:

$$\Delta x = \lambda \frac{f_{\mathrm{obj}}}{a}. \tag{4.11}$$

The first factor here shows that resolution improves at shorter wavelengths. The second factor reflects the construction of the microscope. Here, one defines the **numerical aperture** as the ratio of the aperture radius to the focal length of the objective:

$$\text{Numerical aperture (N.A.)} = \frac{a}{2 f_{\mathrm{obj}}}. \tag{4.12}$$

Note that the spatial resolution is inversely proportional to the numerical aperture. Now we understand why microscopists are so enamored of "high NA" objectives.[2]

---

2. You may wonder why we only talk about the objective lens and not the tube lens. Of course, light scatters at the tube lens as well. However, the range of angles that this produces is proportional to $\lambda / a_{\mathrm{tube}}$, and the diameter a_{tube} of the tube lens tends to be much larger than that of the objective. So diffraction at the objective lens dominates.

4.1.3 The Airy Disk

The arguments in the preceding section apply to an aperture shaped like a slit that extends infinitely in the y-direction perpendicular to the plane of figure 4.2. Real lens apertures, of course, are circular. A round aperture of diameter a can be represented with an aperture function in the (x, y) plane of the objective:

$$A(x, y) = \text{rect}\left(\frac{\sqrt{x^2 + y^2}}{a}\right) = \begin{cases} 1, & \sqrt{x^2 + y^2} < a/2 \\ 0, & \text{otherwise.} \end{cases} \tag{4.13}$$

As in the 1D example considered earlier, a point source in the sample plane will get mapped into a not-quite-parallel beam above the objective. To compute the distribution of angles within that beam, we again integrate over all the little Huygens oscillators within the aperture, which amounts to a **2D Fourier transform of the aperture function**. Finally, we project that distribution of angles back into the (X, Y) coordinates of the sample plane. The result is a two-dimensional (2D) blur function

$$B(X, Y) = \int_{x=-\infty}^{\infty} \int_{y=-\infty}^{\infty} A(x, y) \exp\left(i\frac{2\pi X}{\lambda f_{\text{obj}}} x\right) \exp\left(i\frac{2\pi Y}{\lambda f_{\text{obj}}} y\right) \mathrm{d}x\mathrm{d}y$$

$$\propto \left(\frac{J_1(r)}{r}\right)^2, \tag{4.14}$$

where

$$r = R\frac{\pi a}{\lambda f_{\text{obj}}}$$

and

$$R = \sqrt{X^2 + Y^2}$$

is the distance from the origin in the sample plane, and $J_1(r)$ is the Bessel function of the first kind. A cut through this radial profile is plotted in figure 4.4B. Again, it has a central maximum and a series of concentric rings of decreasing intensity. The first zero of the Bessel function occurs at $r = 1.22$, which defines the radius of the central maximum. This is called the **Airy disk**. The radius of the Airy disk is commonly used as a measure of the spatial resolution of the microscope. It is given by

$$\Delta x = 1.22\frac{\lambda f_{\text{obj}}}{a}. \tag{4.15}$$

Note the close similarity to the 1D results from the slit aperture in equation (4.11).

Suppose now that one could produce an aperture with an arbitrary aperture function $A(x, y)$. For example, this could be a photographic film holding an arbitrary image. By illuminating the film and imaging the transmitted light, one obtains a Fourier transform of the image. It is remarkable that light can perform Fourier transforms so naturally. If the distance to the screen is on the order of 1 meter, light takes only a few nanoseconds to travel this distance and reveal the Fourier transform. Indeed, optical computers have been built on this principle, using a fast spatial light modulator to create the aperture function and a camera to record the result.

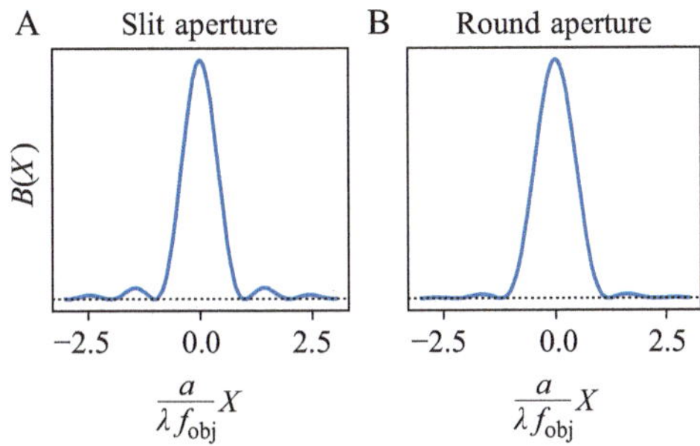

Figure 4.4
Blur profile from diffraction at the objective aperture, translated into the sample plane. A: 1D slit aperture. B: 2D circular aperture. A cut through the blur profile at $Y = 0$. a = aperture diameter, λ = wavelength, f_{obj} = objective focal length, X, Y = coordinates the sample plane.

More generally, the representation of light by a complex phasor is applicable in many circumstances, and thus the calculation of light propagation in optical systems often leads to Fourier transforms (Goodman, 2005).

4.1.4   The Point-Spread Function

In section 4.1.3, we learned that the image of a small point source in the sample is not a point, but rather a blurred spot in the image plane. The profile of this spot is called the **point-spread function (PSF)** of the optical system. As we saw, the minimum size of the PSF is given by the Airy disk corresponding to the numerical aperture of the objective. In practice, the PSF is larger than the Airy disk because of imperfections in the optics.[3]

Now consider an extended object in the sample plane—let's say, of a fluorescence microscope. The sample is a superposition of many point sources, one for each of the fluorophores that emits photons. Each of these sources gets imaged into a blur with the profile of the PSF. Because of linear superposition of light, the resulting image is simply the superposition of all those PSFs.[4] This amounts to a convolution of the sample with the PSF:

$$I(x, y) = S(x, y) * h_{\mathrm{PSF}}(x, y). \tag{4.16}$$

Here, $S(x, y)$ is the sample scaled by the magnification of the instrument, $I(x, y)$ is the generated image, and $h_{\mathrm{PSF}}(x, y)$ is the PSF. Note all three quantities are expressed in the same coordinate system of the image.

---

3. ... and, in most laboratories, because of dirt accumulating on the lenses.

4. A note on superposition: In section 4.1.2, we used the superposition principle for the *electric field*. Because the intensity is the square of the electric field, two light waves can superpose destructively to produce zero intensity. Here, on the other hand, we claim that the *intensities* of different lights superpose. Because the intensity is always positive, no destructive interference can occur. The difference results because the lights emitted by various fluorophores in the sample are not **coherent** and therefore cannot interfere. By contrast, in analyzing diffraction from an aperture, we considered different parts of a coherent light wave emitted from the same point in the sample. Phase contrast microscopy is a variant that also relies intimately on interference effects.

In general, this convolution acts like a low-pass filter that attenuates spatial frequencies higher than the resolution limit of the microscope. Clearly, the point-spread function is one of the most important characteristics of a microscope.

4.2 Crystal Analysis

To analyze the shapes of very small particles like proteins, crystallization is a very powerful tool. Grown from a concentrated solution of the substance in question, the crystal organizes the particles in an array with regular spacing, and all with the same orientation. This regularity offers an opportunity to collect and average the image information from many particles to get a high-resolution view.

For example consider the 2D crystal of the protein cytochrome oxidase in figure 4.5. These transmembrane proteins are confined to the plane of a lipid membrane and packed densely in a periodic array (Frey et al., 1978). They are visualized here by electron microscopy. One can clearly see the periodic structure in the image, but this is imperfect, presumably owing to irregularities in the particle arrangement and noise during the imaging process. One would like to find what all the repeated single-particle images have in common and thus extract some reliable aspect of the particle's shape.

One simple method to extract the average particle shape is as follows: Take a Fourier transform of the image (figure 4.5A) along both dimensions. This yields the Fourier components in 2D frequency space (figure 4.5B). These components are found almost exclusively at discrete, equally spaced spatial frequencies. Also, they get weaker at high frequencies. Measure the complex amplitude at each of the discrete frequencies, and ignore those beyond a certain frequency cutoff. Inverse-transform the result to get an estimate of the average particle shape (figure 4.5C).

Why does this method work? Let us consider first an idealized one-dimensional (1D) crystal: an infinite periodic repeat of the identical small image.

Define the **lattice** as the collection of all lattice points in this periodic array, written as a sum of delta functions at the lattice positions:

$$L(x) = \sum_{j=-\infty}^{+\infty} \delta(x - x_j), \qquad x_j = jd, \tag{4.17}$$

where d is the lattice spacing.

The **unit cell** is the image that is attached to each of those lattice points:

$$U(x), \qquad 0 < x < d. \tag{4.18}$$

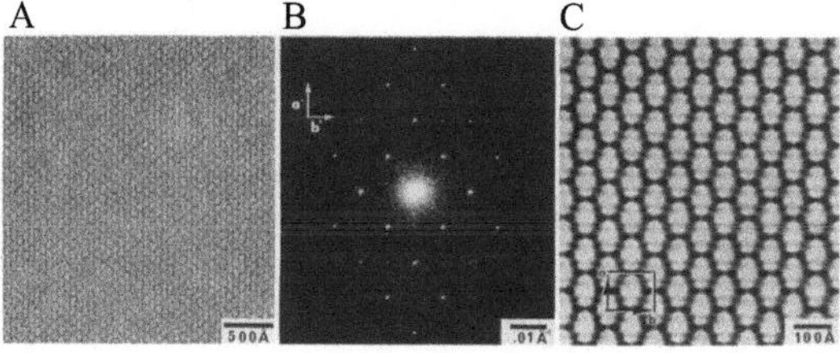

Figure 4.5
Image reconstruction from a 2D protein crystal.

Then the **image** of the entire crystal is the convolution of the lattice with the unit cell

$$I(x) = \sum_{j=-\infty}^{+\infty} U(x - x_j) = \int_{-\infty}^{+\infty} L(x')\, U(x - x')\, dx' = (L * U)(x). \tag{4.19}$$

So the Fourier transform of the crystal is the product of the Fourier transforms of the lattice and the unit cell:

$$\hat{I}(k) = \hat{L}(k) \cdot \hat{U}(k). \tag{4.20}$$

The Fourier transform of the lattice is itself a sum of delta functions at the frequency of the lattice and its multiples:

$$\hat{L}(k) = \int_{-\infty}^{+\infty} L(x)\, e^{-ikx}\, dx = \sum_{j=-\infty}^{+\infty} e^{-ikj \cdot d}. \tag{4.21}$$

Note that this vanishes for all k, except when kd is a multiple of 2π. At those locations, it diverges, and one can show, using equation (1.63), that

$$\hat{L}(k) = \sum_{j=-\infty}^{+\infty} e^{-ikj \cdot d} = \frac{2\pi}{d} \sum_{n=-\infty}^{+\infty} \delta\left(k - n\frac{2\pi}{d}\right). \tag{4.22}$$

So the Fourier transform of the entire crystal is again a series of sharp peaks at multiples of the lattice frequency, $\frac{2\pi}{d}$, but with amplitudes that reflect the Fourier components of the unit cell at those frequencies:

$$\hat{I}(k) = \frac{2\pi}{d} \sum_{n=-\infty}^{+\infty} \delta\left(k - n\frac{2\pi}{d}\right) \cdot \hat{U}\left(n\frac{2\pi}{d}\right). \tag{4.23}$$

With this understanding, we can work backward now: Suppose that a measurement of $\hat{I}(k)$ is available. By looking at the spacing of peaks in $|\hat{I}(k)|$, one can find the lattice frequency $\frac{2\pi}{d}$ of the crystal. Then one can measure the amplitude of each peak by integrating across the peak, which will yield

$$A_n = \frac{2\pi}{d} \cdot \hat{U}\left(n\frac{2\pi}{d}\right). \tag{4.24}$$

Finally, one can derive the shape of the unit cell by inverse transforming (see section 3.2.6):

$$U(x) = \frac{1}{2\pi} \sum_{n=-\infty}^{+\infty} A_n\, e^{2\pi i n \frac{x}{d}}. \tag{4.25}$$

In addition to finding the lattice of the crystal and providing the profile of the unit cell, this method is also very powerful at rejecting noise in the image of the crystal. By selecting only the values of the Fourier transform at multiples of the lattice frequency, we are isolating just those image components that repeat periodically. Any corruption

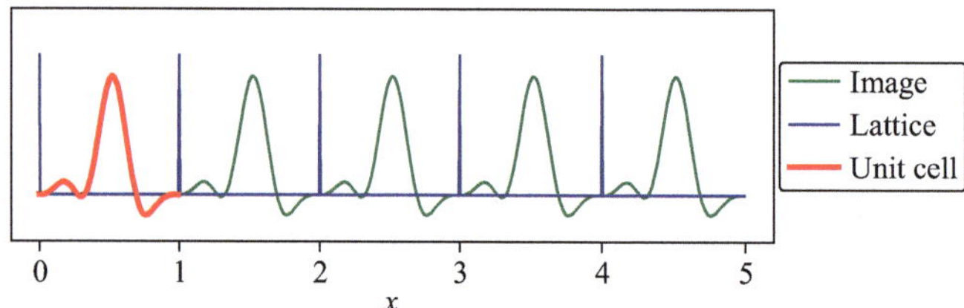

Figure 4.6
A 1D crystal composed of a periodic lattice and repeating unit cell.

that is unrelated to the lattice, such as pixel noise from the image measurement, will get rejected or strongly suppressed. To understand this, consider now a crystal image with added noise:

$$I(x) = (L * U)(x) + N(x) \tag{4.26}$$

with the Fourier transform

$$\hat{I}(k) = \hat{L}(k) \cdot \hat{U}(k) + \hat{N}(k). \tag{4.27}$$

Whereas $\hat{L}(k) \cdot \hat{U}(k)$ is concentrated at the lattice frequencies, the noise spectrum $\hat{N}(k)$ is typically spread throughout. For example, certain forms of imaging noise occur independently in each image pixel, a process known as "white noise." The power spectrum of white noise is distributed equally across all frequencies. Thus, most of this noise power will be rejected if one analyzes only the periodic peaks in $\hat{I}(k)$.

These principles are illustrated in figure 4.7 with the simulation of a 2D crystal. We start with a periodically repeated image to which considerable noise has been added (figure 4.7A). The periodicity is barely visible in the raw image, and even in a zoomed-in view, it is nearly impossible to make out the shape of the unit cell figure. The Fourier transform of the crystal shows sharp peaks at periodic frequencies, as seen in the plot of $|\hat{I}(k)|$ (e.g., the circles in figure 4.7C). From this, we can read off that the crystal has a period of 48 pixels along the x-axis $(1/(0.021 \text{ pix}^{-1}))$ and 32 pixels along the y-axis $(1/(0.031 \text{ pix}^{-1}))$. In addition, one can see a good amount of white noise power distributed at all frequencies. Next, we set all these components to zero, except at multiples of the lattice frequency. Then we inverse transform and obtain a clear estimate of the unit cell (figure 4.7D).

In this reconstruction (in figure 4.7D) most of the noise has already been removed, although one can still see some speckle in the background. This comes from the noise components that are exactly at the lattice frequencies. To reduce this further, one can try to limit which peaks in the Fourier transform are used for the reconstruction. For example, the highest frequencies often contribute less to the image than to the noise. Here, we can compare reconstructions using various low-pass cutoffs for the spatial frequency of the Fourier peaks (figure 4.7E). A cutoff at 0.2 pix^{-1} is too drastic and infringes on the quality of the signal. A cutoff at 0.6 pix^{-1} looks close to optimal in this case.

In practice, of course, one doesn't have the exact noise-free solution for a comparison, and the cutoff must be chosen based on other knowledge of the signal and noise spectra. This is an instance of Wiener filtering (see section 4.7.1).

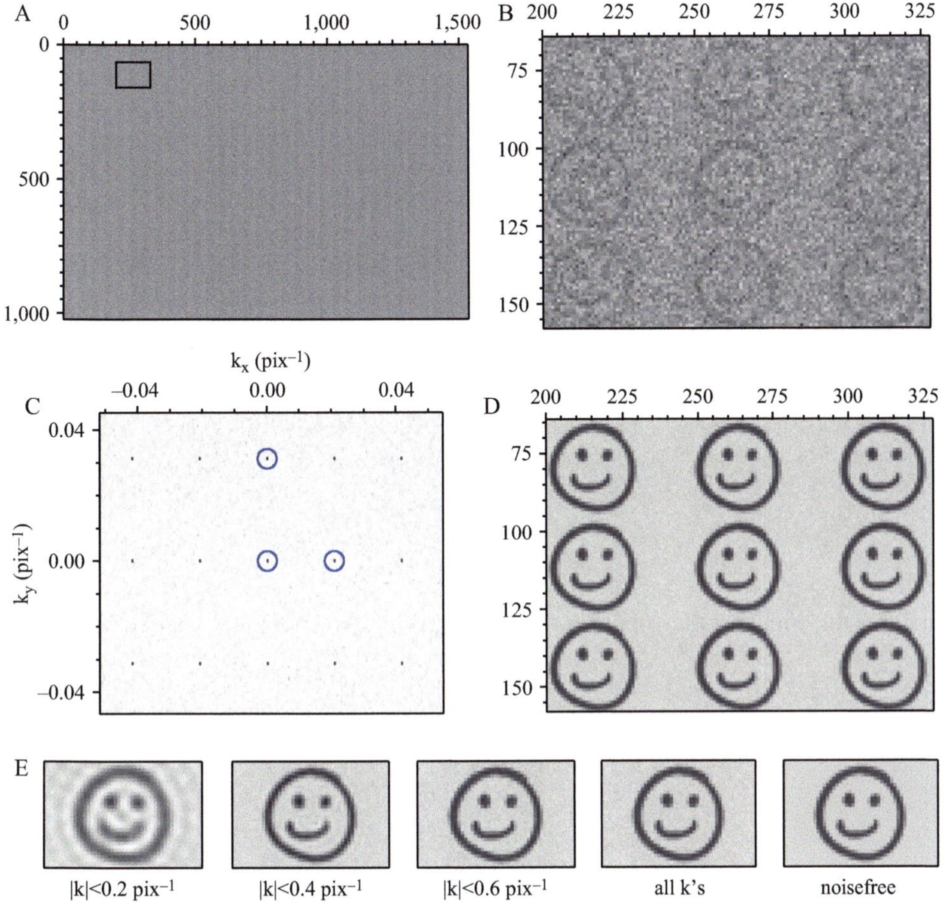

Figure 4.7

Simulation of a 2D crystal with noise. A: Raw image. B: Zoomed-in view of the raw image. C: Fourier transform of the raw image. D: Reconstruction of the unit cell from the Fourier transform. E: Reconstruction of the unit cell after applying various low-pass filters to the Fourier transform.

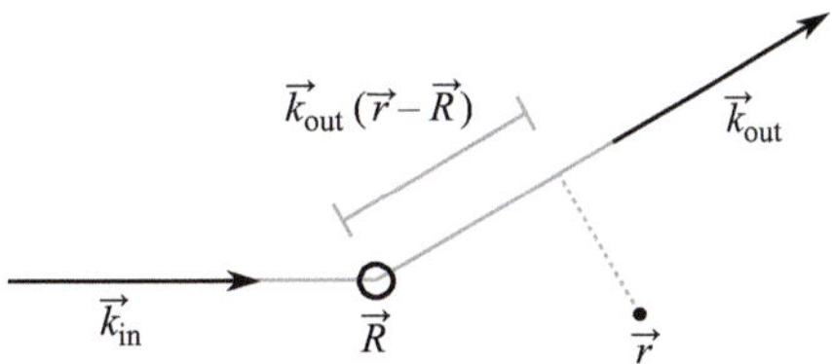

Figure 4.8

Scattering off a single electron.

4.3 X-ray Scattering

Earlier in this chapter, we considered the approach of taking a high-resolution picture of a crystal and then performing the Fourier transform numerically on the image data. A powerful alternative is to perform the Fourier transform on the crystal directly using electromagnetic waves. As we saw in the context of microscope optics (section 4.1), the diffraction of light produces a Fourier transform of the object causing the diffraction, in that case, the objective's aperture. Similarly, the diffraction of X-rays in a crystal produces a Fourier transform of the charge distribution in the crystal. That diffraction pattern can be captured on film or by an X-ray camera and then analyzed by the methods illustrated previously for 2D images.

Here, we present briefly the principles of X-ray crystallography, noting that there are many excellent texts dedicated to this topic (e.g. Drenth (2007)). As discussed previously (section 4.1), a plane electromagnetic wave entering a crystal can be represented by a complex phasor (see section 1.6.4):

$$E(\vec{r}, t) = e^{i(\vec{k}_{\text{in}} \cdot \vec{r} - \omega t)}. \tag{4.28}$$

This reflects the phase of the electric field at location $\vec{r}$ and time t. The wave vector $\vec{k}_{\text{in}}$ indicates the direction of propagation, and the magnitude of $\vec{k}_{\text{in}}$ is related to the wavelength λ of the light

$$|\vec{k}_{\text{in}}| = \frac{2\pi}{\lambda} \tag{4.29}$$

and

$$\omega = \frac{c}{\lambda} \tag{4.30}$$

where c is the speed of light and ω is the frequency of the light.

Consider now a single electron at position $\vec{R}$ (figure 4.8). The incoming light induces that electron to oscillate, with a phasor

$$e^{i\vec{k}_{\text{in}} \cdot \vec{R}}, \tag{4.31}$$

where from now on, we will omit the time dependence $e^{-i\omega t}$. That oscillation in turn radiates a little bit of light of the same frequency in all directions. Looking in some specific direction $\vec{k}_{\text{out}}$, the phasor of the scattered light is

$$e^{i(\vec{r} - \vec{R}) \cdot \vec{k}_{\text{out}}} e^{i\vec{R} \cdot \vec{k}_{\text{in}}} = e^{i\vec{r} \cdot \vec{k}_{\text{out}}} e^{i\vec{R} \cdot (\vec{k}_{\text{in}} - \vec{k}_{\text{out}})}. \tag{4.32}$$

Let us define the scattering vector $\vec{k} = \vec{k}_{\text{out}} - \vec{k}_{\text{in}}$, which represents the change in direction of the scattered light. Then the phasor of the scattered wave is

$$e^{i\vec{r} \cdot \vec{k}_{\text{out}}} e^{-i\vec{R} \cdot \vec{k}}. \tag{4.33}$$

Now suppose there are N electrons located at positions $\vec{R}_j$ ($j = 1, \ldots N$). The phasor describing the linear superposition of their scattered waves is

$$e^{i\vec{r}\cdot\vec{k}_{\text{out}}}\left(\sum_{j=1}^{N}e^{-i\vec{R}_j\cdot\vec{k}}\right).\tag{4.34}$$

The factor

$$F(\vec{k})=\sum_{j}e^{-i\vec{R}_j\cdot\vec{k}}\tag{4.35}$$

is called the **structure factor** for this charge distribution and represents the complex amplitude of the scattered wave in the direction $\vec{k}$.

On the atomic scale, of course, the positions of electrons are not precisely defined, and instead one works with an electron density function $\rho(\vec{R})$. If the amplitude of scattered light from point $\vec{R}$ is proportional to $\rho(\vec{R})$, then the overall amplitude scattered from the object in the direction $\vec{k}$ is proportional to

$$F(\vec{k})=\int\rho(\vec{R})\,e^{-i\vec{k}\cdot\vec{R}}\,dx\,dy\,dz,\tag{4.36}$$

where $\vec{R}=(x,y,z)$. We see that this amplitude is, up to a multiplicative constant, nothing but *the 3D Fourier transform of the electron density function* of the scattering object.

X-ray scattering instruments measure the intensity of this scattered light, for example by projecting it on an image sensor. Thus the scattering vector $\vec{k}$ is converted to a position on the image. However, film and other light sensors only measure the intensity of the light, which is proportional to $|F(\vec{k})|^2$. So by these optical methods, one can only recover the modulus of the Fourier transform, not its phase. This precludes a straightforward inverse transform (equation 3.17) to recover the charge density function $\rho(\vec{R})$. Much of the art of X-ray crystallography is dedicated to this problem of recovering the phases of Fourier components. For example, this involves creative ways of making known perturbations to that function by substituting heavy atoms in certain locations of the molecule.

4.4   Detecting Periodicity

4.4.1   Separating a Periodic Signal from Noise

Fourier transform methods are often used in signal processing to remove noise. When doing so, we typically apply some prior knowledge of what the power spectra of the signal and noise look like and try to implement a method that increases the ratio of the signal to noise. Suppose that an experiment measured the "Signal" trace shown in figure 4.9A. A quick first inspection reveals a large, low-frequency component to the trace, with noise superimposed on it. Let us assume that the only source of noise is a rapid and random fluctuation in the measured signal.

We wish to investigate whether other periodic signals are hidden in the noise. This is not immediately apparent to the naked eye. Again, one wants to decompose the signal into components at all frequencies. We expect that a consistent oscillation at a given frequency, no matter how small, will produce more power at that frequency than random noise does at the other frequencies. Notice that we are already

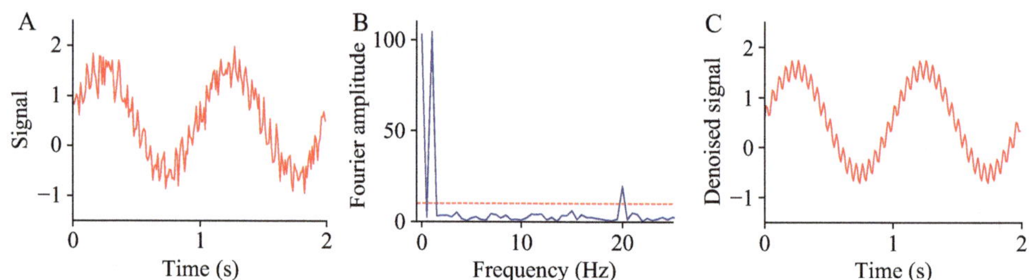

Figure 4.9
A noisy periodic signal. A: Signal observed during a hypothetical experiment. B: Fourier amplitude of that signal. The red dashed line is our threshold to separate the signal from the noise. C: Reconstruction of the signal after denoising.

making two assumptions: that our signal of interest is periodic and that the corrupting noise is not.

We proceed by calculating the Fourier transform $\hat{S}(k)$ of signal $S(t)$ (using the DFT) and then plotting the absolute "Fourier amplitude" $|\hat{S}(k)|$ as a function of the frequency (figure 4.9B).

There are three peaks in the frequency spectrum. The first one appears at frequency 0 and corresponds to the mean of the signal, that is its average offset from zero. The second large peak occurs at a frequency of 1 Hz. But we also notice a smaller, yet significant peak at 20 Hz. This indicates that there is indeed a measurable component to the signal at this frequency. Note that this was obscured in the raw trace (figure 4.9A), but appears clearly above the noise when we look at the signal in the frequency domain (figure 4.9B). We can now threshold the signal in the frequency domain and keep just these three components. If we perform an inverse Fourier transform we obtain the resulting "denoised signal" (figure 4.9C).

4.4.2 Analysis of a Cell Cycle Experiment

In this example, we illustrate the process of exploratory data analysis using Fourier transforms. You can find Python code for this project in the repository associated with this book.

The data set is from a venerable early paper about genomic analysis of the yeast cell cycle (Spellman et al. (1998)). The Stanford server listed in the publication is no longer available (as of November 2023), but the data set has been archived elsewhere, see for example, https://www.yeastgenome.org/dataset/Spellman_1998_PMID_9843569.

We will analyze the data underlying the article's figure 1A, the Alpha series. The starting point is a time series of expression for >6,000 genes. Each time series has 18 measurements, spaced by 7 minutes, and covering 2 cell division cycles. The goal is to replicate what the authors did here:

- Find the genes that are most highly modulated during the cell cycle
- Arrange those genes in order of their phase relative to the cell cycle

To do that, we will Fourier transform the time series of expression for each gene. Then focus on the Fourier component corresponding to the cell cycle (frequency $f = 2$ cycles per experiment). We will sort the genes by the relative strength of that Fourier

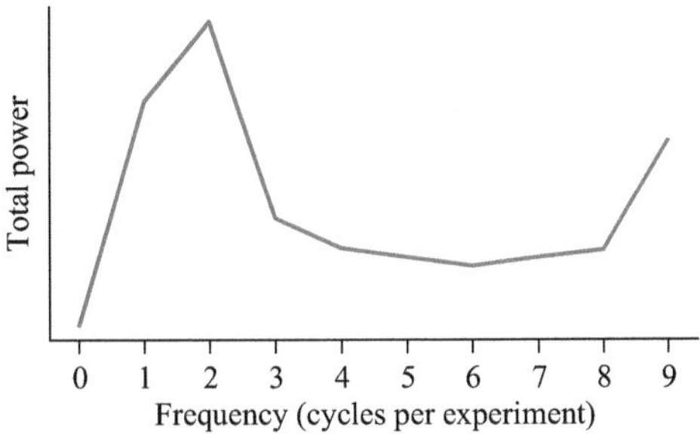

Figure 4.10
Average power spectrum over all genes.

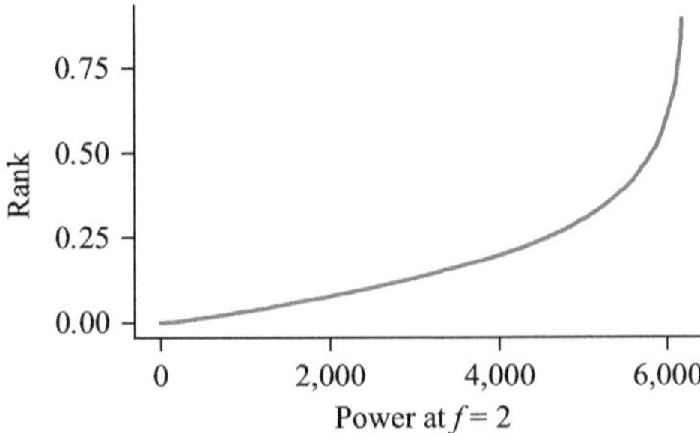

Figure 4.11
Cumulative distribution of power at $f = 2$.

component, pick the top 800 in that list, and then sort those in turn by the phase of the Fourier component at $f = 2$.

For a first look, let's calculate the average power spectrum across all genes. Because the data are real numbers, we can use the Real Fourier Transform (`rfft()` in Python). The power spectrum is the square of the absolute value of the Fourier transform. Then we sum the power over all genes to get the average power spectrum, plotted in figure 4.10.

Here are a few things to note:

- There is zero power at $f = 0$. That's a consequence of the authors' preprocessing of the data: they subtracted the mean from each time series.
- The highest power is at frequency $f = 2$, which corresponds to the cell division cycle, because the experiment spanned two successive cycles.
- The power at $f = 1$ (one cycle per two cell divisions) is also significant.
- And there is substantial power at $f = 9$, which corresponds to periodic alternation on subsequent samples.

Feel free to think harder about the meaning of $f = 1$ and $f = 9$, but here, we will follow the authors and focus on $f = 2$.

Next, let's calculate the relative power at $f = 2$ for each gene and sort them in that order, as plotted in figure 4.11.

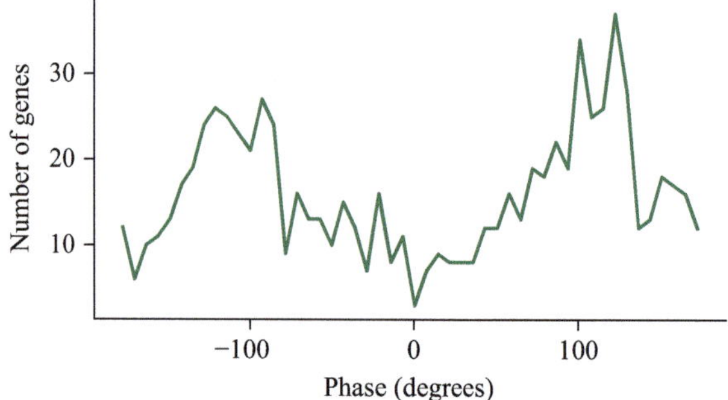

Figure 4.12
Distribution of phases among 800 highly modulated genes.

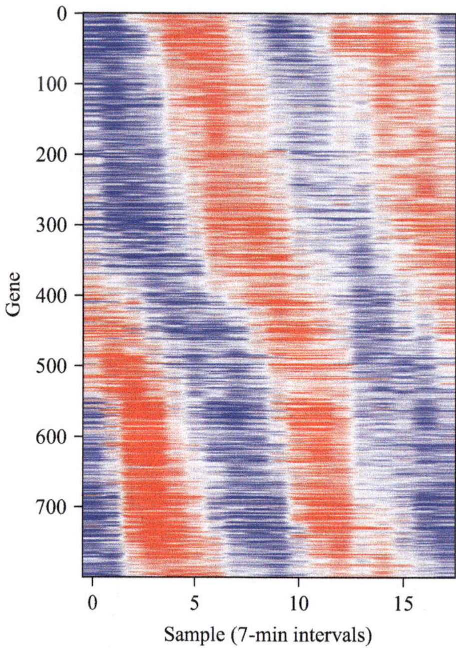

Figure 4.13
Expression time series of highly modulated genes.

It looks like a small number of genes are highly modulated. We will take the top 800 genes fom this list and analyze them further. First, we calculate the phase of each gene—namely, the argument of the complex Fourier coefficient at $f = 2$. Figure 4.12 presents a histogram of those phases.

Clearly, some phases are more common than others.

Now we sort the genes by their phases, in decreasing order, to match the convention in the original paper. Figure 4.13 presents the results, presented as a heat map, with each row representing a gene and each column representing a time point.

Finally, we note the close similarity to the article's figure 1A and pat ourselves and the authors on the back for a successful replication exercise.

4.5 Filtering

In signal processing, one often wants to suppress or enhance certain frequencies in the data, generally with the aim of separating some valuable signal from irritating noise. For example, a microscope camera may be affected by pixel-level noise that obscures the object of interest in the micrograph. Or an electronic recording may contain some high-frequency thermal noise originating in the amplifier. Most often, the noise is at higher frequencies than the signal, so one can improve the data by passing it through a "low-pass" filter. However, other occasions arise where one must remove a low-frequency component, for example because the signal experiences some slow drift unrelated to the phenomenon of interest. In those cases, a "high-pass" filter is appropriate.

A **linear filter** modifies the signal by convolving it with an impulse response, which is called the **kernel** of the filter:

$$g(t) = (f * k)(t) = \int_{-\infty}^{+\infty} f(t')\, k(t - t')\, \mathrm{d}t'. \tag{4.37}$$

The Fourier transform $\hat{k}(\omega)$ of the kernel is called the **transfer function** of the filter. It describes how the filter modifies the amplitude of each frequency component of the signal. Recall that the convolution amounts to multiplication in the frequency domain:

$$\hat{g}(\omega) = \hat{f}(\omega) \cdot \hat{k}(\omega). \tag{4.38}$$

For example, a low-pass filter can be implemented using a Gaussian kernel:

$$k(t) = \frac{1}{\sqrt{2\pi\tau^2}}\, \mathrm{e}^{-t^2/2\tau^2}. \tag{4.39}$$

Effectively, this computes a running average over nearby values of the signal, over a window of width τ, and thus it attenuates all variations that happen faster than τ.

Recall that the Fourier transform of a Gaussian is another Gaussian (see example 3.3), so one can easily visualize how this filter operates in the frequency domain:

$$\hat{k}(\omega) = \mathrm{e}^{-\omega^2\tau^2/2}. \tag{4.40}$$

A kernel of width τ will attenuate components of the signal at frequencies above the **cutoff frequency:**

$$f_c = \frac{\omega_c}{2\pi} = \frac{1}{2\pi\tau}. \tag{4.41}$$

The wider the kernel, the lower the frequency cutoff.

Figure 4.14 illustrates the process of Gaussian filtering for the same signal analyzed in figure 4.9. Note how this eliminates all the corrupting high-frequency noise, as well as the 20-Hz periodic component. The low-frequency component at 1 Hz is preserved.

There are many other options for filter kernels, and entire books have been written on how to design filters for particular purposes. Often it is more convenient to specify the filter in the frequency domain, as was done in the previous crystal analysis example (figure 4.7).

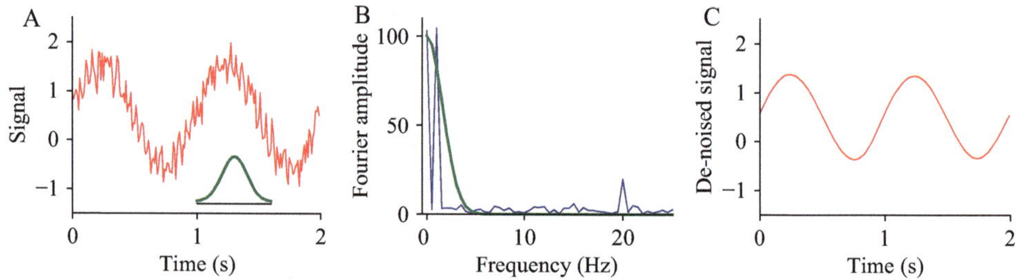

Figure 4.14
A: A periodic signal corrupted by high-frequency noise. Inset: Gaussian kernel function used for a low-pass filter (green). B: Fourier transform amplitude of the signal (blue) and the filter (green). Note that the filter has a high gain at 1 Hz but drops off rapidly beyond that. C: Filtered signal.

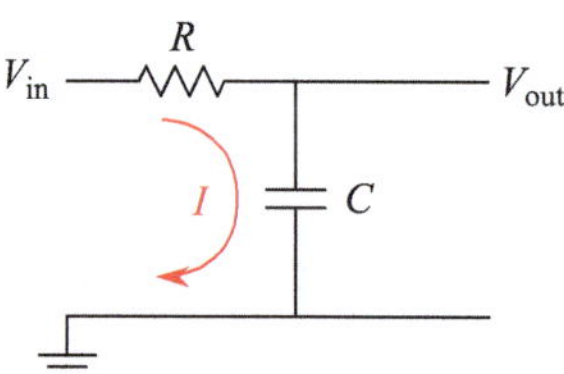

Figure 4.15
An electronic low-pass filter made from a resistor and a capacitor.

4.5.1 RC Filter

A very common motif in electronic circuits is the so-called **low-pass RC filter**. The term "low-pass" indicates that this filter passes low-frequency signals and blocks high-frequency signals. "RC" refers to the fact that it is made of a resistor and a capacitor. This simple circuit also has many biological analogs made of flesh and bone, kinases, or transduction pathways. Engineers will use the terms "low-pass filter" or "RC filter" to refer to any such system, so it is helpful for biologists to know what they mean by it.

This electric circuit (figure 4.15) has a time-varying input voltage $V_{\text{in}}(t)$ and produces an output voltage $V_{\text{out}}(t)$. We want to predict the output signal for any given input signal. To analyze the circuit's operation, we need to understand how the two components work. The resistor follows Ohm's law—the voltage across the resistor is proportional to the current flowing through it:

$$V_{\text{R}}(t) = R\,I(t), \tag{4.42}$$

where R is the resistance. The capacitor follows a different rule—the current through it is proportional to the time-derivative of its voltage:

$$I(t) = C\,\frac{\mathrm{d}}{\mathrm{d}t}V_{\text{C}}(t), \tag{4.43}$$

where C is the capacitance. Because the resistor and capacitor are connected in series, they both experience the same current $I(t)$. And for the same reason, the voltage across both components is the sum of the individual voltages:

$$V_{\text{in}}(t) = V_{\text{R}}(t) + V_{\text{C}}(t).\tag{4.44}$$

Noting that $V_{\text{C}}(t) = V_{\text{out}}(t)$, we can combine equations (4.42) to (4.44) to get

$$RC\frac{\mathrm{d}}{\mathrm{d}t}V_{\text{out}}(t) + V_{\text{out}}(t) = V_{\text{in}}(t).\tag{4.45}$$

Now, this is a fairly simple differential equation, and we could just solve it directly. But instead, let us Fourier-transform it. In section 3.2.4.1, we learned what the Fourier transform does to derivatives. Applying that to equation (4.45), we obtain

$$\widehat{V}_{\text{out}}(\omega) = \frac{1}{1 + i\omega\tau}\,\widehat{V}_{\text{in}}(\omega),\tag{4.46}$$

where

$$\tau = RC\tag{4.47}$$

is called the **time constant** of the RC circuit. Formally, this solves our problem: given any input waveform $V_{\text{in}}(t)$, we use equation (3.18) to compute its Fourier transform, then equation (4.46) to obtain the Fourier transform $\widehat{V}_{\text{out}}(\omega)$ of the output, and then equation (3.17) to transform this back to the time domain, resulting in the desired output waveform $V_{\text{out}}(t)$.

It pays to inspect this system further in the frequency domain. Note that equation (4.46) takes the form

$$\widehat{V}_{\text{out}}(\omega) = \widehat{h}(\omega)\,\widehat{V}_{\text{in}}(\omega),\tag{4.48}$$

where

$$\widehat{h}(\omega) = \frac{1}{1 + i\omega\tau}.\tag{4.49}$$

This means that the RC filter acts as a linear system, and $\widehat{h}(\omega)$ is its transfer function. If we apply a sinusoid input voltage

$$V_{\text{in}}(t) = \sin(\omega t),\tag{4.50}$$

the output will be a sinusoid of the same frequency, but with a different amplitude and phase:

$$V_{\text{out}}(t) = A(\omega)\sin(\omega t + \phi(\omega)).\tag{4.51}$$

The gain $A(\omega)$ and the phase shift $\phi(\omega)$ are given by the magnitude and phase of the transfer function:

$$
\begin{aligned}
A(\omega) &= \left|\widehat{h}(\omega)\right| \\
\phi(\omega) &= \arg\widehat{h}(\omega).
\end{aligned}\tag{4.52}$$

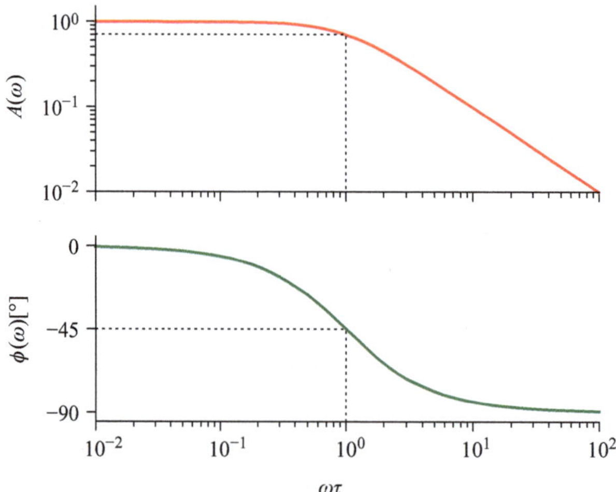

Figure 4.16
Bode plot of an RC low-pass filter.

The transfer function of a linear filter is commonly displayed as a **Bode plot** (figure 4.16) that graphs separately the magnitude and the phase as a function of frequency. Frequency and amplitude are usually plotted on a logarithmic axis. One can read from this plot that the low-pass filter leaves low-frequency signals unaffected: the magnitude is 1 and the phase remains zero. Signals near the **cutoff frequency** of

$$\omega_c = \frac{1}{\tau} = \frac{1}{RC} \tag{4.53}$$

start to eperience attenuation by the filter. At $\omega = \omega_c$ the amplitude is suppressed by $1/\sqrt{2} \approx 0.71$ and the phase of the output signal lags the input by 45 degrees. At much higher frequencies, the output amplitude declines inversely with frequency, as shown by a slope of -1 on the log-log plot of $A(\omega)$, and the phase lag approaches 90 degrees.

Finally, we return to the time domain and ask: What is the impulse response of this linear system? This is the output voltage $V_{\text{out}}(t)$ if the input is presented with a short impulse $V_{\text{in}}(t) = \delta(t)$. From equation (3.41), we know that the transfer function $\widehat{h}(\omega)$ is the Fourier transform of the impulse response $h(t)$, so we obtain $h(t)$ by an inverse Fourier transform:

$$h(t) = \int_{-\infty}^{+\infty} \widehat{h}(\omega)e^{i\omega t}\,d\omega = \frac{1}{2\pi} \int_{-\infty}^{+\infty} \frac{e^{i\omega t}}{1 + i\omega\tau}\,d\omega. \tag{4.54}$$

This integral can be evaluated by performing a contour integration in the complex plane and using Cauchy's residue theorem. The result is plotted in figure 4.17:

$$h(t) = \begin{cases} 0 & t < 0 \\ \frac{1}{\tau}\,e^{-\frac{t}{\tau}} & t \geq 0. \end{cases} \tag{4.55}$$

At negative times, the impulse response is zero because the impulse hasn't happened yet, and the filter is a purely causal system. At time zero, the output voltage jumps up

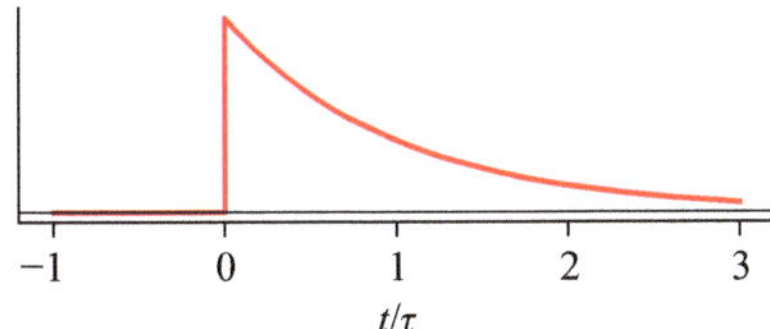

Figure 4.17
Impulse response of an RC low-pass filter.

because the impulse has deposited a charge on the capacitor. At positive times, the voltage decays exponentially with time constant $\tau = RC$ as the capacitor discharges through the resistor. Thus the impulse response of an RC low-pass filter is a decaying exponential. One encounters this behavior in many dynamical systems that are formally equivalent to the RC filter, yet are built from entirely different components.

4.5.2 Resolving Protein Dynamics

There are many instances where the dynamics of a biological system implement something like a low-pass filter. Here is an example that arises in the study of protein dynamics. Protein motors, such as myosin and kinesin, play important roles in the cell for the generation of forces or the shuttling of cargoes. To observe the movement of such a protein, one must make it visible, such as with a fluorescent tag. However, one can gather many more photons, and thus improve the spatial resolution, by attaching a large refractile object like a micron-sized bead. This is how the steplike motion of kinesin on microtubules was discovered (Svoboda et al., 1993).

Consider the situation depicted in figure 4.18: A kinesin molecule moves along a microtubule filament with velocity $v(t)$. It drags along a large bead on an elastic linkage that has length l and spring constant α. We can observe the velocity of the bead $u(t)$. How should one choose the properties of the bead and the linkage to optimize measurement of the kinesin's motion $v(t)$? Intuitively, one would like to make sure that the bead accurately tracks all the motions of the protein. This would be accomplished if the spring is very stiff and the bead is small, so it experiences little drag. Let us see if a more serious analysis supports these predictions.

The motion of the kinesin both extends the spring and drags the bead along:

$$v(t) = u(t) + \frac{\mathrm{d}}{\mathrm{d}t} l. \tag{4.56}$$

When the bead moves at velocity $u(t)$, it experiences a frictional drag force equal to

$$F_{\text{fric}} = -cu(t). \tag{4.57}$$

where c is the bead's frictional drag coefficient. The bead also experiences a force proportional to the extension of the spring:

$$F_{\text{spring}} = \alpha l. \tag{4.58}$$

Since no other forces act on the bead, we have $F_{\text{fric}} + F_{\text{spring}} = 0$, and thus

$$l = \frac{c\,u(t)}{\alpha}. \tag{4.59}$$

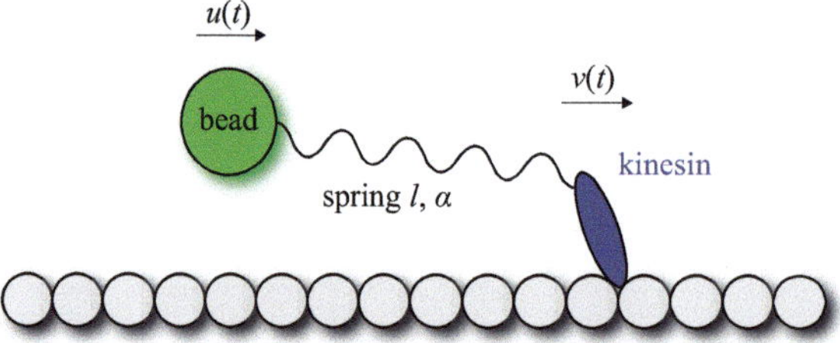

Figure 4.18
Kinesin stepper: The motion of a kinesin molecule (blue), which is too small to be directly observable can be followed by tagging it with a larger bead (green).

By substituting this into equation (4.56), we obtain

$$v(t) = u(t) + \frac{c}{\alpha}\frac{d}{dt}u. \tag{4.60}$$

Now we can Fourier-transform this with respect to time. The time derivative just introduces a multiplicative factor $i\omega$:

$$\hat{v}(\omega) = \hat{u}(\omega) + \frac{c}{\alpha}i\omega\hat{u}(\omega)$$
$$= \left(1 + \frac{i\omega c}{\alpha}\right)\hat{u}(\omega). \tag{4.61}$$

Here, we recognize that the bead+linkage acts as a linear system. The kinesin velocity $v(t)$ is the input signal, and the bead velocity $u(t)$ is the output signal. In the Fourier domain, input and output are simply related by the transfer function, $\hat{u}(\omega) = \hat{h}(\omega)\hat{v}(\omega)$. It follows from equation (4.61) that the transfer function $\hat{h}(\omega)$ is

$$\hat{h}(\omega) = \frac{1}{1 + \frac{i\omega c}{\alpha}}. \tag{4.62}$$

We observe that this transfer function resembles that of the electric low-pass filter of section 4.5.1. In fact, this system too acts like a low-pass filter with cutoff frequency $\omega_{\text{cutoff}} = \alpha/c$. If the kinesin molecule steps more often than this, these steps will be difficult to detect because the signal gets attenuated and may drop below the resolution of the optical instrument. One therefore wants this cutoff frequency to be as high as possible, which corresponds to a stiff spring with large α and a small bead with small c, just as we predicted. In practice, of course, there are compromises: An excessively large bead will impose a large force on the stepper, and a stiff molecular linkage may become bulky and interfere with the motion. Furthermore, a complete treatment of the process will need to include measurement noise, as discussed elsewhere in this chapter.

4.6 Sampling

Sampling is the critical step in data acquisition when a continuous signal gets turned into a discrete series of numbers. Most commonly, the continuous signal is a voltage trace coming out of some instrument, and it gets sampled and digitized at regular time

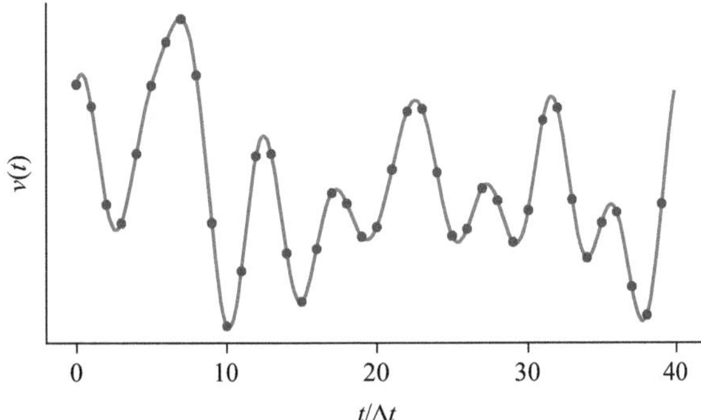

Figure 4.19
Sampling. A continuous voltage signal (green) gets sampled at periodic time intervals (blue dots).

intervals by an electronic circuit called an "analog-to-digital converter" (figure 4.19).[5] So a continuous real-world signal $v(t)$ gets reduced to a set of discrete samples $v_i = v(i\Delta t)$ taken at intervals of Δt.

Naively, one would think that sampling incurs a huge loss of information. After all, the continuous voltage curve has "infinitely many" points on it, but after sampling, we are left with observing only a finite number of points. The infinite number of points between the samples, therefore, must be lost.

Here is where the **Sampling theorem** makes a remarkable statement: **Under suitable conditions, the discrete samples contain all the information needed to perfectly reconstruct the continuous voltage function that produced them.** In that case, there is no information loss from sampling.

Those suitable conditions are simple: The input signal should vary slowly enough that the sequence of samples captures it well. **Specifically, the signal must have no sinusoidal components at frequencies above the so-called Nyquist frequency,** meaning that its power spectrum $P(f)$ must be zero for $f > f_{\text{Nyquist}}$.

The Nyquist frequency is defined as the highest frequency that can be reconstructed from the samples. It is equal to half the sampling frequency:

$$f_{\text{Nyquist}} = \frac{1}{2} f_{\text{sample}}. \tag{4.63}$$

For example, if the digitizer takes samples at a rate of 10,000 Samples/s, the input signal should have no power at frequencies above 5,000 Hz.

4.6.1 Aliasing

What happens if the input signal varies faster than the Nyquist frequency? Those frequency components cannot be reconstructed from the samples because their origin is perfectly ambiguous (figure 4.20).

---

5. A digital camera performs discrete sampling in space, when the continuous optical image gets digitized at discrete pixel locations. The considerations for spatial sampling are analogous to temporal sampling discussed here.

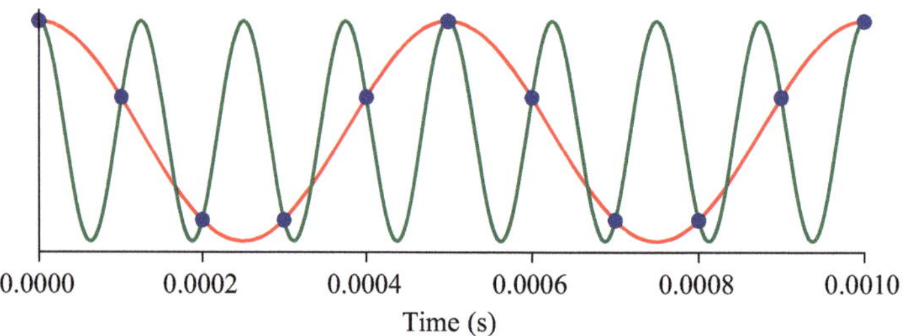

Figure 4.20
Aliasing. A sine wave with a frequency of 2 kHz (red) gets sampled at a rate of 10,000 samples/s (blue dots). Another sine wave with frequency 8 kHz (green) produces the exact same sequence of samples. So after sampling, one cannot reconstruct whether the original signal was at 2 kHz or at 8 kHz.

The same set of samples may have resulted from a signal below the Nyquist frequency or from any number of possible signals above the Nyquist frequency. This phenomenon is called "aliasing" because a high-frequency signal can masquerade in the sampled waveform as a low-frequency signal. If no other constraints are known about the input signal, then aliasing leads to information loss because the input can no longer be reconstructed from the samples.

How can we avoid aliasing? Simply make sure to eliminate frequency components above the Nyquist rate from the analog signal before the sampling step. Most commonly, this is done with an electronic low-pass filter that allows low frequencies through but blocks high frequencies (see section 4.5.1). A great deal of engineering expertise has gone into designing these filters because of their ubiquitous applications. So the usual design process for an acquisition system is this:

1. Determine the bandwidth of the signal that you want to observe, namely, the highest-frequency component that you will want to reconstruct for your purpose. This is the so-called cutoff frequency f_{cutoff} of your system.
2. Send the signal through a low-pass filter that removes all frequencies above f_{cutoff}.
3. Sample the output of the filter at a sampling rate:

$$f_{\text{sample}} \geq 2 f_{\text{cutoff}}. \tag{4.64}$$

A simple way to remember this rule is that **one needs at least two samples per cycle at the highest frequency present in the signal.**

4.6.2 Reconstruction

After sampling the function $v(t)$, we only have the discrete samples $v_i = v(i\Delta t)$ available. The Sampling theorem states that no information was lost in the process. That means that we should be able to reconstruct the full function $v(t)$ by interpolating somehow between those samples. Indeed the correct method for interpolation is the sinc function: Multiply each sample value by a sinc-shaped pulse centered on the sampling point and add the results:

$$v(t) = \sum_i v_i \operatorname{sinc}\left(\pi\left(\frac{t}{\Delta t} - i\right)\right), \tag{4.65}$$

where

$$\text{sinc}(x) = \frac{\sin x}{x}. \tag{4.66}$$

The sinc function is scaled so it has a value of 1 at the origin and of zero at multiples of Δt. Thus the reconstructed value at a sampling point, where we know the original signal exactly, is not contaminated by any of the other samples.

Note that we encountered the sinc function previously as the Fourier transform of a rectangular pulse as discussed in section (4.7). A dive into the proof of the Sampling theorem shows that it appears here for the same reason.

4.7   Optimal Estimation

4.7.1   Wiener Filter

Suppose that we want to measure a signal $s(t)$ that is corrupted by some additive noise source $n(t)$ to produce the observable $r(t)$, as shown in figure 4.21. How should one process $r(t)$ into an estimate $s_{\text{est}}(t)$ that comes as close as possible to reconstructing $s(t)$? Obviously, deciding this requires that we know something about the properties of the signal and the noise. Suppose in particular that we have their power spectra, $P_s(\omega)$ and $P_n(\omega)$, respectively. This is not uncommon because one usually knows what sort of process generated the signal and the noise. For example, $s(t)$ might be an electrical recording from nerve cells whose action potentials have a well-defined shape but an unknown occurrence time, and $n(t)$ might be electronic white noise with a uniform spectrum.

The simplest approach to estimating the input is to try a linear filter $f(t)$ applied to $r(t)$:

$$s_{\text{est}}(t) = \int r(t')f(t-t')\, dt'. \tag{4.67}$$

We want to choose $f(t)$ to minimize the discrepancy between the estimated $s_{\text{est}}(t)$ and the true $s(t)$, as measured by the mean squared error:

$$\chi^2 = \left\langle \int |s_{\text{est}}(t) - s(t)|^2 dt \right\rangle. \tag{4.68}$$

Here, the expectation value (section 6.3.3) is taken over many repeated versions of the experiment, each with a different instantiation of signal and noise.

In the Fourier domain, we have

$$\hat{s}_{\text{est}}(\omega) = \hat{r}(\omega)\hat{f}(\omega) \tag{4.69}$$

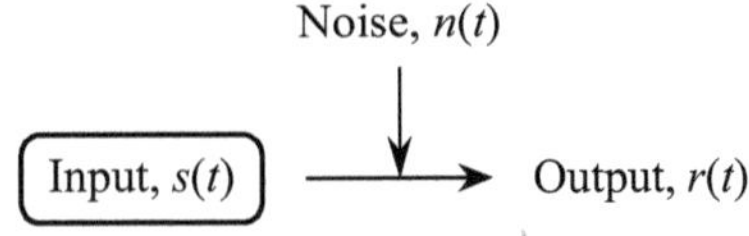

Figure 4.21
The Wiener filter: Estimating the input after corruption by additive noise.

and, from Parseval's theorem (section 3.2.3),

$$
\begin{aligned}
\chi^2 &= \frac{1}{2\pi} \left\langle \int |\hat{s}_{\text{est}}(\omega) - \hat{s}(\omega)|^2 d\omega \right\rangle \\
&= \frac{1}{2\pi} \left\langle \int | \left(\hat{s}(\omega) + \hat{n}(\omega)\right) \hat{f}(\omega) - \hat{s}(\omega)|^2 d\omega \right\rangle.
\end{aligned}
\tag{4.70}
$$

The minimization can be done separately at each frequency ω. After omitting the hats and ω for clarity and moving the average inside the integral, the integrand is

$$
\begin{aligned}
\left\langle |(s+n)f - s|^2 \right\rangle &= \left\langle |s|^2 \right\rangle |f - 1|^2 + \left\langle |n|^2 \right\rangle |f|^2 \\
&= P_s|f - 1|^2 + P_n|f|^2,
\end{aligned}
\tag{4.71}
$$

where we have used the following definition of the power spectrum (see section 3.2.3):

$$
P_s(\omega) = \left\langle |s(\omega)|^2 \right\rangle, \quad P_n(\omega) = \left\langle |n(\omega)|^2 \right\rangle
\tag{4.72}
$$

and the fact that signal and noise are uncorrelated, so

$$
\left\langle s^\star(\omega) n(\omega) \right\rangle = 0.
\tag{4.73}
$$

Because P_s and P_n are real and positive, one can see that the minimum value of equation (4.71) occurs for f on the real axis somewhere between 0 and 1. One finds the minimum by differentiating

$$
\begin{aligned}
0 &= \frac{\partial}{\partial f} \left(P_s(f - 1)^2 + P_n f^2 \right) \\
&= 2 \left(P_s(f - 1) + P_n f \right),
\end{aligned}
\tag{4.74}
$$

so the optimal filter function is given in the frequency domain by

$$
\hat{f}(\omega) = \frac{P_s(\omega)}{P_s(\omega) + P_n(\omega)}.
\tag{4.75}
$$

This particular version of noise reduction is called the **Wiener filter**. Depending on the particular goals of signal processing, one may end up with different optimal filters. For example, Wiener's classic text (Wiener, 1949) also discusses the optimal causal filter (no impulse response at negative times) or the optimal filter with an impulse response of specified length. This is relevant in the context of real-time signal processing with minimal delay, where one cannot look into the future for more signal values to average. For after-the-fact data processing, these concerns do not apply, and one may optimize the filter by different criteria. Thick books have been written on the theory and practice of filter design.

4.7.2 Deconvolution Microscopy

We saw in section 4.1 that a light microscope maps a single point in the sample into a small blur circle in the image plane, called the **point spread function** (PSF). This is the impulse response of the linear system. The output image $r(x)$ is the convolution of the sample's intensity profile $s(x)$ with the PSF $h(x)$: [6]

$$r(x) = \int s(x')h(x - x')\, dx'. \tag{4.76}$$

That PSF can be estimated analytically (see section 4.1.3) or measured experimentally by imaging a point-like sample. So if the PSF $h(x)$ is known, can we not simply undo the convolution in equation (4.76) and restore $s(x)$ to its full glory? Such image-processing steps are known as **deconvolution** (figure 4.22).

In the frequency domain,

$$\hat{r}(k) = \hat{s}(k)\hat{h}(k), \tag{4.77}$$

so the deconvolution offers itself simply as

$$\hat{u}(k) = \frac{\hat{r}(k)}{\hat{h}(k)}. \tag{4.78}$$

One concern is immediately apparent: this expression diverges when $\hat{h}(k) = 0$. At frequencies that are not represented at all in the impulse response, the image information is simply lost and cannot be restored. For optical PSFs, those are generally the high spatial frequencies.

However, even if $\hat{h}(k)$ is small but not strictly zero, we encounter a problem. Realistic measurements are usually corrupted by some degree of noise (figure 4.22). For example, a digital camera has pixel noise deriving from limited photon counts or the electronics of the image sensor. So a more realistic model of the image includes the noise as follows:

$$\hat{r}(k) = \hat{s}(k)\hat{h}(k) + \hat{n}(k), \tag{4.79}$$

where $\hat{n}(k)$ is the Fourier transform of the image noise. Now the brute-force deconvolution produces

$$\hat{u}(k) = \frac{\hat{r}(k)}{\hat{h}(k)} - \frac{\hat{n}(k)}{\hat{h}(k)}. \tag{4.80}$$

At frequencies where $\hat{h}(k)$ is small, the second term dominates, and the reconstruction will be almost all noise. The signal portion of the image, $\hat{s}(k)\hat{h}(k)$, is generally band-limited, because it got filtered through the low-pass PSF, whereas the noise portion $\hat{n}(k)$ can extend over all frequencies. For example, photon-counting noise is independent in each pixel, and thus it has equal power at all spatial frequencies. So the second problem is that brute-force deconvolution amplifies noise at high frequencies in the image. How can this be alleviated?

One solution is related to the Wiener filter (section 4.7.1). We want to process the image with a deconvolution filter $\hat{f}(k)$ that produces a reconstruction $\hat{s}_{\text{est}}(k)$ of the

---

6. For simplicity, we treat the image as 1D, but the results extend in obvious ways to 2D images.

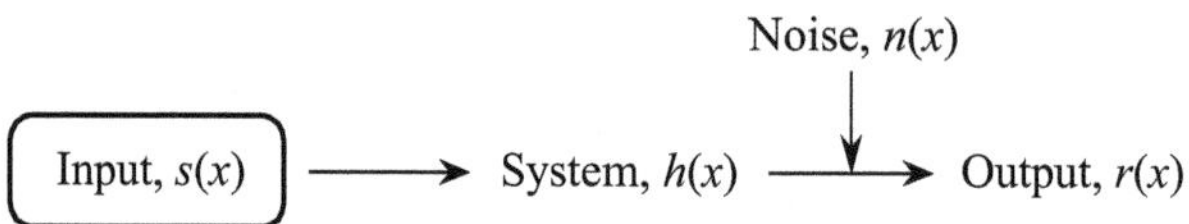

Figure 4.22
Deconvolution: An input $s(x)$ is processed by a linear system with impulse response function $h(x)$. It is then corrupted by additive noise $n(x)$ before being measured as $r(x)$. What linear filter $f(x)$ should we use to reconstruct the input $s(x)$?

sample

$$\hat{s}_{\text{est}}(k) = \hat{f}(k)\hat{r}(k), \tag{4.81}$$

which on average comes as close as possible to the real sample $\hat{s}(k)$. As before, we try to minimize the mean squared error between the estimate and the true sample, as shown in equation (4.70). In addition, we need to provide some knowledge about the statistics of signal and noise, particularly their power spectra $P_s(k)$ and $P_n(k)$. Following the same derivation as in equations (4.71) to (4.75), one finds the optimal reconstruction filter as follows:

$$\hat{f}(k) = \frac{\hat{h}^{\star}(k) \cdot P_s(k)}{|\hat{h}(k)|^2 P_s(k) + P_n(k)}. \tag{4.82}$$

Note that in the limit of very low noise P_n, this filter truly divides by the transfer function. If the impulse response is low-pass, as is often the case, the deconvolution filter in this regime is high-pass. In the limit of high noise, on the other hand, the filter becomes low-pass and furthermore is greatly attenuated.

Figure 4.23 illustrates this process. Here, the sample is a micrograph of microtubules obtained by super resolution imaging (Roubinet et al., 2018), which offers a much finer resolution than conventional light microscopy. To simulate the blurring by a confocal microscope, we convolved the image with an Airy disk. Then we added various amounts of photon-counting noise, as might be obtained from different exposure times. The reconstructions were computed with the Wiener filter, as in equation (4.82).

Note that for low amounts of noise, the Wiener filter recovers a good portion of fine detail that is not apparent in the blurred image. As the noise increases, it corrupts the information about high spatial frequencies, and the reconstruction adjusts by emphasizing the low frequencies. Note that the brute-force deconvolution in equation (4.78) fails entirely under all these noise conditions.

For many applications, this is a useful approach to deconvolution. Even when the power spectra are not well known, an educated guess at their shape will produce decent results. Real deconvolution microscopy software includes several other tricks, which often go beyond linear filtering, for example, see the Lucy-Richardson algorithm.

Because the objective information about high-frequency features of the sample is weak, the reconstruction algorithms must make educated guesses about those features based on all kinds of prior knowledge. This combination of data and prior expectations is often called **regularization**.

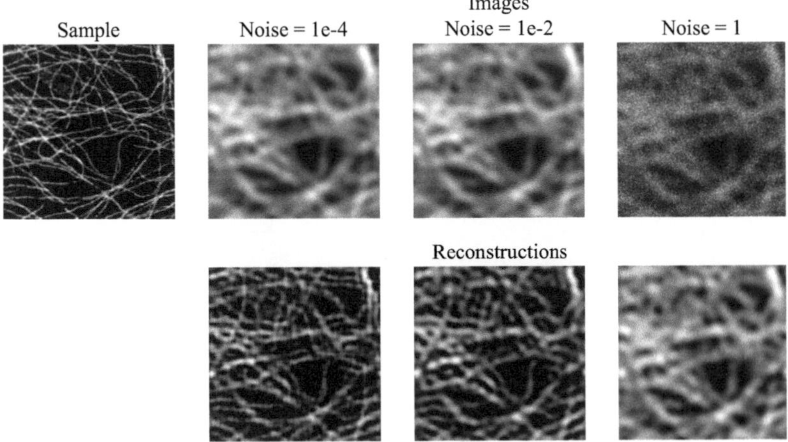

Figure 4.23
Reconstructions (bottom row) of a sample (top left, image width $= 10\ \mu$m) from a blurred image with varying amounts of noise (rest of top row).

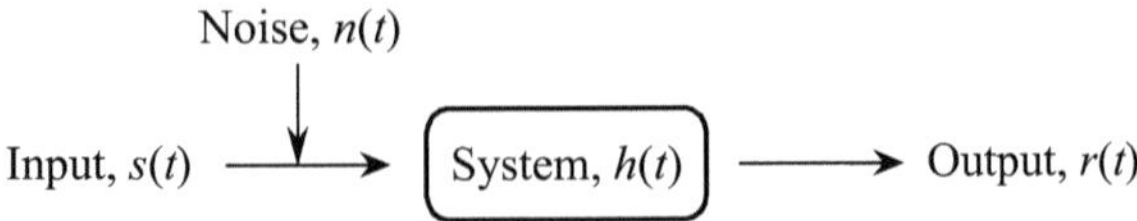

Figure 4.24
System identification: What is the transfer function $h(t)$?

4.7.3   System Identification and White Noise Analysis

In yet another version of linear systems analysis, the object of interest is the system itself. Here, the properties of the system are unknown, but we have a sense that it can be usefully approximated as linear. We want to determine the system's impulse response $h(t)$. This is called **system identification** (figure 4.24).

For example, we may be studying signal transduction in a cell, where an external stimulus gets transduced into a physiological response, or the dynamics of an organism's motor system where commands to the muscles get transformed into movements; or we may want to measure the spatial transfer function of a new optical imaging system.

The task then is to probe the system with various kinds of inputs, measure the resulting outputs, and infer the system's transfer function from the results. For the input function, there are many possible choices, and figure 4.25 illustrates some common ones.

A brief impulse seems very convenient because the resulting output directly reflects the system's impulse response (figure 4.25). Sinusoids are a common alternative (figure 4.25): By varying the frequency of the sinusoid and measuring amplitude and phase of the output, one obtains the Fourier transform of the impulse response (see section 3.2.5). Another popular option is "white noise"; namely, a random input stimulus as in figure 4.25. This may be constructed by setting each sample of the input waveform $s(t_i)$ to an independent random number.

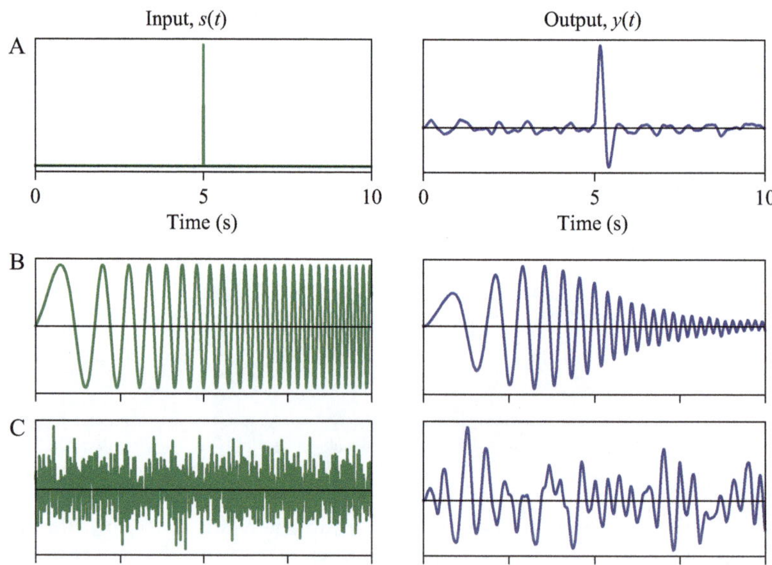

Figure 4.25
System identification: What is the transfer function $h(t)$? Some popular probe stimuli are impulses (A), sinusoids (B), and random noise (C).

How can we choose among these options? That decision is part of the experimenter's art. It may depend on what kinds of input waveforms are easy to produce, and what tools are available to measure the output. For example, if the input signal is the concentration of a ligand in water, it may be difficult to generate a white-noise waveform. A more fundamental consideration relates to the assumption of linearity. Many biological systems are intrinsically nonlinear, but they can be approximated as linear over some narrow regime of operation, such as by using a Taylor series (see sections 1.2.8 and 1.5.2.4). In that case, it matters what stimulus environment the experiment creates, since that defines the set point around which we approximate the system. An impulse stimulus can be delivered on a background of complete silence, whereas a white-noise stimulus creates a background that is broadly distributed in both time and frequency.

Suppose now that all those choices have been made, and the experiment performed, and we know the input $s(t)$ and the output $r(t)$.[7] We want to derive the impulse response $h(t)$ of the system. The noise $n(t)$ will limit the accuracy of our estimate, but suppose that we know a little bit about this noise, namely, its power spectrum $P_n(\omega)$. What is the best estimate of $h(t)$?

This problem is entirely analogous to deconvolution (section 4.7.2). In one case, the impulse response is known and we seek to reconstruct the input signal; in the other case, the input signal is known, and indeed under experimental control, but we seek the impulse response (compare figures 4.22 and 4.24).

As always, the problem is best formulated in the frequency domain:

$$\hat{r}(\omega) = \hat{h}(\omega)\left(\hat{s}(\omega) + \hat{n}(\omega)\right), \tag{4.83}$$

7. As always, these could be functions of space instead of time.

where s is the input stimulus used to probe the system, r is the resulting response, h is the system's impulse response, and n is a noise variable that corrupts the input signal and is statistically independent of the input. We seek a linear filter f that can be applied to the measured output r to estimate h:

$$\hat{h}_{\text{est}}(\omega) = \hat{f}(\omega)\hat{r}(\omega),\tag{4.84}$$

so as to minimize the mean squared discrepancy from the real $\hat{h}$:

$$\chi^2 = \left\langle \int |\hat{h}_e(\omega) - \hat{h}(\omega)|^2 \mathrm{d}\omega \right\rangle.\tag{4.85}$$

Again, one can minimize this expression with respect to $\hat{f}(\omega)$. Following the same logic as deriving the Wiener filter as in equations (4.71) to (4.75), one finds that the optimal estimate of the impulse response, under the conditions stated here, is

$$\hat{h}_e(\omega) = \frac{\hat{r}(\omega)}{\hat{s}(\omega)} \frac{1}{(1 + P_n(\omega)/P_s(\omega))}.\tag{4.86}$$

Practical white noise analysis usually involves additional steps of regularization to extract a realistic impulse response. For example, one typically has some expectation about the duration of the impulse response based on prior knowledge of the system's dynamics, whereas this treatment allows the impulse response to be as long as the entire duration of the experiment. By restricting the duration of the impulse response, one can substantially improve the rejection of noise in the measurement.

Exercise 5.1 (Matrix warm-up)

a) Compute the following products:

$$
\begin{bmatrix} 4 & 0 & 1 \\ 0 & 1 & 0 \\ 4 & 0 & 1 \end{bmatrix} \cdot \begin{bmatrix} 3 \\ 4 \\ 7 \end{bmatrix}
$$

$$
\begin{bmatrix} 1 & 0 & 0 \\ 0 & 1 & 0 \\ 0 & 0 & 1 \end{bmatrix} \cdot \begin{bmatrix} 5 \\ -2 \\ 3 \end{bmatrix}
$$

$$
\begin{bmatrix} 2 & 0 \\ 1 & 3 \end{bmatrix} \cdot \begin{bmatrix} 4 & 5 \\ 7 & 6 \end{bmatrix}
$$

b) Write the 3×3 matrices $\mathbf{A}$ and $\mathbf{B}$, which have the entries $a_{ij} = i + j$ and $b_{ij} = (-1)^{i+j}$.

Exercise 5.2 (Mechanics of matrix multiplication)

a) How many multiplications and additions are required to multiply two $n \times n$ matrices? Is it possible to reduce the number of operations significantly from the brute-force algorithm? Hint: Online research is allowed if necessary.
b) The transpose of a matrix is obtained by reflecting all the elements about the diagonal; that is, $\mathbf{A} = [a_{ij}]$, then $\mathbf{A}^{\top} = [a_{ji}]$. Show that $(\mathbf{AB})^{\top} = \mathbf{B}^{\top}\mathbf{A}^{\top}$.

Exercise 5.3 (Matrix puzzles) By trial and error, find examples of 2×2 matrices with real coefficients, such that

a) $\mathbf{A}^2 = -\mathbf{I}$
b) $\mathbf{B}^2 = 0$, whereas $\mathbf{B} \neq 0$

Exercise 5.4 (Differentiation is a linear operator) On a computer, a continuous function is approximated by a set of discrete values. For example, if the function $f(x)$ is defined over $0 \leq x \leq L$, one divides the range into n steps:

$$
x_i = i\frac{L}{n}, \quad i = 0, \ldots, n
$$

and represents the function by the set of values that it takes at those discrete points:

$$f_i = f(x_i).$$

Suppose that we want to calculate the derivative of the function

$$g(x) = \frac{d}{dx} f(x).$$

One approach is to approximate

$$g_i = g(x_i) \approx \frac{f_{i+1} - f_{i-1}}{2L/n}.$$

a) Show that this is a reasonable approximation. For example, check whether it works for $f(x) = x^2$. Does it work this well for all functions $f(x)$? Under what conditions will it fail seriously?

b) Some special treatment is needed at the "edges"; namely, for $i=0$ and $i=n$. Recommend a useful definition for g_0 and g_n.

c) Think of the set of values f_i as a column vector, and do the same for g_i:

$$\mathbf{f} = \begin{bmatrix} f_0 \\ \vdots \\ f_n \end{bmatrix}, \quad \mathbf{g} = \begin{bmatrix} g_0 \\ \vdots \\ g_n \end{bmatrix}$$

Show that this calculation of the derivative in parts (a) and (b) can be written in matrix form:

$$\begin{bmatrix} g_0 \\ \vdots \\ g_n \end{bmatrix} = \begin{bmatrix} a_{00} & \cdots & a_{0n} \\ \vdots & \ddots & \vdots \\ a_{n0} & \cdots & a_{nn} \end{bmatrix} \cdot \begin{bmatrix} f_0 \\ \vdots \\ f_n \end{bmatrix}.$$

Give the values of the matrix $\mathbf{A} = [a_{ij}]$.

d) Does the matrix $\mathbf{A}$ have an inverse? Hint: Is differentiation an invertible operation; that is, given $g(x)$, is there a unique $f(x)$ with $g(x) = \frac{d}{dx} f(x)$?

Exercise 5.5 (Solving equation systems) Write code for a function that allows you to solve a system of linear equations: Given an $n \times n$ matrix $\mathbf{A}$, and n-dimensional vector $\mathbf{y}$, find $\mathbf{x}$ to solve

$$\mathbf{A} \cdot \mathbf{x} = \mathbf{y}.$$

If your programming language has a built-in function with that exact feature, then write your code without using that.

Use your function to solve the following system of linear equations:

$$2x_1 - x_2 - 3x_3 = -21.25$$

$$-2x_1 + 5x_2 + x_3 = 46.5833$$

$$x_1 + 4x_2 + 6x_3 = 35$$

Exercise 5.6 (Change of basis) Consider a linear operator and a vector that are represented in a certain basis by

$$A = \begin{bmatrix} 2 & 1 & 0 \\ 1 & 2 & 0 \\ 0 & 0 & 5 \end{bmatrix} \quad \text{and } \mathbf{x} = \begin{bmatrix} 1 \\ 2 \\ 3 \end{bmatrix}.$$

Transform the operator and the vector to a new basis, whose basis vectors are

$$\mathbf{e}_1' = \begin{bmatrix} 1 \\ 1 \\ 0 \end{bmatrix}, \quad \mathbf{e}_2' = \begin{bmatrix} 1 \\ -1 \\ 0 \end{bmatrix}, \quad \mathbf{e}_3' = \begin{bmatrix} 0 \\ 0 \\ 1 \end{bmatrix}.$$

Hint: A "block-diagonal" matrix has elements that are zero except in square blocks arranged along the diagonal:

$$\mathbf{M} = \begin{bmatrix} \mathbf{A} & & \\ & \mathbf{B} & \\ & & \mathbf{C} \end{bmatrix}.$$

In such a case, the inverse is found by simply inverting the individual blocks:

$$\mathbf{M}^{-1} = \begin{bmatrix} \mathbf{A}^{-1} & & \\ & \mathbf{B}^{-1} & \\ & & \mathbf{C}^{-1} \end{bmatrix}.$$

Bonus question: Why?

Exercise 5.7 (Eigenvalues of an idempotent matrix) A matrix $\mathbf{A}$ is said to be idempotent if $\mathbf{A} \cdot \mathbf{A} = \mathbf{A}$.

a) Explain why all eigenvalues of $\mathbf{A}$ must be 0 or 1.
b) Show that

$$A = \begin{bmatrix} 1 & 0 & 0 \\ 0 & 3 & 6 \\ 0 & -1 & -2 \end{bmatrix}$$

is idempotent. Find its eigenvalues and eigenvectors and confirm the result in (a).

Exercise 5.8 (Diagonalization) Transform the matrix $\mathbf{A}$ and vector $\mathbf{x}$ to a basis in which $\mathbf{A}$ is diagonal:

$$A = \begin{bmatrix} 0 & -i \\ i & 0 \end{bmatrix}, \quad \mathbf{x} = \begin{bmatrix} 1 \\ a \end{bmatrix}.$$

Exercise 5.9 (Concatenation of two linear systems) Consider the situation depicted in figure 5.1: the concatenation of two linear systems $\mathcal{L}_1$ and $\mathcal{L}_2$, with impulse response functions $h_1(t)$ and $h_2(t)$, respectively. The output of $\mathcal{L}_1$ is fed directly into $\mathcal{L}_2$. Show

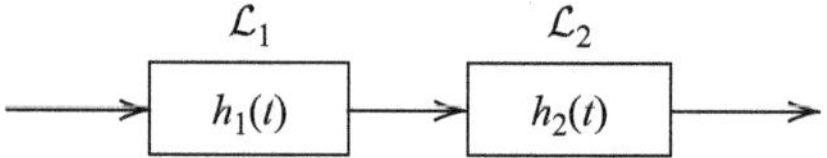

Figure 5.1
The concatenation of two linear systems.

that the concatenated system, from the input to $\mathcal{L}_1$ to the output from $\mathcal{L}_2$, is also a linear system. What is the impulse response function of this combined linear system?

Exercise 5.10 (Time translation as a linear system) Show that the operation of shifting a function in time,

$$f(t) \rightarrow f(t+s),$$

is a linear system. What is its impulse response? What is the transfer function?

Exercise 5.11 (Differentiating by convolution) Show that the operation of differentiating a function,

$$f(t) \rightarrow \frac{df}{dt},$$

is a linear system. What is its impulse response? What is the transfer function?

Exercise 5.12 (Shifting and scaling) Evaluate the Fourier transforms of the following in terms of the Fourier transform $\widehat{f}(\omega)$ of $f(t)$:

a) $\quad f(-t)$

b) $\quad f(at)$

c) $\quad f(t+a)$

d) $\quad e^{ia}f(t)$

where a is a constant.

Exercise 5.13 (Convolution commutes) Prove that the convolution in equation (3.11) is commutative. That is,

$$f * g = g * f.$$

Exercise 5.14 (Exponential representation of the delta function) The Fourier theorem relies intimately on the identity expressed in equation (1.62), which represents the delta function as an integral of complex exponentials:

$$\int \delta(x) = \frac{1}{2\pi} \int_{k=-\infty}^{\infty} e^{ikx}\, dk. \tag{5.1}$$

Try to derive this identity using methods of elementary calculus.

Hint: One trick is to "regularize" the integral with a factor that decays nicely as k goes to $\pm\infty$:

$$\int_{k=-\infty}^{\infty} e^{ikx}\,dk = \lim_{a\to 0}\int_{k=-\infty}^{\infty} e^{ikx}\,e^{-a|k|}\,dk. \tag{5.2}$$

Exercise 5.15 (Forward and inverse transform) Prove that the inverse transform in equation (3.17) really does invert the Fourier transform in equation (3.18).

Hint: You may have use for the identity in equation (1.62), which links the complex exponentials to the delta function.

Exercise 5.16 (Forward and inverse discrete Fourier transform) Verify the relations in equations (3.59) and (3.60), which define the discrete Fourier transform (DFT).

Exercise 5.17 (A proof of Parseval's theorem) Show that the integral of the power spectrum is equal to the total power in the signal waveform, as stated in equation (3.27).

Hint: Start with the right side and continue operating on it until it turns into the left side.

Exercise 5.18 (The Fourier transform turns differential equations into algebraic equations) Consider the following differential equation:

$$\frac{d^2 f}{dt^2} + \Omega^2 f = 0. \tag{5.3}$$

a) Show that the Fourier transform of $f(t)$ satisfies the following algebraic equation:

$$\left(-\omega^2 + \Omega^2\right)\hat{f}(\omega) = 0. \tag{5.4}$$

b) Solve equation (5.4) for $\hat{f}(\omega)$.
c) Use the inverse Fourier transform to find $f(t)$.

Hint: The solution is a linear combination of two complex exponentials.

Exercise 5.19 (Phase of the Fourier transform) What is the Fourier transform of $f(t) = \sin(t)$?

Hint: There are at least two ways to do this: (a) straightforward integration; (b) notice that the sine is a shifted version of the cosine and recall the effect of time shift on the Fourier transform.

Exercise 5.20 (Fourier transform of a Gaussian) What is the Fourier transform of $f(x) = e^{-x^2/2a^2}$?

In example 3.3, we derived this using contour integration in the complex plane, and in fact, this is a common way to solve Fourier-transform integrals. Now suppose that you had not studied complex analysis. Try to derive this Fourier transform using conventional tools from real analysis.

Hint: Use the symmetry of the function $f(x)$ to write the Fourier transform as an integral over real numbers. Then apply the "Feynman trick" of differentiating under the integral.

Exercise 5.21 (Fourier series of an even function) Show that an even function has a Fourier series with exclusively real coefficients, as in the example 3.5.

Exercise 5.22 (Sawtooth function) Find the general formula for the Fourier coefficients of the following 2π periodic function:

$$f(t) = t \qquad (-\pi < t \le \pi).$$

Exercise 5.23 (Eigenfunctions) Show that the complex exponentials e^{ikx} are eigenfunctions of the derivative operator $\frac{d}{dx}$. What are the corresponding eigenvalues?

Exercise 5.24 (Double transform) What is the Fourier transform of the Fourier transform of $f(x)$?

Exercise 5.25 (Circular convolution and padding) Using your favorite programming package, perform a convolution of two arrays in three ways:

a) Circular convolution in the time domain, using the definition in equation (3.69).
b) Circular convolution using the fast Fourier transform (FFT), based on equation (3.70). Check that the results are the same as in question (a).
c) Pad each of the arrays with zeros to twice the length, and perform a circular convolution using the FFT. Then truncate the result back to the original length. Explain how the result is related to the result of question (b).

Exercise 5.26 (Linkage for tracker beads) The discussion of the kinesin stepping experiments in section 4.5.2 concluded that the bead linker to a protein should be as small as possible and the linkage should be as stiff as possible. What are the possible drawbacks of a small bead? What about a stiff linkage? Discuss qualitatively how these drawbacks will affect the optimal choice for those two parameters. What would you need to know to make an informed quantitative choice?

Exercise 5.27 (Image deblurring) Use your camera to take an image with high-resolution information, such as text. Blur it with an Airy disk to simulate imaging though a microscope. Corrupt that with additive noise to simulate limited photon counts. Deconvolve. First, use a naive method (i.e., divide by the transfer function). Then use regularized deconvolution, as in section 4.7.2. Compare the results, such as by checking whether text is legible. You can also compare to a canned deblurring routine from your favorite programming language.

Exercise 5.28 (Convolutions in your eye) This exercise concerns transduction in photoreceptor cells of the retina, where photons of light are converted to electrical signals that are then further processed and interpreted by the brain (Schnapf et al., 1990) (see also sections 9.4.1 and 9.4.2). The input of the linear system is the light intensity as a function of time. The output is electrical current as a function of time.

When a photon gets absorbed by a photoreceptor neuron in your retina, it triggers a cascade of intracellular signaling reactions that ultimately leads to a change in the electric current across the cell membrane. This current, called the *single-photon response*, lasts for a few tenths of a second and then returns to zero. The photoreceptor cell is remarkably linear: when several photons get absorbed, the resulting current is simply the sum of all their single photon responses.

Consider the following questions:

a) Say that the waveform of the single-photon response is $h(t)$. What will be the photoreceptor response $R(t)$ to an arbitrary time-varying light stimulus, in which the rate of photon absorptions varies in time as $I(t)$, where

$$I(t)\, dt = \text{number of photons absorbed in short interval } dt?$$

b) A group of experimenters wanted to test the linearity of phototransduction. First, they measured the flash response of a photoreceptor by delivering a very brief flash of light containing N photons. The resulting current had the following waveform:

$$R_{\text{flash}}(t) = \begin{cases} 0 & t < 0 \\ A \sin(\omega t) e^{-\alpha t} & t \geq 0. \end{cases}$$

Given this measurement, and assuming linearity, what is the single-photon response $h(t)$?

c) The experimenters then applied a long step of light with the waveform

$$I(t) = \begin{cases} 0 & t < 0 \\ M & t \geq 0. \end{cases}$$

Assuming linearity, what current response did they expect to get?

6 Basics of Probability and Statistics

We see that the theory of probability is in essence only common sense reduced to calculation; it makes us appreciate with exactitude what reasonable minds feel by a sort of instinct, often without being able to account for it ... The most important questions in life are, for the most part, really only problems of probability.

Pierre Simon Laplace, *Théorie Analytique des Probabilités*, 1812

6.1 Motivation

The prolific scholar and occasional gambler Pierre Simon Laplace (1749–1827) is credited with extending the theory of probability from a calculation of odds in games of chance to broader applications in other domains. He may not have anticipated the profound relevance that probability has to not only life, but also the *life sciences*. The modern biologist encounters questions of probability calculus every step of the way; for example:

- At the most fundamental level, the one true theory of biology, namely, evolution, relies intimately on a process of random chance: mutations in the genome. Starting from there, many of the arguments of population genetics are fundamentally probabilistic (see section 9.3).
- The basic unit of biological function, the cell, is microscopic enough to be governed by the law of small numbers. There are typically one or two copies of every gene. Many important proteins and ribonucleic acid (RNA) species exist in only a handful of copies. Movement of molecules across the cell is governed by random processes like Brownian motion. In many of its operations, the cell is a stochastic machine, not a deterministic one.
- Biological measurements tend to exert much weaker constraints over the system under study than those in physics or engineering. Many of the variables that may affect the outcome of experiments remain uncontrolled and thus inject uncertainty into the interpretation. This puts a premium on experimental design and statistical analysis of the results, again drawing on concepts in probability.

In this and the following chapters, we will present the basic concepts of probability theory with an eye toward the applications encountered by the biologist. We will also honor Laplace's other observation, namely that much of this is simply formalized common sense, and stick to a practical rather than an abstract formulation of the ideas.

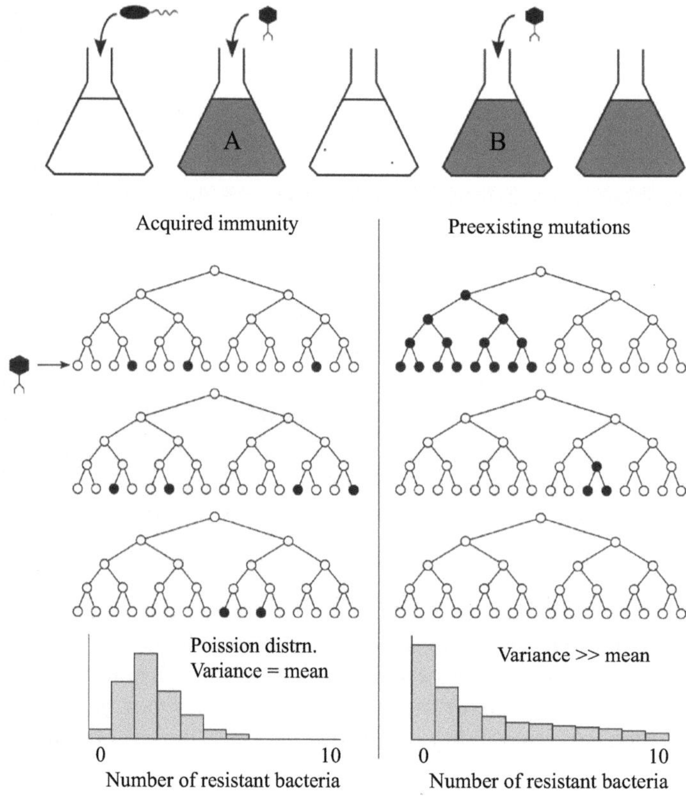

Figure 6.1
The Luria-Delbrück fluctuation test of acquired immunity. Top: Growth of bacterial culture A in a flask. Adding a bacteriophage kills almost all the cells. The survivors form a new culture B, which is resistant to the phage. Bottom: Growth of culture A through successive cell divisions. Mutations (black) could arise either from exposure to the phage (left) or spontaneously throughout the growth (right).

6.1.1 Luria and Delbrück's Fluctuation Test

In a classic 1943 paper, Salvador Luria and Max Delbrück reached a deep biological conclusion based on a purely probabilistic argument. They grew a large culture (A) of bacteria starting from a single cell (figure 6.1), then exposed it to a virus that preys on bacteria (phage). The phage killed almost all the bacteria in the culture except for a small number that survived. When a second culture (B) was grown from such a survivor cell, the phage could not kill it. Apparently, a mutation had occurred that made the surviving bacteria resistant to the phage. The question that arose was whether the phage had caused the mutation (the "acquired immunity" model) or whether the mutation had already been present among the bacteria in culture A before they were exposed to the phage (the "preexisting mutation" model).

Luria and Delbrück considered how mutations would arise under the two models. In the "acquired immunity" model, the phage acts on all the cells in the fully grown culture, and each cell has a very small chance of mutating, independent of all the others. The number of mutated bacteria is impossible to specify with certainty. But if one performed the same experiment many times, the number of mutants should follow the so-called Poisson probability distribution. Under the "preexisting mutations" model, the resistance mutation could have arisen at any stage during the growth of culture A, prior to exposure to the phage. If the mutation happened early, it would be passed to all the descendants of the lucky cell and produce a large number of mutants in the final population. If it happened late, there would be a much smaller number.

Common sense suggests (and a bit of calculation confirms) that the number of mutants in the final population should vary a lot more under the second model than under the first model if the experiment was repeated many times. This is indeed what happened: By simply counting mutant bacteria and noting the scatter of the results across otherwise identical experiments, the authors unequivocally concluded that mutations happened before and independent of attacks by the virus. They go on to remark that the "apparent lack of reproducibility of results... lies in the very nature of the problem and is an essential element for its analysis." In the next few sections, we will develop the concepts underlying their reasoning.

6.2 Events and Probabilities

An **event** is a proposition that can be true or false, typically about the outcome of a measurement. For example, "there are seven mutant bacteria in this culture" or "the top card is an ace."

The **probability** associated with an event is the fraction of times that we expect this event to happen if we performed the same measurement many times. That fraction ranges from 0 (never) to 1 (always). For example, in a shuffled deck of fifty-two cards, the event "the top card is an ace" will be true a fraction 1/13 of the time.[1]

$$A = \texttt{"the top card is an ace"}$$
$$P(A) = 1/13. \tag{6.1}$$

The **complement** A' of an event A is defined by the logical negation of the proposition for A. The probabilities of two complementary events sum to 1. For example:

$$R = \texttt{"the top card is red"}$$
$$R' = \texttt{"the top card is not red"} \tag{6.2}$$
$$P(R) + P(R') = 1.$$

6.2.1 Combined Events

The **union** of two events is defined by the disjunction of the two statements, for example

$$A \cup R = \texttt{"the top card is an ace or red"}. \tag{6.3}$$

The **intersection** of two events is defined by the conjunction of their statements, for example

$$A \cap R = \texttt{"the top card is a red ace"}. \tag{6.4}$$

---

1. There are alternative ways of defining probability, and philosophical battles are fought about them. However, none of that matters for the practical purposes of a research scientist.

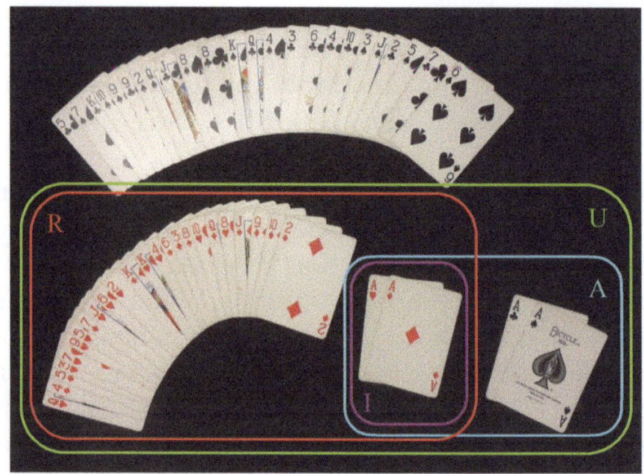

Figure 6.2
The union (U) and intersection (I) of the events R="the card is red" and A="the card is an ace".

The probabilities of these combined events are related by

$$P(A \cup R) = P(A) + P(R) - P(A \cap R).$$ (6.5)

This is understood easily by looking at a Venn diagram of all the possible outcomes of the measurement (figure 6.2).

6.2.2 Statistically Independent Events

Two events are **statistically independent** if the outcomes cannot influence each other and they do not share any common influences. In that case, the probability of the intersection—also called the "joint probability"—is the product of the individual probabilities:

$$P(A \cap R) = P(A) \cdot P(R), \quad \text{if } A \text{ and } R \text{ are statistically independent.}$$ (6.6)

This holds for the events A and R in the example (6.3) above:

$$P(\texttt{"red ace"}) = P(\texttt{"red"}) \cdot P(\texttt{"ace"}) = \frac{1}{2} \cdot \frac{1}{13} = \frac{1}{26}.$$ (6.7)

6.2.3 Conditional Probability

In the more general case, observing event A may offer some information about event B. In that case, knowing that A occurred changes the probability for B. One defines the **conditional probability** $P(B|A)$ as the probability of B given the knowledge of A.

Conditional and joint probabilities are related by **Bayes' rule**:

$$P(A \cap B) = P(A) \cdot P(B|A) = P(B) \cdot P(A|B).$$ (6.8)

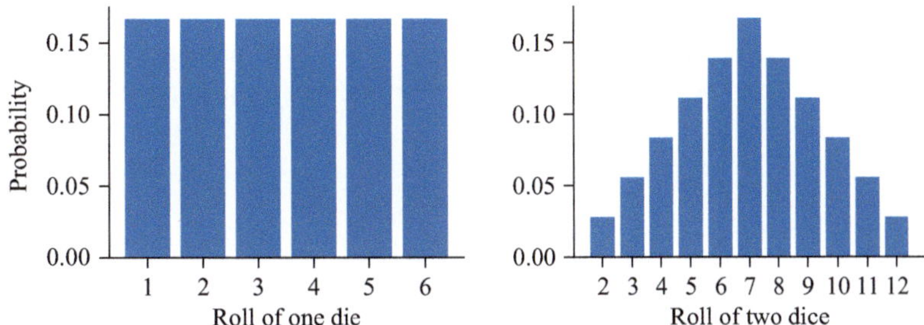

Figure 6.3
Probability distributions of the roll of one die (left) and the roll of two dice added together (right).

As Laplace said, this is just formalizing common sense: the fraction of time that A and B happen together is the fraction of time when A happens times the fraction of those instances when B also happens.[2]

6.3 Discrete Random Variables

A **random variable** is a measurement whose outcome can take on different values. A **discrete random variable** has a discrete set of possible outcomes. Examples are the number of mutant bacteria in a culture, or the number of molecules of a protein in the cell. A **continuous random variable** can take on a continuous range of values, such as the position of a molecule on a gel, or the light intensity of a fluorophore. We begin by covering discrete random variables before moving to the continuous case.

Consider a discrete random variable X that can take $x_1, \ldots, x_n$ as possible values. Each of these outcomes is associated with a probability $P(x_i)$, such that

$$P(x_i) = \text{probability that } X \text{ is equal to } x_i. \tag{6.9}$$

$P(x_i)$ is called the **probability distribution** of X. For discrete random variables, one sometimes uses the term **probability mass function**.

6.3.1 Display
Probability distributions are often displayed as a bar graph, with the outcomes x_i listed on the horizontal axis, and the associated probabilities drawn as bars of length $P(x_i)$, as shown in figure 6.3.

---

2. A word of warning: Notation in probability texts can be sloppy at times. For example, in the statement of Bayes' rule, a naive reader might think that $P(B|A)$ is the same function as $P(A)$, just evaluated at a different argument. Instead, one needs to read the argument list carefully because it defines a different function. Also, many writers seem unaware that the alphabet has letters other then P for denoting functions. Fixing these problems in this book would give the student a false sense of security, and thus we will follow the convention for sloppy notation.

6.3.2 Normalization

The various outcomes x_i are all mutually exclusive: only one of them can happen. They are also exhaustive: one of them must happen. Therefore, their probabilities must sum to 1:

$$\sum_i P(x_i) = 1. \tag{6.10}$$

One says that the probability distribution is **normalized**.

6.3.3 Mean of a Distribution

If we measure a random variable X many times, drawing from the same probability distribution over and over, what is the average value we will get? The answer to this question is called the expectation value of the random variable and it is given by the mean of its probability distribution. One obtains the mean by averaging all the possible outcomes of the random variable, weighted by the probability of each outcome:

$$\begin{aligned} \mathrm{E}\,[X] &= \text{mean of } X \\ &= \sum_i P\,(x_i) \cdot x_i. \end{aligned} \tag{6.11}$$

There are many equivalent notations for the mean, including

$$\mathrm{E}\,[X] \equiv \langle X \rangle \equiv \bar{X}. \tag{6.12}$$

6.3.4 Variance and Standard Deviation

The second-most-frequently asked question is "How wide is the distribution?" We begin by calculating the average squared deviation from the mean,[3] which is called the **variance**:

$$\begin{aligned} \mathrm{Var}\,[X] &= \text{variance of } X \\ &= \sum_i P\,(x_i) \cdot (x_i - \mathrm{E}[X])^2. \end{aligned} \tag{6.13}$$

If the most frequent outcomes all cluster close to the mean, then the squared terms $(x_i - \mathrm{E}[X])^2$ will be small and the variance is low. By contrast, if the outcomes scatter widely, both below and above the mean, they will contribute a large variance.

The variance can also be expressed as follows (see exercise 10.6):

$$\begin{aligned} \mathrm{Var}\,[X] &= \mathrm{E}[X^2] - \mathrm{E}[X]^2 \\ &= \text{"the mean of the square minus the square of the mean."} \end{aligned} \tag{6.14}$$

Because of the squaring operation, the variance $\mathrm{Var}\,[X]$ has different units from the random variable X itself. To obtain a measure of width on the same scale as X one

---

3. If we did not square this distance, positive and negative values would cancel each other out, such that a distribution that is symmetric around its mean would automatically give zero, regardless of its width.

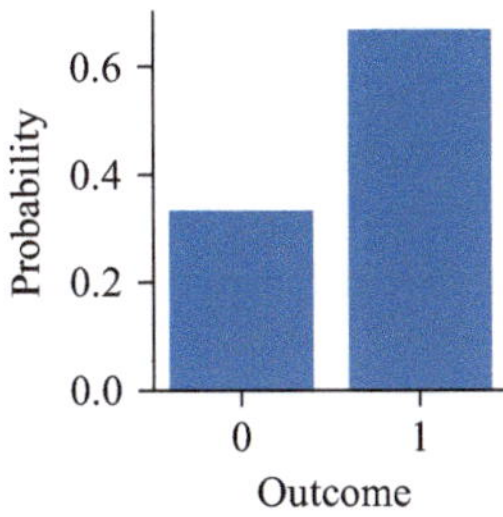

Figure 6.4
Distribution for a Bernoulli trial that delivers "1" with probability $p = 2/3$.

therefore takes the square root of the variance, which is called the **standard deviation**:

$$\text{Std}\,[X] = \text{standard deviation of } X$$
$$= \sqrt{\text{Var}\,[X]}. \tag{6.15}$$

6.3.5 Bernoulli Distribution

This is the simplest possible random variable, but as we will see, it gives rise to many interesting processes. A **Bernoulli trial** is the outcome of a single coin toss, where we allow the coin to be biased in favor of one outcome or the other[4]. One generally labels the two possible outcomes as "0" and "1." The probability of obtaining "1" is defined as p. Obviously, the probability of getting "0" is $1 - p$. An example is shown in figure 6.4.

Formally, if the random variable X follows the Bernoulli distribution, one writes

$$X \sim \text{Bern}(p), \tag{6.16}$$

which means

$$P(X = 1) = p$$
$$P(X = 0) = 1 - p. \tag{6.17}$$

6.3.5.1 Mean and variance If $X \sim \text{Bern}(p)$, then

$$\text{E}\,[X] = p \cdot 1 + (1 - p) \cdot 0 = p$$
$$\text{Var}\,[X] = p \cdot (1 - p)^2 + (1 - p) \cdot (0 - p)^2 = p\,(1 - p). \tag{6.18}$$

6.3.6 Binomial Distribution

The binomial distribution governs a process that involves many independent Bernoulli events. For example, in a sequence of 10 coin tosses, what is the probability of getting 7 "heads"? As it turns out, many processes in the natural world involve the accumulation of binary events, each of which may happen or not. The cumulative effect of these is governed by the binomial distribution.

---

4. You might think that this object is so trivial that it doesn't deserve the name of a famous mathematician, but you would be wrong. The "Rademacher distribution" is reserved for the special case of the Bernoulli distribution where $p = 1/2$, literally the toss of an unbiased coin.

Consider a series of n **identically and independently distributed (i.i.d.)** Bernoulli events. Here, "identically" means that all the trials use the same probability p, and "independently" means that the outcome of one trial is not influenced by any other trial. We want to know the probability distribution of the number of "1"s that occurred in those n trials. This random variable follows the **binomial distribution**:

$$X \sim \text{Bin}(n, p), \tag{6.19}$$

with[5]

$$P(X = m; n, p) = \binom{n}{m} p^m (1 - p)^{n-m}, \tag{6.20}$$

where

$$\binom{n}{m} = \frac{n!}{m! \, (n - m)!} \tag{6.21}$$

and

$$n! = n \cdot (n - 1) \cdot \ldots \cdot 2 \cdot 1. \tag{6.22}$$

To understand this result, we first ask how many distinct sequences of n outcomes there are with exactly m "1"s. The answer is $\binom{n}{m}$, pronounced "n choose m", namely, the number of combinations of m things of one type and $n - m$ things of another[6]. Second, for any one of these sequences, what is the probability that it will happen? On each trial in the sequence, "1" occurs with probability p and "0" with probability $1 - p$. Because the trials are statistically independent, the probability of the entire sequence is the product of the individual probabilities. Therefore, all sequences with m "1"s and $n - m$ "0"s occur with the same probability: $p^m(1 - p)^{n-m}$. Finally, the event "exactly m 1's" is the union of all those sequences, so we have to add their individual probabilities, leading to equation (6.20).

Figure 6.5 plots some examples of the binomial distribution. Here, we abandoned the conventional bar graph in favor of a line graph, which accomplishes the same thing with less ink and allows the plotting of multiple distributions in the same panel. One can make several qualitative observations right away:

- In each case, the distribution has a bell shape.
- For higher p, the distribution is shifted farther to the right, as expected by common sense.
- For higher n, the bell shape is narrower, at least when measured relative to the range of possible outcomes. This is perhaps less obvious from common sense.

---

5. The notation $P(X = m; n, p)$ in (6.20) should be understood as the probability that the random variable X is equal to m, where the parameters n and p are given.

6. Many problems in probability require some kind of approach to counting sequences or patterns. The field of **combinatorics** is devoted to these methods.

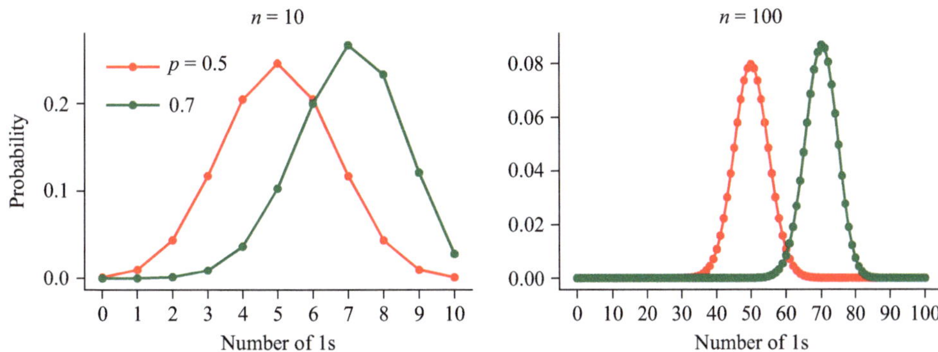

Figure 6.5
Binomial distribution for $n = 10$ (left) or $n = 100$ (right) and two values of p.

6.3.6.1 Mean and variance If $X \sim \text{Bin}(n, p)$, then

$$E[X] = np$$
$$\text{Var}[X] = np(1 - p).$$

(6.23)

This follows from the results for a Bernoulli random variable $\text{Bin}(n, p)$ shown in equation (6.18) and the fact that a binomial random variable is just a sum of n independent Bernoulli random variables. As we will see in section 6.5.4, the mean and variance of sums of independent random variables are just the sums of the means and variances.

6.3.7 Poisson Distribution
The **Poisson distribution**

$$X \sim \text{Pois}(\mu)$$

(6.24)

is an approximate expression for the binomial distribution $\text{Bin}(n, p)$ in the limiting case where n is very large and p is very small, but $n \times p = \mu$ is some reasonable number. Under those conditions,

$$P(X = m) = e^{-\mu} \frac{\mu^m}{m!}.$$

(6.25)

In general, this applies when there is a large number of independent binary trials, each of which has only a tiny probability of success. For example:

- X is the number of mutant bacteria growing on a Petri dish. There may be $n = 10^9$ bacteria spread on the dish, and each has a tiny chance of $p = 10^{-8}$ of having acquired the right mutation that allows it to grow. The number of mutant bacteria will follow the distribution $X \sim \text{Pois}(10)$. This was what Luria and Delbrück (1943) predicted under the hypothesis of "acquired immunity" (as discussed in section 6.1.1).
- X is the number of photons absorbed by a single cone photoreceptor in your eye in the next second. Maybe 1 trillion photons entered your eye in that second, but each had only a one-in-a-billion chance of getting absorbed by that particular cone photoreceptor. The number of absorbed photons will be distributed like $X \sim \text{Pois}(1000)$.

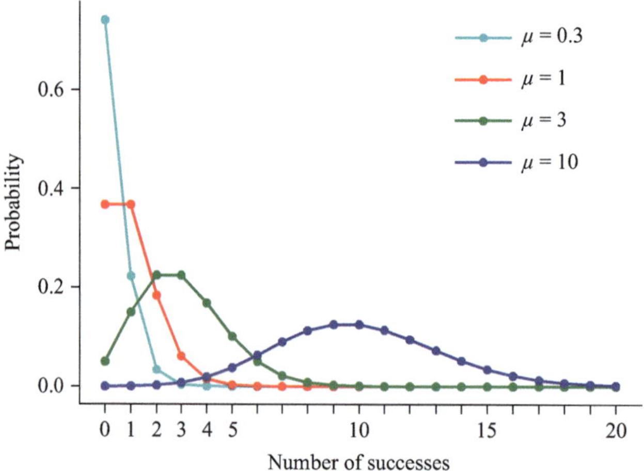

Figure 6.6
Poisson distribution for various values of the mean μ.

6.3.7.1 Mean and variance Note that the Poisson distribution has a single parameter, μ. Conveniently, that parameter is also the mean and the variance of the distribution.
If $X \sim \mathrm{Pois}(\mu)$, then

$$
\begin{aligned}
\mathrm{E}\,[X] &= \mu \\
\mathrm{Var}\,[X] &= \mu.
\end{aligned}
\tag{6.26}
$$

Both of these relations follow immediately from equation (6.23) in the limit where

$$
p \to 0, \quad np = \mu.
\tag{6.27}
$$

6.3.7.2 Display Figure 6.6 shows some examples of the Poisson distribution for different values of the mean parameter μ.

6.3.7.3 Comparison to binomial Figure 6.7 shows that the Poisson approximation to the binomial distribution works quite well, even at moderate values of n.

6.3.8 Gaussian Distribution

Another interesting limit to the binomial distribution $\mathrm{Bin}\,(n,p)$ arises when n is very large and p is something reasonable, such that the typical number of successes m is also very large. In that case,

$$
X \sim \mathcal{N}\,(\mu, \sigma)
\tag{6.28}
$$

follows the **Gaussian distribution**–namely,

$$
P\,(X = m) = \frac{1}{\sqrt{2\pi}\,\sigma}\, e^{-\frac{(m-\mu)^2}{2\sigma^2}}.
\tag{6.29}
$$

This is often called the **normal distribution**, hence the symbol $\mathcal{N}$. Many random variables in the natural world follow this normal distribution for reasons that we will understand soon (section 6.6).

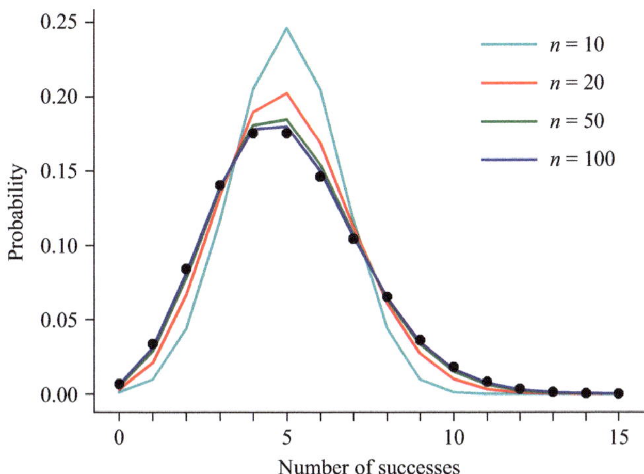

Figure 6.7
Poisson distribution (dots) for $\mu = 5$, along with the corresponding binomial distribution Bin(n, p) (lines) for various values of n and $p = \mu/n$.

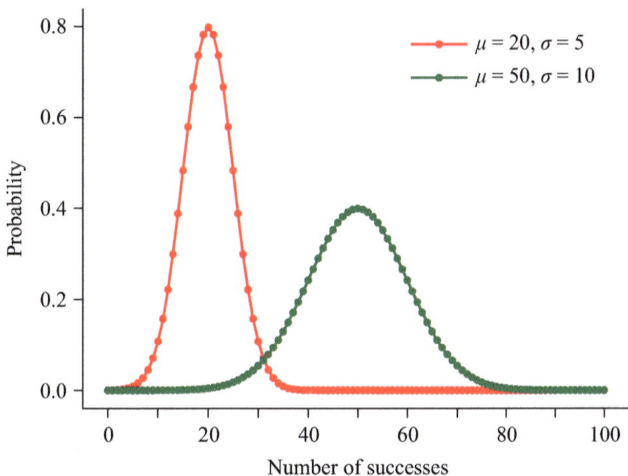

Figure 6.8
Gaussian distribution for various values of the mean μ and standard deviation σ.

6.3.8.1 Display The Gaussian distribution is a classic bell curve (figure 6.8). Its maximal value occurs at $m = \mu$. Note that the bell is symmetric around μ because $P(X = m)$ is the same for $m = \mu - d$ and $m = \mu + d$. The decline on either side of the maximum is governed by the value of σ: it drops by a factor of $e^{-1/2} \approx 0.606$ when m is one σ away from the maximum.

6.3.8.2 Mean and variance The two parameters of the Gaussian distribution, μ and σ, correspond to its mean and standard deviation: if $X \sim \mathcal{N}(\mu, \sigma)$, then

$$E[X] = \mu$$

$$\text{Var}[X] = \sigma^2. \tag{6.30}$$

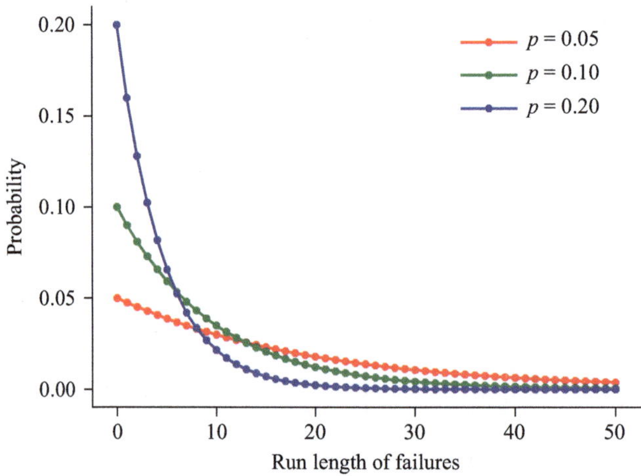

Figure 6.9
Geometric distribution for various values of success probability p.

6.3.9 Geometric Distribution

Consider a sequence of independent Bernoulli trials with probability of success p. What is the probability that the first success occurs after a run of k failures? In the context of coin tossing, one might ask about the probability of a run of k "tails" before the first "heads" occurs. This run-length variable k is distributed according to the **geometric distribution**:

$$X \sim \text{Geom}(p) \tag{6.31}$$

with

$$P(X = k) = (1 - p)^k p. \tag{6.32}$$

One can understand this result directly from the fact that the Bernoulli trials are statistically independent, and that the compound event requires exactly k failures, each with probability $1 - p$, followed by 1 success, with probability p.

6.3.9.1 Display The geometric distribution (figure 6.9) has a maximum at $k = 0$: the most probable run of failures is "no failure." From there, it declines like an exponential function.

6.3.9.2 Mean and variance If $X \sim \text{Geom}(p)$, then

$$\begin{aligned} \text{E}[X] &= \frac{1 - p}{p} \\[2mm] \text{Var}[X] &= \frac{1 - p}{p^2}. \end{aligned} \tag{6.33}$$

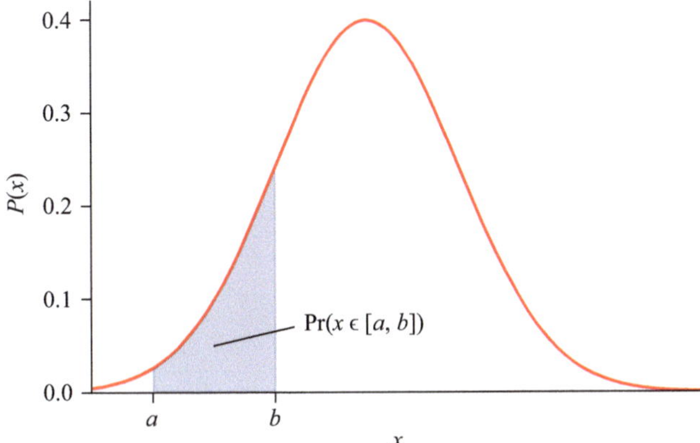

Figure 6.10
The probability density function of a continuous random variable.

6.4 Continuous Random Variables

A **continuous random variable** can take on a continuous range of values. Some obvious examples are the time of an event, the spatial position of an event, the light intensity of a fluorophore, the membrane voltage of a neuron, and many others. While it is true that any scientific measurement of such variables yields a digitized—and therefore discrete—version (see section 9.2.1), it still helps to be able to reason about the underlying continuous variables.

6.4.1 Continuous Probability Distribution

For a continuous variable X, the probability that it takes on any specific real number $X = a$ is zero except in trivial cases, like a particle being locked in to one position in space. However, one can ask about the probability of finding X inside some interval $[a, b]$. This leads to the definition of the **probability distribution function** $P(x)$ for a continuous random variable:

$$\int_a^b P(x)\, dx = \text{probability that } x \in [a, b]. \tag{6.34}$$

In graphical terms (figure 6.10), the area under the curve of $P(x)$ between a and b is the probability of finding x in $[a, b]$.

If one makes the interval infinitesimally small, say $[a, a + dx]$, then the area under the curve is simply the width of the interval times the height of the curve:

$$P(a)\, dx = \text{probability that } x \in [a, a + dx]. \tag{6.35}$$

This suggests an interpretation of $P(a)$ as a density of "probability per unit x." In fact, a continuous probability distribution is often called a **probability density function**.

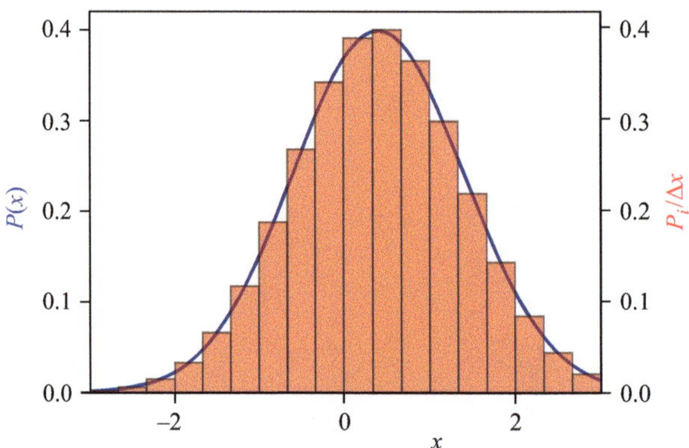

Figure 6.11
A continuous distribution as the limiting case of a discrete distribution.

6.4.2 Relation to Discrete Distributions
Imagine dividing the x-axis into many small bins of width Δx: $x_i = i \cdot \Delta x$. Then consider the probability that x will fall into bin i. That is a discrete distribution:

$$P_i = \Pr(x \in [x_i, x_i + \Delta x]). \tag{6.36}$$

As illustrated in figure 6.11, as the bin width Δx goes to zero, the discrete distribution P_i will approach the continuous distribution $P(x)$, namely

$$\frac{P_i}{\Delta x} \xrightarrow[\Delta x \to 0]{} P(x). \tag{6.37}$$

6.4.3 Normalization
From equation (6.34), it follows that the probability density must be positive or zero:

$$P(x) \geq 0. \tag{6.38}$$

Also, because x must take on *some* value, the area under the entire curve equals 1:

$$\int_{-\infty}^{\infty} P(x)\, dx = 1. \tag{6.39}$$

6.4.4 Mean, Variance, and Higher Moments
For a continuous random variable, the expectation value is computed by the weighted integral:

$$E[X] \equiv \langle X \rangle = \int_{-\infty}^{\infty} xP(x)\, dx. \tag{6.40}$$

Similarly, the second and higher moments are:

$$\langle X^2 \rangle = \int_{-\infty}^{\infty} x^2 \cdot P(x)\, dx$$

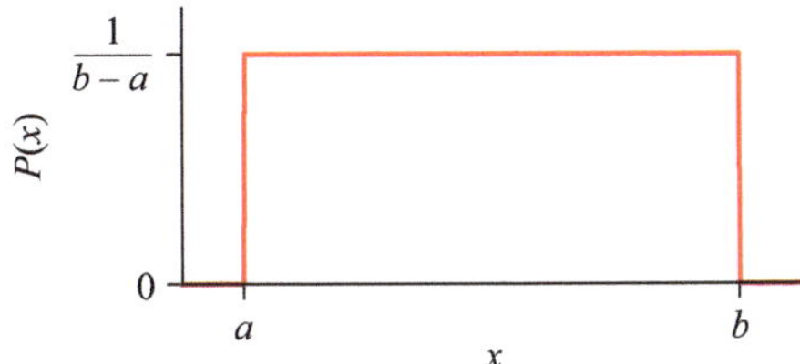

Figure 6.12
The uniform probability density in the interval $x \in [a, b]$.

$$\langle X^n \rangle = \int_{-\infty}^{\infty} x^n \cdot P(x)\, dx. \tag{6.41}$$

As for discrete random variables, the variance is defined as

$$\text{Var}[X] = \langle X^2 \rangle - \langle X \rangle^2. \tag{6.42}$$

6.4.5 Uniform Distribution

Perhaps the simplest continuous distribution is the **uniform distribution**, plotted in figure 6.12:

$$X \sim \text{Unif}(a, b) \tag{6.43}$$

with an equal probability density over some interval $x \in [a, b]$:

$$P(x) = \begin{cases} \frac{1}{b-a}, & x \in [a, b] \\ 0, & \text{otherwise} \end{cases}. \tag{6.44}$$

6.4.5.1 Mean and variance:

$$\begin{aligned} E[X] &= \frac{a+b}{2} \\ \text{Var}[X] &= \frac{(b-a)^2}{12}. \end{aligned} \tag{6.45}$$

6.4.6 Exponential Distribution

The **exponential distribution**

$$T \sim \text{Exp}(\tau) \tag{6.46}$$

arises commonly when "waiting" for random events (figure 6.13). For example, what is the amount of time until a cosmic ray strikes the chromosome to cause a mutation? Generally, the random variable is limited to positive values:

$$P(t) = \frac{1}{\tau} e^{-\frac{t}{\tau}}, \quad t \geqslant 0. \tag{6.47}$$

Note that the most probable value (meaning the highest density) is at $t = 0$. The **mean** and **variance** are:

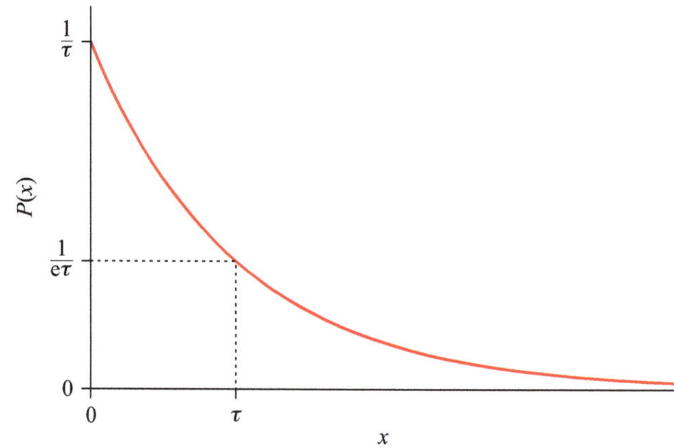

Figure 6.13
The exponential probability density.

$$E[T] = \tau$$
$$\text{Var}[T] = \tau^2. \tag{6.48}$$

Sometimes the distribution is parameterized instead by the "rate parameter" $1/\tau$, which has the inverse units of t. Although we used time t here as the random variable, the exponential distribution can apply to spatial and other random variables as well.

The exponential distribution is closely related to the geometric distribution for discrete variables. Imagine dividing the time axis into short bins of width Δt. In each bin, the probability of the event happening is $p = \Delta t/\tau$, proportional to the bin width. What is the probability that the first event will happen in the interval $[t, t + \Delta t]$? That requires

$$k = t/\Delta t \tag{6.49}$$

bins without an event, and according to equation (6.32),

$$P_k = (1 - p)^k \cdot p. \tag{6.50}$$

By allowing the time bins to get infinitesimally small, we make the transition to the continuous probability density expressed in equation (6.37):

$$P(t) = \lim_{\Delta t \to 0} \frac{P_k}{\Delta t} = \frac{1}{\tau} \lim_{\Delta t \to 0} \left(1 - \frac{\Delta t}{\tau}\right)^{\frac{t}{\Delta t}} = \frac{1}{\tau}\left(\lim_{\varepsilon \to 0} (1 - \varepsilon)^{1/\varepsilon}\right)^{\frac{t}{\tau}}. \tag{6.51}$$

This last expression has an interesting limit–namely,

$$\lim_{\varepsilon \to 0} (1 - \varepsilon)^{1/\varepsilon} = \frac{1}{e}, \tag{6.52}$$

which leads to the result shown in equation (6.47).

The exponential distribution with parameter λ corresponds to the distribution of waiting times between events that occur randomly with a constant rate λ. A process that generates these random events is called a "Poisson process." Is there any

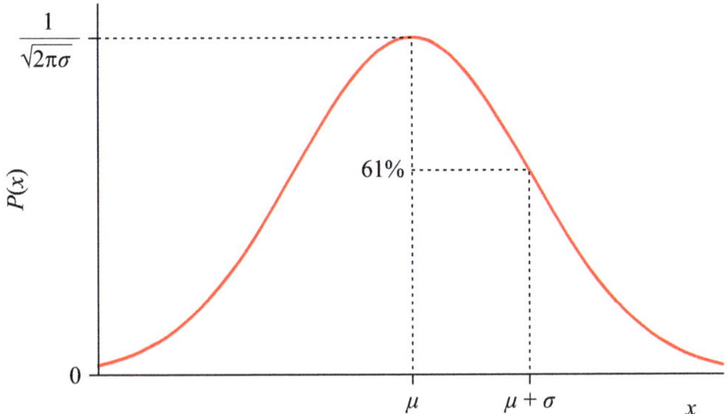

Figure 6.14
The Gaussian probability density.

relation between this stochastic process and the discrete Poisson distribution that we encountered before? In fact, there is. In a Poisson process, we may be interested in asking: How many events do we expect to observe in a time window ΔT? If the rate is λ, the average number of events will be $\lambda \Delta T$. The distribution of the number of events in fact will be given by a Poisson distribution with this mean. So while an exponential distribution describes the waiting times between random events, the Poisson distribution describes the number of random events that occur in a finite time window.

6.4.7 Gaussian Distribution

The Gaussian distribution for a continuous variable

$$X \sim \mathcal{N}(\mu, \sigma) \tag{6.53}$$

takes the form

$$P(x) = \frac{1}{\sqrt{2\pi}\sigma} e^{-\frac{(x-\mu)^2}{2\sigma^2}}. \tag{6.54}$$

This produces the classic bell shape plotted in figure 6.14. The maximal value is also the **mean**:

$$\mathrm{E}[X] = \mu, \tag{6.55}$$

The **variance** is

$$\mathrm{Var}[X] = \sigma^2 \tag{6.56}$$

and the **standard deviation** is

$$\mathrm{Std}[X] = \sigma. \tag{6.57}$$

At 1 standard deviation on either side of the mean, the density drops by $e^{-1/2} \approx 0.61$.

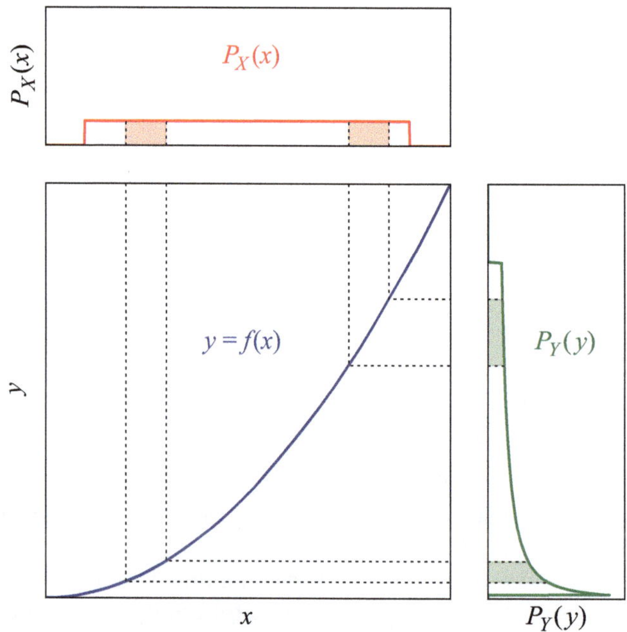

Figure 6.15
Probability density of the function of a random variable.

6.4.8 Function of a Random Variable

Suppose that X is distributed according to some probability density $P_X(X)$ and another random variable Y is defined through the function $y = f(x)$. Then what is the probability density $P_Y(Y)$?[7]

Figure 6.15 illustrates this problem. Here, the variable X is uniformly distributed. The variable Y results from squaring X: $y = f(x) = x^2$. As a result, Y is not uniformly distributed. In regions where $f(x)$ is steep, the range of x gets spread out over a wider range of y, so the probability density of y is low. The opposite is the case in regions where $f(x)$ is shallow. If a small interval $[x, x + dx]$ gets mapped into the interval $[y, y + dy]$, then we must have

$$P_X(x)dx = P_Y(y)dy$$

$$P_Y(y) = P_X(x)\left|\frac{dx}{dy}\right| = P_X(x)\left|\frac{d}{dy}f^{-1}(y)\right|. \tag{6.58}$$

Note that the scaling between $P_X(x)$ and $P_Y(y)$ depends only on the absolute value of the slope of f.

6.4.9 Arbitrary Probability Density

One can use this insight to create random variables with any desired distribution. Suppose that we want to sample a variable Y from some arbitrary distribution $P(y)$. Then we should:

---

7. Note: The subscript here reminds us that P_X and P_Y are two different functions. Some authors will omit the subscript and expect the reader to figure out that the P in $P(X)$ and $P(Y)$ stands for different functions.

- Compute the cumulative of P–namely,

$$g(y) = \int_{-\infty}^{y} P(y')\, dy'. \tag{6.59}$$

- Sample a random variable X from a uniform distribution on $[0, 1]$:

$$X \sim \mathrm{Unif}\,(0, 1). \tag{6.60}$$

- Transform X into Y according to

$$y = g^{-1}(x). \tag{6.61}$$

This gives the desired result:

$$Y \sim P. \tag{6.62}$$

6.5 Multiple Random Variables

The most elementary experiment involves at least two variables: you control one thing, X, and measure another thing, Y. So one is immediately led to thinking about the joint distribution of two random variables. Many modern technologies in biology aim at collecting vast numbers of measurements on the state of a system. Any image that you capture has millions of pixels, each a separate measurement. Sequencing methods produce fantastic volumes of data, such as the expression level of every gene in every cell type. Neuroscience tools allow both the control and the monitoring of thousands of nerve cells in parallel. A subdiscipline of computational biology is forming around the big data challenges posed by experiments with such global ambitions. Here, we introduce some of the basic concepts surrounding the analysis of multiple random variables.

6.5.1 Joint Distribution
Given two random variables X and Y, the joint probability distribution is defined by

$$P(x, y)\, dx\, dy = \text{Probability that } X \in [x, x + dx] \text{ and } Y \in [y, y + dy]. \tag{6.63}$$

One can visualize this function as a surface $z = P(x, y)$ above the (x, y)-plane as shown in figure 6.16. By analogy to equation (6.34), the probability of finding (x, y) inside some region A of the (x, y)-plane is equal to the volume under the surface and enclosed by A.

$$\int_{A} P(x, y)\, dy\, dx = \text{Probability that } (x, y) \text{ is in region } A. \tag{6.64}$$

Because some combination of (x, y) has to occur, the joint distribution is **normalized**:

$$\int_{x=-\infty}^{\infty} \int_{y=-\infty}^{\infty} P(x, y)\, dy\, dx = 1. \tag{6.65}$$

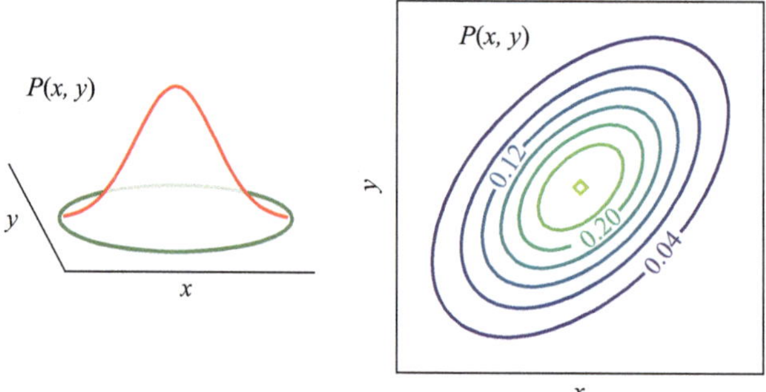

Figure 6.16
A bell-shaped joint probability density function $P(x, y)$ of two random variables. Left: Sketched as a surface in 3D. Right: Displayed as a contour plot.

We will explore some aspects of joint probability distributions using the simplest case of just two random variables X and Y, but all these ideas extend in obvious ways to higher dimensions.

6.5.2 Marginal Distribution

One defines the **marginal** distribution of one of the random variables, X, as the distribution $P_X(x)$ of that variable without regard to the others. This is obtained from the joint distribution by "integrating over" all the unneeded variables. For the case of two variables X and Y,

$$P_X(x) = \int_{y=-\infty}^{\infty} P(x, y)\, dy$$

$$P_Y(y) = \int_{x=-\infty}^{\infty} P(x, y)\, dx. \tag{6.66}$$

As is fitting, these marginal distributions are often drawn in the margins of the joint probability plot (figure 6.17).

6.5.3 Moments of the Joint Distribution

The **mean** of a multivariate distribution is obtained by averaging the random variables weighted by their probability:

$$\langle X \rangle = \int_{x=-\infty}^{\infty} \int_{y=-\infty}^{\infty} x \cdot P(x, y)\, dy\, dx = \int_{x=-\infty}^{\infty} x \cdot P_X(x)\, dx$$

$$\langle Y \rangle = \int_{x=-\infty}^{\infty} \int_{y=-\infty}^{\infty} y \cdot P(x, y)\, dy\, dx = \int_{y=-\infty}^{\infty} y \cdot P_Y(y)\, dy. \tag{6.67}$$

Note that the mean of any of the variables, like $\langle X \rangle$, is also the average of its marginal distribution, $P_X(x)$.

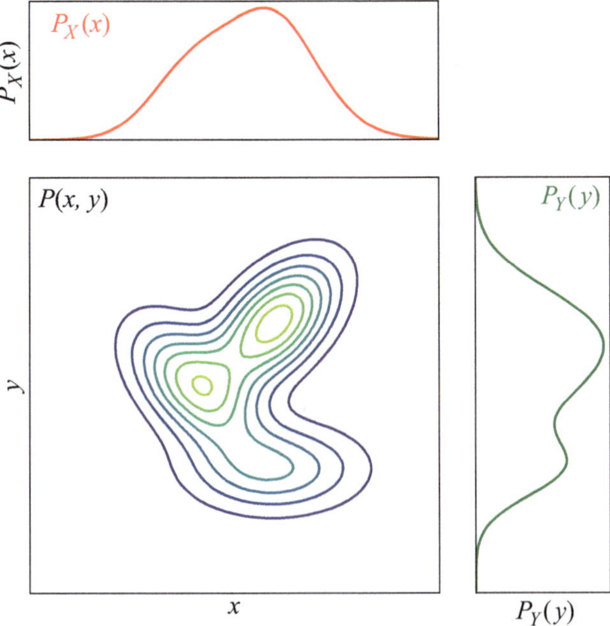

Figure 6.17
A joint distribution and its marginal distributions.

Similarly, the **moments** of a multivariate distribution are obtained by averaging the appropriate powers of the variables:

$$\langle X^m Y^n \rangle = \int\limits_{x=-\infty}^{\infty} \int\limits_{y=-\infty}^{\infty} x^m y^n \cdot P(x, y)\, \mathrm{d}y \mathrm{d}x. \tag{6.68}$$

Of particular interest is the **covariance** $\mathrm{Cov}\,[X, Y]$ of two random variables X and Y, defined as

$$\mathrm{Cov}\,[X, Y] = \langle (X - \langle X \rangle)(Y - \langle Y \rangle) \rangle = \langle XY \rangle - \langle X \rangle \langle Y \rangle. \tag{6.69}$$

As expected, this is a measure of how much X and Y covary. For example, if X tends to be larger than its mean when Y is also larger than its mean, the covariance will be positive.

A related measure is the **correlation coefficient** $\mathrm{Cor}\,[X, Y]$, which normalizes the covariance by the amount that X and Y vary individually:

$$\mathrm{Cor}\,[X, Y] = \frac{\mathrm{Cov}\,[X, Y]}{\sqrt{\mathrm{Var}\,[X] \cdot \mathrm{Var}\,[Y]}}. \tag{6.70}$$

6.5.4 Independent Random Variables

If two variables are **statistically independent**, then their probability distribution factors into the product of the marginal distributions—namely,

$$P(x, y) = P_X(x) \cdot P_Y(y). \tag{6.71}$$

For independent variables, the mixed moments are particularly simple because

$$\langle X^m Y^n \rangle = \int\limits_{x=-\infty}^{\infty} \int\limits_{y=-\infty}^{\infty} x^m y^n \cdot P(x, y) \, dy \, dx$$

$$= \int\limits_{x=-\infty}^{\infty} \int\limits_{y=-\infty}^{\infty} x^m y^n \cdot P_X(x) \cdot P_Y(y) \, dy \, dx \tag{6.72}$$

$$= \int\limits_{x=-\infty}^{\infty} x^m P_X(x) \, dx \cdot \int\limits_{y=-\infty}^{\infty} y^n P_Y(y) \, dy = \langle X^m \rangle \langle Y^n \rangle.$$

In particular,

$$\langle XY \rangle = \langle X \rangle \langle Y \rangle. \tag{6.73}$$

Therefore, **independent variables have zero covariance and correlation:**

$$\mathrm{Cov}[X, Y] = \langle XY \rangle - \langle X \rangle \langle Y \rangle = 0. \tag{6.74}$$

Note that the opposite is not true: $\mathrm{Cov}[X, Y] = 0$ does not imply statistical independence.

6.5.5 The Multivariate Gaussian Distribution

Recall the one-dimensional (1D) Gaussian distribution:

$$P(X; \mu, \sigma) = \frac{1}{\sqrt{2\pi}\,\sigma} e^{-\frac{(x-\mu)^2}{2\sigma^2}}. \tag{6.75}$$

It has a peak at the expectation value $x = \mu$ and falls off symmetrically on either side with a bell shape, whose width is characterized by the standard deviation σ.

Now suppose that there are n random variables $x_1, \ldots, x_n$. Define the vector

$$\mathbf{x} = \begin{bmatrix} x_1 \\ \vdots \\ x_n \end{bmatrix} \tag{6.76}$$

Then the **multivariate Gaussian distribution** is

$$P(\mathbf{x}; \boldsymbol{\mu}, \mathbf{C}) = \frac{1}{\sqrt{2\pi}^n \sqrt{\det \mathbf{C}}} \exp\left(-\frac{1}{2}(\mathbf{x} - \boldsymbol{\mu})^\top \mathbf{C}^{-1} (\mathbf{x} - \boldsymbol{\mu})\right), \tag{6.77}$$

where

$$\boldsymbol{\mu} = \mathrm{mean} = \begin{bmatrix} \mu_1 \\ \vdots \\ \mu_n \end{bmatrix} = \begin{bmatrix} \langle x_1 \rangle \\ \vdots \\ \langle x_n \rangle \end{bmatrix} \tag{6.78}$$

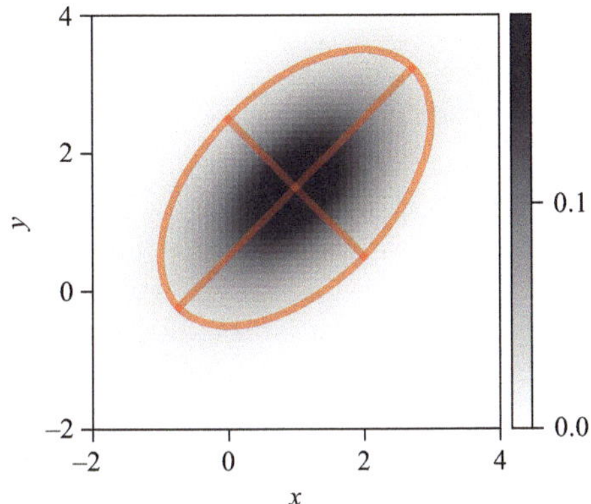

Figure 6.18

A two-dimensional (2D) Gaussian distribution, plotted in gray scale, with mean $\boldsymbol{\mu} = \begin{bmatrix} 1.0 \\ 1.5 \end{bmatrix}$ and covariance $\mathbf{C} = \begin{bmatrix} 1.0 & 0.5 \\ 0.5 & 1.0 \end{bmatrix}$. A contour ellipse at 2 SD from the mean is drawn, along with its principal axes.

is the vector containing all the means, and

$$\mathbf{C} = \text{covariance matrix} = \begin{bmatrix} \text{Cov}\,[X_1, X_1] & \cdots & \text{Cov}\,[X_1, X_n] \\ \vdots & \ddots & \vdots \\ \text{Cov}\,[X_n, X_1] & \cdots & \text{Cov}\,[X_n, X_n] \end{bmatrix} \tag{6.79}$$

is the matrix containing all the pairwise covariances between the random variables.

As in the 1D case, the density peaks at the mean $\boldsymbol{\mu}$ and then falls off in all directions as $e^{-\frac{1}{2}(\text{distance from the mean})^2}$, where the distance d is defined by

$$d^2 = (\mathbf{x} - \boldsymbol{\mu})^\top \mathbf{C}^{-1} (\mathbf{x} - \boldsymbol{\mu}). \tag{6.80}$$

The right side of equation (6.80) is a quadratic form in the variable x_i. Because the covariance matrix is positive definite, equation (6.80) defines an ellipsoid. So in the case $n = 2$, the contour lines of $P(\mathbf{x})$ are ellipses centered on $\boldsymbol{\mu}$ as shown in figure 6.18; for higher n, the hypersurfaces of constant density are ellipsoids.

Recall that an ellipse is defined by its principal axes, which form a set of orthogonal directions in the n-dimensional space of the $\mathbf{x}$. These directions are given by the eigenvectors of the matrix $\mathbf{C}$ and one suspects that a basis transform into this eigenbasis could be useful. The matrix $\mathbf{C}$ is symmetric and positive definite, which implies that all the eigenvectors are real and positive. This implies that the matrix function $\mathbf{C}^{-1/2}$ exists, as discussed in section (2.11.2), so one can define a new set of random variables u_i by the linear transformation

$$\mathbf{u} = \mathbf{C}^{-1/2} (\mathbf{x} - \boldsymbol{\mu}). \tag{6.81}$$

Then the distribution of the u_i is

$$P(\mathbf{u}) = \frac{1}{\sqrt{2\pi}^n} e^{-\frac{1}{2}\mathbf{u}^\top \mathbf{u}}$$

$$= \prod_{i=1}^{n} \frac{1}{\sqrt{2\pi}} e^{-\frac{1}{2}u_i^2} = \prod_{i=1}^{n} \mathcal{N}(u_i; \mu = 0, \sigma = 1). \tag{6.82}$$

Now all the u_i are independent random variables, distributed according to a 1D Gaussian with mean 0 and variance 1. Obviously, that transformation to i.i.id. random variables greatly simplifies any further treatment of the problem.

This also offers an effective way for generating samples x from a multivariate Gaussian distribution:

- Decide what should be the mean $\boldsymbol{\mu}$ and covariance matrix C.
- Draw n numbers u_i independently from a normal distribution $\mathcal{N}(0, 1)$.
- Transform $\mathbf{x} = \boldsymbol{\mu} + \mathbf{C}^{1/2}\mathbf{u}$.

6.5.6 The Sum of Random Variables

Many situations arise where we have to deal with the sum of two random variables. Suppose that X and Y are distributed according to $P(X, Y)$, and define Z as their sum:

$$Z = X + Y. \tag{6.83}$$

What is the probability distribution of Z? The general answer is

$$P_Z(z) = \int_{x=-\infty}^{\infty} P(x, z - x)\, dx = \int_{y=-\infty}^{\infty} P(z - y, y)\, dy. \tag{6.84}$$

A few characteristics of this distribution are worth remembering by heart.

The **mean of the sum of random variables** is equal to the sum of the means:

$$\langle Z \rangle = \langle X + Y \rangle = \langle X \rangle + \langle Y \rangle. \tag{6.85}$$

More generally, this relationship applies to any linear mixture of two variables:

$$\langle \alpha X + \beta Y \rangle = \alpha \langle X \rangle + \beta \langle Y \rangle. \tag{6.86}$$

This is because "compute the mean" is a linear operation on the probability distribution P. As a result, it obeys the superposition principle of all linear systems, as shown in section 3.1.2.

The **variance of the sum of random variables** is equal to the sum of the variances plus twice the covariance:

$$\mathrm{Var}\,[X + Y] = \left(\langle X^2 \rangle - \langle X \rangle^2\right) + 2\left(\langle XY \rangle - \langle X \rangle \langle Y \rangle\right) + \left(\langle Y^2 \rangle - \langle Y \rangle^2\right)$$

$$= \mathrm{Var}\,[X] + 2\,\mathrm{Cov}\,[X, Y] + \mathrm{Var}\,[Y]. \tag{6.87}$$

For **independent variables**, the covariance vanishes, so the variance of the sum is equal to the sum of the variances:

$$\text{Var}\,[X+Y] = \text{Var}\,[X] + \text{Var}\,[Y]\,, \text{ if } X \text{ and } Y \text{ are independent.} \tag{6.88}$$

6.5.7 Averaging Repeated Measurements

Perhaps the most common operation in empirical science is to repeat the same measurement several times and average the results to obtain a more reliable estimate. Suppose that we are interested in some parameter X whose measurement is affected by measuring errors, leading to a probability density $P\,(x)$ with mean

$$\text{E}\,[X] = \mu \tag{6.89}$$

and variance

$$\text{Var}\,[X] = \sigma^2. \tag{6.90}$$

We take n successive measurements of that same random variable:

$$X_1, \ldots, X_n. \tag{6.91}$$

Then we average the n measurements to arrive at the variable

$$Y = \frac{1}{n} \sum_{i=1}^{n} X_i, \tag{6.92}$$

which is called the **sample mean**.

Based on the results for i.i.d. variables, this average has moments

$$
\begin{aligned}
\text{E}\,[Y] &= \frac{1}{n} \sum_{i=1}^{n} \text{E}\,[X_i] = \mu \\[2mm]
\text{Var}\,[Y] &= \frac{1}{n^2} \sum_{i=1}^{n} \text{Var}\,[X_i] = \frac{1}{n}\sigma^2.
\end{aligned}
\tag{6.93}
$$

The standard deviation of Y is less than that of X by a factor of $\frac{1}{\sqrt{n}}$:

$$\text{Std}\,[Y] = \frac{1}{\sqrt{n}} \text{Std}\,[X]\,. \tag{6.94}$$

This, fundamentally, is the motivation for averaging independent measurements: the result is distributed more sharply than an individual measurement, leaving less uncertainty about the underlying true value of X. Note however that averaging offers diminishing returns: you need 10 measurements to reduce the uncertainty by a factor of 3, but 100 measurements to get another factor of 3.

6.5.8 The Sum of Independent Gaussian Random Variables

The sum of two independent Gaussian random variables is again distributed according to a Gaussian. Also, given the above results, we can immediately spell out what the mean and standard deviation of that Gaussian are. If

$$X_1 \sim \mathcal{N}(\mu_1, \sigma_1) \text{ and } X_2 \sim \mathcal{N}(\mu_2, \sigma_2), \tag{6.95}$$

then

$$X_1 + X_2 \sim \mathcal{N}\left(\mu_1 + \mu_2, \sqrt{\sigma_1^2 + \sigma_2^2}\right). \tag{6.96}$$

In higher dimensions, if one vector variable $\mathbf{X}_1$ is Gaussian with mean $\boldsymbol{\mu}_1$ and covariance matrix $\mathbf{C}_1$, and another Gaussian variable $\mathbf{X}_2$ has mean $\boldsymbol{\mu}_2$ and covariance matrix $\mathbf{C}_2$, then the sum $\mathbf{X}_1 + \mathbf{X}_2$ is Gaussian again, with mean $\boldsymbol{\mu}_1 + \boldsymbol{\mu}_2$ and covariance matrix $\mathbf{C}_1 + \mathbf{C}_2$.

6.6 The Central Limit Theorem

One frequently encounters situations where **a variable of interest is the result of many small and independent additive contributions.** For example, a polygenic trait, like the height of a human, may be influenced by many independent loci in the genome. The position of a protein particle on a gel is the consequence of many independent microscopic interactions with the solvent and the gel matrix. The measurement error of an instrument may be the sum of many independent sources of noise. In those cases, one can make a strong statement about the resulting random variable: it will have a Gaussian probability distribution.

This **central limit theorem** is one of the most consequential results in probability theory. The proper mathematical statement comes with the usual caveats, but for the normal user, it goes like this: Suppose that

$$Y = X_1 + \ldots + X_n \tag{6.97}$$

is the sum of n independent random variables. The X_i can have different distributions, with means and variances

$$\mathrm{E}\,[X_i] = \mu_i$$
$$\mathrm{Var}\,[X_i] = \sigma_i^2. \tag{6.98}$$

Then, as n gets large, the distribution of Y approximates a Gaussian distribution:

$$Y \underset{n \to \infty}{\sim} \mathcal{N}\left(\mu_1 + \ldots + \mu_n, \sqrt{\sigma_1^2 + \ldots + \sigma_n^2}\right). \tag{6.99}$$

Note that this is a much stronger statement than that in section 6.5.6 because it specifies not just the first two moments, but the entire shape of the distribution.

6.6.1 Works Reasonably Well Even for Small n

The central limit theorem can provide a pretty reasonable approximation when summing even a small number of random variables. Figure 6.19 shows an example using five variables that all have different distributions. The predicted Gaussian is off by just a few percent at the peak of the distribution.

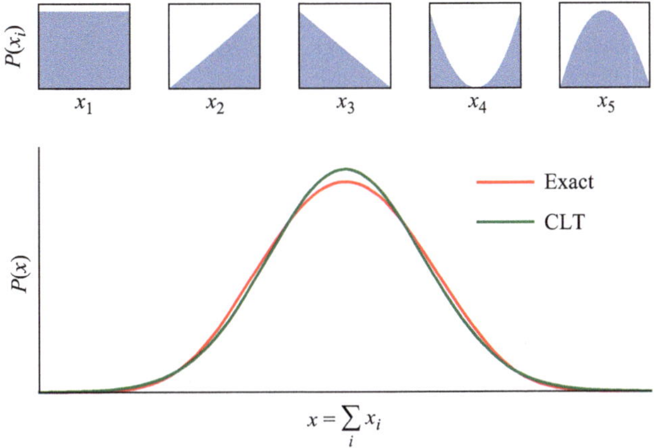

Figure 6.19
The sum of these five random variables (top row, each with a different probability distribution) is approximately a Gaussian variable. CLT = prediction from the central limit theorem.

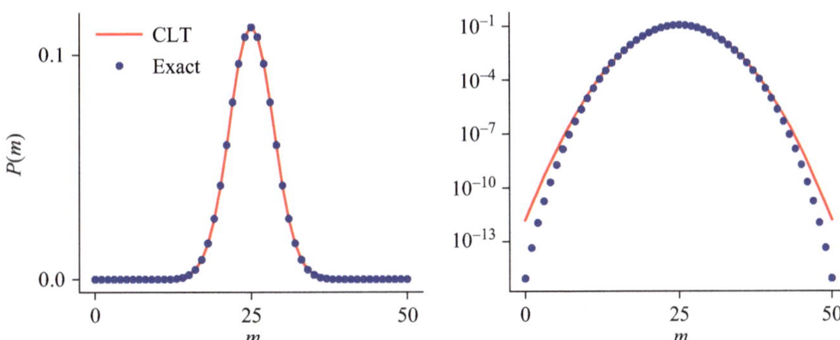

Figure 6.20
The sum m (this is the label of the x-axis) of 50 coin tosses has a binomial distribution (Exact), shown with its Gaussian approximation (CLT). The probability is plotted on a linear scale (left) and a log scale (right).

6.6.2 Works Best Near the Center of the Distribution

This is where the central limit theorem gets its name: The approximation works really well for central values, meaning near the mean. However, it cannot be trusted far out in the tails. Figure 6.20 shows an example using the binomial distribution—namely, the sum of 50 coin tosses.

On an absolute scale, the Binomial is very well approximated by a Gaussian, as expected from the central limit theorem. Nevertheless, a log-scale plot shows that the relative error in the tails of the distribution is enormous: many orders of magnitude, in fact. So if the phenomenon of interest depends on these rare events far from the mean of the distribution, then one cannot rely on the central limit theorem to estimate their frequency.

6.6.3 Multiplicative Contributions Lead to a Log-Normal Distribution

Sometimes the various random variables X_i combine not by addition, but by multiplication. For example, a population of bacteria may grow exponentially, multiplying by a factor X_i every hour. After n hours, the number of bacteria is proportional to the product of the X_i values:

$$Y = X_1 \cdot \ldots \cdot X_n = \prod_{i=1}^{n} X_i. \tag{6.100}$$

In such a case, one can apply the central limit theorem to the logarithms:

$$\ln Y = \sum_{i=1}^{n} \ln X_i. \tag{6.101}$$

Therefore, $\ln Y$ will be distributed according to a Gaussian. This is called a **log-normal distribution**. This distribution occurs very frequently in nature. Some people claim that in biology and chemistry, the log-normal is a better default approximation than the normal because multiplicative influences are so common. For example, a series of enzymatic reactions depends on the product of the rates at each stage.

Inference is the process by which conclusions can be reached from the evidence observed and thinking logically about it. Fundamentally, this is where data get converted to knowledge. A commonly useful framework starts with some model of how the world works. That model includes one or more parameters that are unknown. To learn about that parameter, we design experiments whose measurements are sensitive to the parameter. Once the measurements are done, we assess what we have learned about the parameter of the model. This reduction in uncertainty about the model increases our knowledge.

To illustrate this process, it helps to return to the example of coin tossing.[1] Suppose that we are handed a mystery coin of unknown provenance. Our "model" of this system says that tossing the coin will yield Heads with probability p and Tails with probability $1 - p$. Further, the model says that subsequent tosses do not influence each other. We are nevertheless uncertain about the value of the bias p: it could be anything between 0 and 1.

We design an experiment on this system, which is to toss the coin $n = 10$ times and count the number of Heads. The model predicts that the outcome follows the binomial distribution (see section 6.3.6), such that the probability of getting m Heads in n tosses is

$$P(m; n, p) = \binom{n}{m} p^m (1 - p)^{n-m}. \tag{7.1}$$

Now suppose that we perform the experiment and observe $m = 7$ heads. What have we learned about p from that result? In particular, one might ask: Given the data, what is the best estimate of p? And what is the uncertainty in that estimate?

7.1   Maximum Likelihood Estimation

There are multiple approaches to this problem of parameter estimation, but methods based on the likelihood function are particularly useful, with a good theoretical basis and broad applicability. In this framework, one reinterprets the probability distribution $P(m|n, p)$ in equation (7.1) as a **likelihood function** for the parameter p. The values of n and m are now known, one from experimental design, and the other from observed data. The independent variable is p, and the likelihood function spells out how likely

---

1. The literal tossing of coins is not a common pursuit in biological research. However, by now the reader will understand that the simple Bernoulli process, with two possible outcomes, is indeed ubiquitous: mutated or not, dead or alive, photon or no photon, etc.

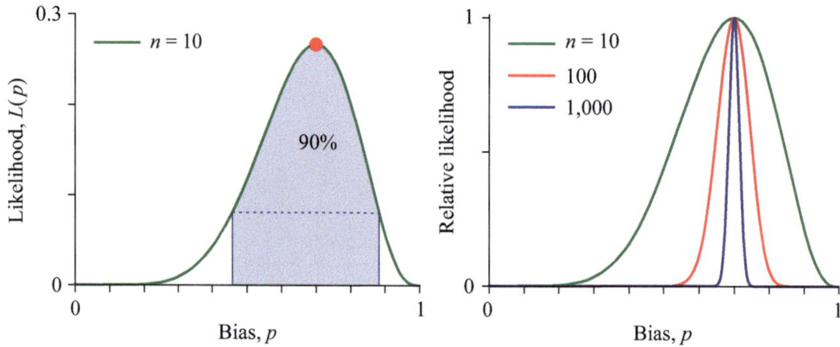

Figure 7.1
Left: The likelihood function for the bias p after observing $m = 7$ Heads in $n = 10$ coin tosses. The MLE is $\hat{p} = 0.7$ (red) and the 90% confidence limits are [0.46, 0.88]. Right: The relative likelihood function (normalized to a peak of 1) for increasing number of coin tosses n, of which $m = 0.7\,n$ returned Heads.

it is to get the observed data if the bias has value p:

$$L(p|n, m) = P(m|n, p) = \binom{n}{m} p^m (1 - p)^{n-m}. \tag{7.2}$$

This likelihood function (figure 7.1) summarizes what we have learned about the parameter p. It is identical to the probability density function, but seen as a function of the parameters instead of the observations. Clearly there is still some uncertainty about p but some values are much more likely than others. The best estimate of p is the value with the largest likelihood, called the **maximum likelihood estimator (MLE)**:

$$\hat{p} = \arg\max_p L(p|n, m). \tag{7.3}$$

In this particular case of the binomial distribution, one can show (see exercise 10.13) that the likelihood peaks at

$$\hat{p} = \frac{m}{n}, \tag{7.4}$$

which corresponds to the observed fraction of Heads in the data.

What about the remaining uncertainty regarding p? Again, the full knowledge is embodied in the likelihood function. The more tightly peaked the function, the smaller the uncertainty. One often quotes the uncertainty in terms of **confidence limits**. For example, we might find the interval of p that accounts for 90 percent of the likelihood, and quote the bounds of that interval as 90 percent confidence limits. In a report, one can summarize the experimental result as

$$p = 0.70\ [0.46,\ 0.88],\ 90\%\ \text{CI}. \tag{7.5}$$

To reduce the uncertainty about the bias parameter p, we could of course perform more coin tosses. Figure 7.1 illustrates how the likelihood function changes as the number of samples n increases. As expected, the relative likelihood function gets sharper and sharper, so the confidence region of our estimate $\hat{p}$ gets progressively narrower. As discussed next, the shape of the likelihood function tends to a Gaussian, with a width that shrinks approximately as $1/\sqrt{n}$.

7.1.1 Likelihood Function with Many Samples

Suppose that a model with parameter θ predicts that the distribution for an experimental outcome X is $f(X|\theta)$. We want to estimate the true value θ_0 of the model parameter. We perform a repeated series of measurements that gives outcomes $X_1, \ldots, X_n$. Assuming that these experimental samples are independent, the likelihood function for θ is

$$L(\theta|\{X_i\}) = \prod_{i=1}^{n} f(X_i|\theta). \tag{7.6}$$

An important result from estimation theory says that when the number of samples n is large, the likelihood function becomes a Gaussian centered on the true parameter θ_0, and in particular,

$$L(\theta) \sim \exp\left[(\theta - \theta_0)^2 \, \frac{nI(\theta_0)}{2}\right] \tag{7.7}$$

where

$$I(\theta) = -\left\langle \frac{\partial^2}{\partial \theta^2} \log f(X|\theta) \right\rangle_X \tag{7.8}$$

and the expectation value is taken over the possible values of X. Therefore, the variance declines as $1/n$ and the **uncertainty about** θ **declines as** $1/\sqrt{n}$. The quantity $I(\theta)$ is called the **Fisher information.**

For illustration, let us return to the exercise of tossing a coin n times and observing m Heads. In section 7.1, we discussed this as drawing a single sample m from the binomial distribution, $m \sim \text{Bin}(n, p)$. Instead, we can regard the single coin toss as a sample from a bernoulli distribution, where the two possible outcomes are $X \in \{0, 1\}$ with probabilities

$$f(X|\theta) = \begin{cases} \theta & \text{if } X = 1 \\ 1 - \theta & \text{if } X = 0. \end{cases} \tag{7.9}$$

Here, the unknown model parameter θ is the coin bias p. Now we perform n independent draws from this distribution, and want to infer θ from the results. As derived in section 7.1, the MLE for θ is

$$\hat{\theta} = \frac{m}{n}. \tag{7.10}$$

What is the distribution of the true θ about that value of $\hat{\theta}$? The Fisher information is found to be

$$I(\theta) = -\left[f(1|\theta) \frac{\partial^2}{\partial \theta^2} \log f(1|\theta) + f(0|\theta) \frac{\partial^2}{\partial \theta^2} \log f(0|\theta) \right]$$

$$= -\left[\theta \frac{\partial^2}{\partial \theta^2} \log \theta + (1 - \theta) \frac{\partial^2}{\partial \theta^2} \log(1 - \theta) \right]$$

$$= -\left[\theta \frac{-1}{\theta^2} + (1-\theta)\frac{-1}{(1-\theta)^2}\right]$$

$$= \frac{1}{\theta(1-\theta)} \tag{7.11}$$

and the variance of the likelihood distribution becomes

$$\frac{1}{nI(\hat{\theta})} = \frac{\hat{\theta}(1-\hat{\theta})}{n}. \tag{7.12}$$

In conclusion, for large n, the relative likelihood of θ is a normal distribution with mean $\hat{\theta} = m/n$ and standard deviation $\sqrt{\hat{\theta}(1-\hat{\theta})/n}$:

$$\theta \sim \mathcal{N}\left(\hat{\theta}, \frac{\hat{\theta}(1-\hat{\theta})}{n}\right). \tag{7.13}$$

7.1.2 Maximum Likelihood Estimation for the Gaussian Distribution

Suppose that the data come from a Gaussian distribution with unknown mean and variance:

$$X \sim \mathcal{N}(\mu, \sigma^2)$$

$$f(X|\mu, \sigma) = \frac{1}{\sqrt{2\pi}\sigma} e^{-\frac{(X-\mu)^2}{2\sigma^2}}. \tag{7.14}$$

Then the likelihood function for the data set $\{X_i\}$ is

$$L(\mu, \sigma|\{X_i\}) = \prod_{i=1}^{n} f(X_i|\mu, \sigma). \tag{7.15}$$

Often, it is convenient to work with the logarithm of the likelihood:

$$\log L(\mu, \sigma|\{X_i\}) = \sum_{i=1}^{n} \log f(X_i|\mu, \sigma) = -\frac{1}{2\sigma^2}\sum_{i=1}^{n}(X_i - \mu)^2 - n\log\sigma - n\log\sqrt{2\pi}. \tag{7.16}$$

Because the logarithm is a monotonic function, maximizing the log-likelihood is the same as maximizing the likelihood. Differentiating with respect to the parameters

$$\frac{\partial}{\partial\mu} \log L(\mu, \sigma|\{X_i\}) = \frac{\partial}{\partial\sigma} \log L(\mu, \sigma|\{X_i\}) = 0 \tag{7.17}$$

leads to the maximum likelihood estimators for mean and variance:

$$\hat{\mu} = \bar{X} = \frac{1}{n}\sum_{i=1}^{n} X_i$$

$$\hat{\sigma}^2 = S = \frac{1}{n}\sum_{i=1}^{n}(X_i - \bar{X})^2. \tag{7.18}$$

To establish confidence limits on these estimators, we also compute their large-sample variance using the Fisher information defined in equation (7.7):

$$\text{Var}\,[\mu] = \frac{1}{nI(\mu)}, \tag{7.19}$$

where

$$I(\mu) = -\left\langle \frac{\partial^2}{\partial\mu^2} \log f(X|\mu,\sigma) \right\rangle_X = -\left\langle \frac{-1}{\sigma^2} \right\rangle_X = \frac{1}{\sigma^2}. \tag{7.20}$$

So

$$\text{Var}\,[\mu] = \frac{\sigma^2}{n}. \tag{7.21}$$

After replacing σ^2 with its MLE, this becomes

$$\text{Var}\,[\mu] = \frac{\hat{\sigma^2}}{n} = \frac{S}{n}. \tag{7.22}$$

The square root of this variance is often quoted as the **standard error of the parameter estimate**, so one might write in a report

$$\hat{\mu} = \bar{X} \pm \sqrt{\frac{S}{n}} \quad \text{(standard error)}. \tag{7.23}$$

Similarly,

$$\text{Var}\,[\sigma^2] = \frac{1}{nI(\sigma^2)} = \frac{1}{-n\left\langle \frac{\partial^2}{\partial(\sigma^2)^2} \log f(X|\mu,\sigma) \right\rangle_X}$$

$$= \frac{1}{-n\left\langle \frac{-1}{2(\sigma^2)^2} \right\rangle_X} \tag{7.24}$$

$$= \frac{2\sigma^4}{n}.$$

So the MLEs of the variance and its standard error are

$$\hat{\sigma^2} = S \pm \sqrt{\frac{2}{n}}S \quad \text{(standard error)}. \tag{7.25}$$

7.1.3 Sample Mean and Sample Variance

The previous derivation relied strongly on two statistics of the data sample $\{X_1, \ldots, X_n\}$:

The **sample mean** is the average of the data:

$$\bar{X} = \frac{1}{n} \sum_{i=1}^{n} X_i. \tag{7.26}$$

The **sample variance** is the average squared deviation from the sample mean:

$$S^2 = \frac{1}{n} \sum_{i=1}^{n} (X_i - \bar{X})^2. \tag{7.27}$$

Loosely speaking, these reflect the center of the data distribution and its width, respectively. We will see that they play important roles in many aspects of estimation from data.

7.1.4 Maximum Likelihood Estimators for Various Distributions

By the same method, one can derive MLEs for the parameters of other distributions and the standard errors of their estimates. Here is a partial table:

Distribution	Probability function	Parameter	MLE	SE
$\mathcal{N}(\mu, \sigma^2)$	$\frac{1}{\sqrt{2\pi}\sigma} \exp\frac{(X-\mu)^2}{2\sigma^2}$	μ	$\bar{X}$	$S/\sqrt{n}$
		σ^2	S^2	$S^2\sqrt{2/n}$
$\mathrm{Exp}(\lambda)$	$\frac{1}{\lambda}\exp\left(-\frac{X}{\lambda}\right)$	λ	$\bar{X}$	$S/\sqrt{n}$
$\mathrm{Poiss}(\nu)$	$\frac{\nu^X}{X!}\exp(-\nu)$	ν	$\bar{X}$	$S/\sqrt{n}$
$\mathrm{Binom}(\theta)$	$X\theta + (1-X)(1-\theta)$	θ	$\bar{X}$	$\sqrt{\bar{X}(1-\bar{X})/n}$

Note that all these MLEs are related to the sample mean and sample variance.

7.2 Bayesian Estimation

On many occasions, we have some prior knowledge about the model parameters. They may be restricted to a certain domain because of the nature of the model, such as a parameter corresponding to a drug concentration cannot be negative. Or a prior series of studies may have measured the parameter already to some accuracy. One can incorporate such prior knowledge using **Bayesian estimation**. Suppose that before our experiments we believe that the parameter θ has a probability distribution $P_\theta(\theta)$. This is called the **prior distribution**. Then the experiment produces measurements X. Treating both X and θ as random variables, Bayes rule, given by equation (6.8), says that

$$P_{\theta|X}(\theta|X) = \frac{P_{X|\theta}(X|\theta)P_\theta(\theta)}{P_X(X)}, \tag{7.28}$$

where the subscripts serve to clearly distinguish all the different probability distributions.

$P_{\theta|X}(\theta|X)$ is called the **posterior probability** of θ, so called because it arises **after** the measurement of X. When comparing this posterior probability for different values of θ, the data X always remain the same, so one can ignore the term $P_X(X)$ in the denominator of equation (7.28). We are left with

$$P_{\theta|X}(\theta|X) \propto P_{X|\theta}(X|\theta)P_\theta(\theta). \tag{7.29}$$

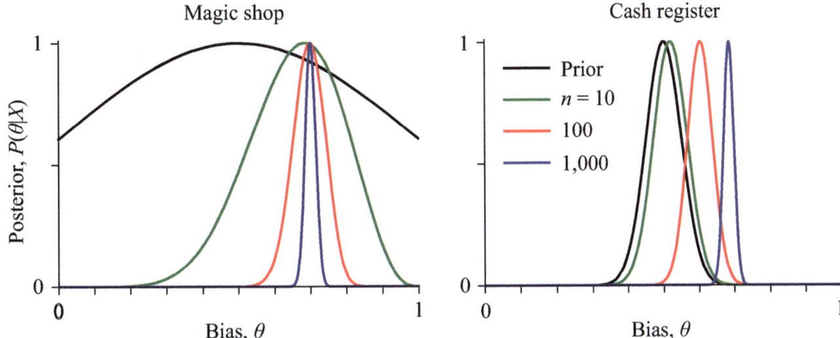

Figure 7.2
The posterior distribution for the bias θ after observing 70% Heads in n coin tosses. Left: With a weak prior belief. Right: With a strong prior belief that the coin is fair.

The **posterior probability is proportional to the product of likelihood and prior probability**. This posterior function comprises the new knowledge that we have about the parameter θ given the knowledge we started with and the additional evidence from the newly observed data. All subsequent analysis can be based on this posterior distribution.

Returning to the coin-tossing example, what is the prior knowledge of the parameter θ? Suppose that we know that the coin just came out of a cash register, then it is most likely a standard coin. Given our experience with coins and some common sense, it would be highly improbable for such a coin to have a strong bias away from $\theta = 0.5$. Conservatively, we might assign a prior distribution that allows for say 5 percent bias either way, for example a normal distribution

$$P_\theta(\theta) \sim \mathcal{N}(0.5, 0.05). \tag{7.30}$$

On the other hand, if the coin was just purchased at a magic shop, we might feel much less confident about its behavior during coin tosses. Still, extreme biases should be less plausible than $\theta = 0.5$, so let us propose a broad normal distribution like

$$P_\theta(\theta) \sim \mathcal{N}(0.5, 0.5). \tag{7.31}$$

What are the consequences of these prior beliefs? Figure 7.2 shows the resulting posterior distributions, again based on the data $X =$ "7 Heads in 10 tosses." The "magic shop" case looks similar to the situation we already treated in figure 7.1: because the prior distribution varies only weakly with θ, the posterior is almost equal to the likelihood function. In the "cash register" case, on the other hand, the posterior distribution is strongly dominated by the prior belief. The data after 10 coin tosses barely alter the knowledge about the coin and only shift the posterior ever so slightly to the right. The 99 percent confidence limits on θ are [0.40,0.64]. With a much larger number of coin tosses, the likelihood function gets sufficiently steep to make a difference. For example, with $n = 1,000, m = 700$ the confidence limits on the bias are [0.65,0.72], which no longer includes the fair coin bias of 0.5. At that point, we could declare with 99 percent certainty that something is wrong with the coin, despite the fact that it came out of a cash register.

7.2.1 Pros and Cons of Bayesian Estimation

Some scientists argue against the use of Bayesian priors in evaluating data because it may be difficult to quantify the amount of prior knowledge. If an experiment is the first attempt at a particular measurement, then there may be no prior knowledge about the target parameters, other than constraints from natural laws. On the other hand, if the experiment follows on prior work, especially from the same researchers, then it seems necessary to include the expectations that come from that prior work. Some studies want to claim that the prior work is somehow faulty and led to the wrong conclusions: In such a case, one obviously does not want to incorporate those conclusions in the new analysis. This situation falls under the rubric of "hypothesis testing," to be discussed next.

7.3 Hypothesis Testing

7.3.1 Simple Hypotheses

Suppose that hypothesis H_1 makes a concrete prediction about the experimental outcome, in the form of a probability distribution for the data $P_1(X|H_1)$. This is called a **simple hypothesis**. If we want to compare multiple such hypotheses $H_1, H_2, \ldots$ their relative likelihood is given simply by the ratio of the probabilities that they predict for the data.

$$l_i = \text{relative likelihood of hypothesis } H_i = \frac{P_i(X)}{\sum_j P_j(X)}. \tag{7.32}$$

Optionally, one may again include a Bayesian prior for each of the hypotheses, as discussed previously.

7.3.2 Composite Hypotheses

More often, the hypotheses still have some unknown parameters to them. For example, we may ask whether a given data sample resulted from a Gaussian or a Poisson distribution, each with unknown parameters. In that case, one generally finds for each hypothesis the parameters that maximize the likelihood, and then computes the likelihood ratio. Note that this optimal parameter depends on the data X. For example, if H_1 and H_2 predict data distributions $P_1(X|\theta_1)$ and $P_2(X|\theta_2)$, with unspecified θ_1 and θ_2, then one defines the likelihood ratio as

$$l_i = \text{relative likelihood of hypothesis } H_i = \frac{P_i(X|\hat{\theta}_i)}{\sum_j P_j(X|\hat{\theta}_j)} \tag{7.33}$$

where

$$\hat{\theta}_i = \text{MLE of } \theta_i. \tag{7.34}$$

7.3.3 Null Hypothesis Significance Testing

This approach based on likelihood can evaluate multiple hypotheses and assign them relative plausibilities based on measurements. In one conception of scientific research, workers keep these alternative explanations in mind, each weighted with some posterior probability, and plan further studies on that landscape of potential explanations. However, much of the arsenal of traditional statistics is instead aimed at killing off hypotheses in a permanent fashion. In this framework, the researcher proposes a statement to explain the data, called the **null hypothesis**, and then sets out to disprove that statement with a categorical decision. Then the research community goes on to pursue alternative hypotheses, never to look back again on the discredited null hypothesis.

This traditional approach to hypothesis testing has value in certain situations where a binary decision must be made, such as whether to approve a drug for distribution to patients. However, we believe that its value in day-to-day research is overrated, and we will advocate for alternative ways of drawing inference from data. Because the reader will be confronted with many instances of null hypothesis testing, and because (at least as of this writing) the editors of biological journals are overly enamored with the resulting statistical significance tests, we present here an introduction to that discipline.

In this framework, a particular claim is singled out as the **null hypothesis**. Typically, this is a conservative claim based on conventional wisdom, such as that "this treatment has zero effect." The goal of the inference method is to test whether one can reject this null hypothesis. One starts by formulating an alternative hypothesis, which may be rather vague, such as "the treatment has nonzero effect." Then one decides what measurements to take that would reveal a departure from the null hypothesis. Further, one identifies a **statistic**, X, that will be computed as a function of the observed data. Under the null hypothesis, this statistic would obey a certain probability distribution, $P_0(X)$, called the **null distribution**. If the statistic takes on an extreme value within that distribution, this is taken as evidence against the null hypothesis. What counts as extreme? The analyst must decide on a rejection criterion. This is a set of values of the statistic X that will decisively count against the null hypothesis. Of course, there is some probability that the statistic will land in that region "by chance," even if the null hypothesis is correct. This probability that the null hypothesis will be falsely rejected even if it is true is called the **significance level**, α. It is also known as the **false positive rate** because a rejection of the null hypothesis is often thought of as a positive finding. Conversely, the **power of the test**, $(1 - \beta)$, is the probability that the test will reject the null hypothesis when it is false. Thus, β is the **false negative rate**. A good statistical test would have low α and low β.

All these design considerations, including the choice of significance level and rejection region, are supposed to be completed before any measurements are taken. Finally, the analyst looks at the measurements and if the statistic X falls in the rejection region, the analyst declares that "the null hypothesis is rejected at significance level α." In a follow-up stage, one can try to decrease α in the same test. At some point, the rejection region becomes so small that the measured value of X falls outside it. That critical value of α is called the p-value: the lowest level of significance at which the null hypothesis can be rejected.

7.4   The z-test

To illustrate the process of hypothesis testing, let us consider a simple situation. Suppose that the analyst has decided on computing a certain statistic X whose distribution under the null hypothesis is a normal distribution with mean 0 and variance 1:[2]

---

2. In fact, many statistical tests lead to a test statistic with an approximately Normal null distribution.

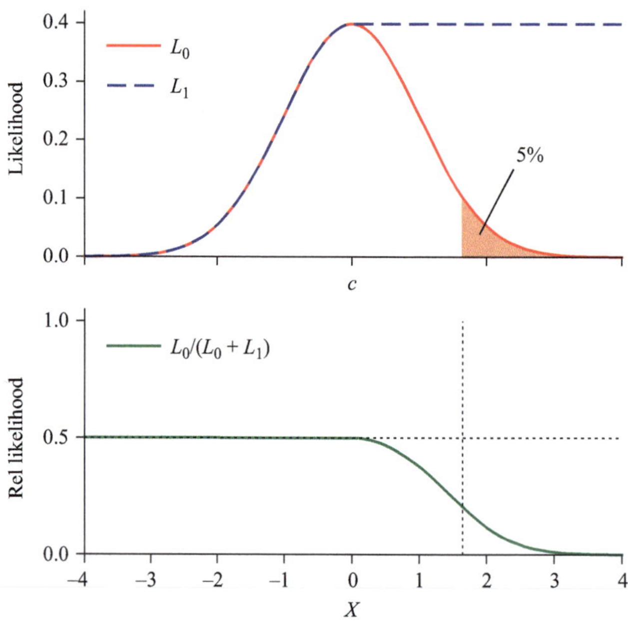

Figure 7.3
Illustration of a one-sided z-test. The null hypothesis has likelihood $L_0(X)$ (red), and the alternative hypothesis $L_1(X)$ (blue). The relative likelihood is $L_0/(L_0 + L_1)$ (green) and declines with increasing X. For significance level $\alpha = 0.05$, the rejection region is $X > 1.65$ (shaded red).

$$P_0(X) = \frac{1}{\sqrt{2\pi}} \exp\left(-\frac{X^2}{2}\right). \tag{7.35}$$

Where in that distribution should one place the rejection region? Presumably, very large values of $X \gg 1$ would raise suspicions whether X really was drawn from the null distribution. What about large negative values $X \ll -1$? Or very small values, like $X = 0.00001$? This depends entirely on what the alternative hypotheses are. Once an alternative hypothesis is specified, one can often make a reasoned argument for a specific rejection region based on likelihood ratios.

For example, consider the alternative hypothesis H_1 that "the treatment had a positive effect on X" (figure 7.3). One might predict the corresponding alternative probability distribution $P_1(X)$ to be normal with variance 1 and a positive mean $\mu > 0$ whose value is otherwise unknown:

$$P_1(X, \mu) = \frac{1}{\sqrt{2\pi}} \exp\left(-\frac{(X - \mu)^2}{2}\right), \quad \mu > 0. \tag{7.36}$$

This is a composite hypothesis (see section 7.3.2), so in evaluating the likelihood of hypothesis H_1, we choose the value of μ that maximizes the likelihood. Because of the constraint $\mu > 0$, that means

$$\mu = \max(0, X). \tag{7.37}$$

Thus the likelihood ratio becomes

$$\frac{L_0}{L_1} = \frac{P_0(X)}{\max\limits_{\mu>0} P_1(X,\mu)} = \begin{cases} 1, & \text{if } X < 0 \\ \exp\left(-\frac{X^2}{2}\right), & \text{if } X > 0. \end{cases} \tag{7.38}$$

We will want to reject values of X where the likelihood ratio is small. Clearly, that corresponds to the right tail of the distribution at positive X (figure 7.3). How much of that tail should be the rejection region? That depends on the desired significance level, α. The cutoff value c for X should be such that values greater than c happen by chance a fraction α of the time. In other words,

$$\int_c^\infty P_0(X)\mathrm{d}X = \alpha. \tag{7.39}$$

Because $P_0(X)$ is the normal distribution, one can obtain the desired cutoff c in terms of the inverse complementary error function:

$$c = \sqrt{2}\,\mathrm{erfc}^{-1}(2\alpha), \tag{7.40}$$

where

$$\mathrm{erfc}(z) \equiv 1 - \mathrm{erf}(z) \equiv 1 - \frac{2}{\sqrt{\pi}} \int_0^z e^{-t^2}\mathrm{d}t. \tag{7.41}$$

For example, for a significance level of $\alpha = 0.05$, this recommends $c = 1.65$. If $X > c$, we can declare that "the null hypothesis can be rejected at the 5 percent significance level." This specific statistical procedure is called a **one-sided z-test**.

If, instead, we posit the alternative hypothesis H_1 that "the treatment favors *negative* values of X," then the constraint on the mean becomes $\mu < 0$, the likelihood ratio L_0/L_1 declines for negative values of X, and thus one would place the rejection region in the left tail of the null distribution at $X < -c$.

If the alternative hypothesis H_1 is "the treatment has some nonzero effect on X," then there is no constraint on μ. To maximize L_1, one can adjust $\mu = X$ for any value of X, which leads to $L_1(X) = 1/\sqrt{2\pi}$. Now the relative likelihood of L_0 declines for both large positive and large negative values of X, and one divides the rejection region with area α equally between the two tails of the distribution $P_0(X)$ (figure 7.4). This is called a **two-sided z-test**.

Some caution is warranted because under certain alternative hypotheses, the framework leads to absurd results. For example, suppose that the alternative hypothesis H_1 is "the treatment tends to increase X, with an average effect greater than 5." The corresponding probability distribution is

$$P_1(X,\mu) = \frac{1}{\sqrt{2\pi}} \exp\left(-\frac{(X-\mu)^2}{2}\right), \quad \mu > 5. \tag{7.42}$$

Note that this is rather similar to the assumption $\mu > 0$ discussed previously. Following the same recipe, we find that the relative likelihood again decreases monotonically with increasing X, so we are led to choose a rejection region at large X (figure 7.5). With a significance level of $\alpha = 0.05$, all values $X > 1.65$ lead to the rejection

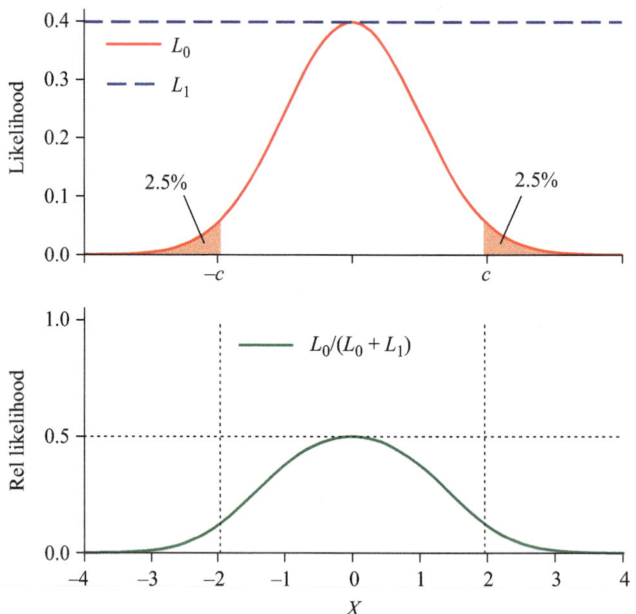

Figure 7.4
Illustration of a two-sided z-test. Display as in figure 7.3. For significance level $\alpha = 0.05$, the rejection region is $|X| > 1.96$ (shaded red).

of the null hypothesis. However, over most of that region, the null hypothesis is in fact much more likely than the alternative hypothesis. In this case, the recipe for null hypothesis significance testing produces misleading results.

These examples illustrate that even in the simplest cases, the criteria for rejecting the null hypothesis depend entirely on what one contemplates as the alternative hypotheses. One weakness of this logical framework is that it puts all the emphasis asymmetrically on the null hypothesis, and unsuspecting practitioners may forget all the assumptions involved about alternative explanations for the data.

7.5   The t-test

This is undoubtedly the statistical test that accounts for the largest number of error bars and p-values in biological research reports. Typically, one has taken a set of replicate measurements X and wants to know whether they are significantly different from zero or from some other set of measurements Y taken under different conditions.

7.5.1   One-Sample t-test

Suppose that we have taken m independent measurements of a random variable X. We also have reason to believe that X comes from a normal distribution. Both the mean μ and variance σ^2 of that distribution are unknown.[3] We want to establish confidence limits on μ. Or alternatively, we may want to test the null hypothesis that $\mu = 0$.

---

3. ... unlike for the z-test, where σ^2 is known.

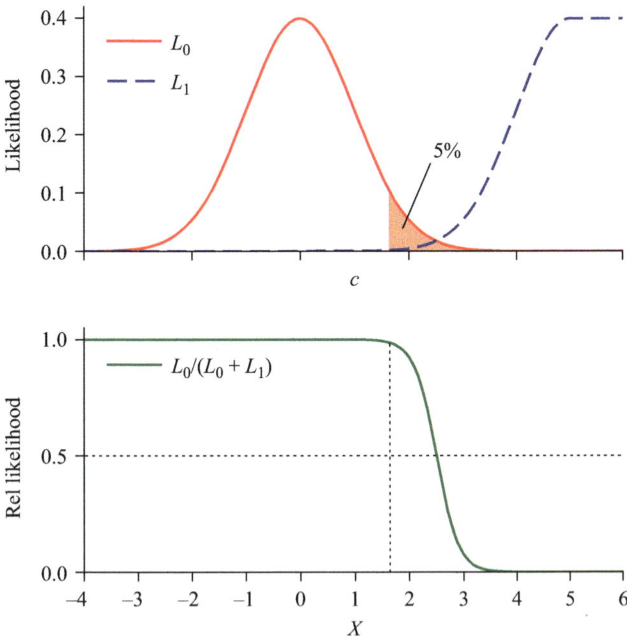

Figure 7.5
A failure of null hypothesis testing. Display as in figure 7.3. For significance level $\alpha = 0.05$, the rejection region is $X > 1.65$. But for most of those values, $L_0/(L_0 + L_1) > 0.5$, so the null hypothesis is much more likely than the alternative.

One can treat this problem by the method of likelihood functions introduced previously. After finding the likelihood function for μ, one gets the maximum-likelihood estimator $\hat{\mu}$. From the width of the likelihood function, one establishes confidence intervals (CIs) for that estimator. Alternatively, one can use the likelihood ratio to define a rejection region for the null hypothesis. We skip the mathematical derivation (see, e.g., Rice (2006)) and present the results.

From the data, one first computes the **sample mean**:

$$\bar{X} = \frac{1}{m} \sum_{i=1}^{m} X_i \tag{7.43}$$

and the **sample variance:**[4]

$$S^2 = \frac{1}{m-1} \sum_{i=1}^{m} (X_i - \bar{X})^2. \tag{7.44}$$

---

4. Note that the denominator in equation (7.44) is $(m-1)$, rather than m as in equation (7.27). This relates to a subtle difference between the maximum-likelihood estimator and the so-called unbiased estimator for the variance. As a rule, one should not fret too much about that difference. As a wise scientist once said: "If you really have to worry about the difference between m and $m-1$, you probably don't have enough data."

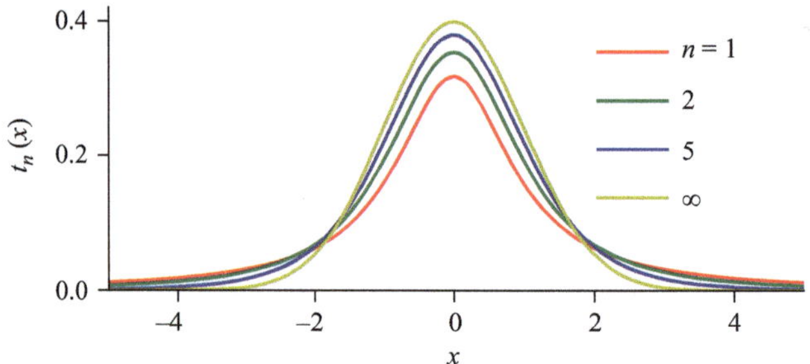

Figure 7.6
The *t*-distribution for various degrees of freedom.

Then one computes the **t-statistic** as follows:

$$t = \frac{\bar{X}}{S/\sqrt{m}}. \tag{7.45}$$

The key result is that the statistic $t - \mu$ follows the **t-distribution** with $m - 1$ degrees of freedom:

$$t - \mu \sim t_{m-1}. \tag{7.46}$$

The *t*-distribution with n degrees of freedom has a symmetric bell shape around zero, given by

$$t_n(z) = \frac{\Gamma\left(\frac{n+1}{2}\right)}{\sqrt{n\pi}\,\Gamma\left(\frac{n}{2}\right)} \left(1 + \frac{z^2}{n}\right)^{-(n+1)/2}. \tag{7.47}$$

where $\Gamma(z)$ denotes the Gamma function. For large n, it tends to the normal distribution.[5]

To quote confidence limits on μ at significance level α, we find the $(1 - \alpha/2)$ quantile of the *t*-distribution $t_{m-1}(z)$—namely, that value of z that has a fraction $\alpha/2$ of the area above it. For this purpose, let us define for any distribution f the function

$$C(f, x) = \text{ inverse cumulative of } f \text{ evaluated at } 1 - x, \tag{7.48}$$

which means that the area under f to the right of $C(f, x)$ equals x:

$$\int_{z=C(f,x)}^{\infty} f(z)\mathrm{d}z = x. \tag{7.49}$$

Returning to the present case, the confidence limits on μ at confidence level α are

$$\mu = \bar{X} \pm C(t_{m-1}, \alpha/2). \tag{7.50}$$

---

5. This follows from the central limit theorem, which states that for large samples, the sample mean $\bar{X}$ will be normally distributed.

Alternatively, one can evaluate the null hypothesis, $H_0 : \mu = 0$ against the alternative hypothesis, $H_1 : \mu \neq 0$. This leads to a two-tailed t-test, where the null hypothesis is rejected if the statistic t falls into either of the two tails of the distribution; namely, $|t| > C(t_{m-1}, \alpha/2)$. If, on the other hand, the alternative hypothesis is $H_1 : \mu > 0$, then one performs a one-sided test and rejects H_0 if t is large and positive $t > C(t_{m-1}, \alpha)$.

7.5.2 Two-Sample t-test

In a closely related situation we have taken m independent measurements of random variable X, and another n measurements of variable Y. Typically, X and Y are observed under two different conditions, such as "control" and "treatment." We have reason to believe that both distributions are normal, with means μ_X and μ_Y and identical variance σ^2. The question is whether the two means μ_X and μ_Y are significantly different. The procedure is very similar to the one-sample case.

First, evaluate the sample means, $\bar{X}$ and $\bar{Y}$, and the sample variances, S_X^2 and S_Y^2, as in equations (7.43) and (7.44). Compute the **pooled sample variance**:

$$S_P^2 = \frac{(m-1)S_X^2 + (n-1)S_Y^2}{m+n-2}. \tag{7.51}$$

Then compute the t-statistic:

$$t = \frac{\bar{Y} - \bar{X}}{S_P \sqrt{\frac{1}{m} + \frac{1}{n}}}. \tag{7.52}$$

Then the statistic $t - (\mu_Y - \mu_X)$ follows a t-distribution with $n + m - 2$ degrees of freedom:

$$t - (\mu_Y - \mu_X) \sim t_{m+n-2}. \tag{7.53}$$

Based on this, one finds the $(1 - \alpha)$-confidence limits as

$$\mu_Y - \mu_X = (\bar{Y} - \bar{X}) \pm C(t_{m+n-2}, \alpha/2). \tag{7.54}$$

Alternatively, one can test the null hypothesis $H_0 : \mu_Y = \mu_X$ by a one-tailed or two-tailed test on t, rejecting H_0 if $t > C(t_{m+n-2}, \alpha)$ or $|t| > C(t_{m+n-2}, \alpha/2)$, respectively.

7.5.3 Paired-Sample t-test

In this case, the data points come in pairs—namely, $\{(X_1, Y_1), \ldots, (X_n, Y_n)\}$. For example, (X_i, Y_i) may be before and after measurements on the same subject. We suspect that there may be some variability across subjects, but we want to know whether within a subject, Y is systematically different from X. Here, one simply computes the differences:

$$D_i = Y_i - X_i \tag{7.55}$$

and then proceeds with a regular one-sample t-test on the $\{D_1, \ldots, D_n\}$, as before, in order to compare these data to the null expectation of zero effect.

7.6 Goodness of Fit to a Distribution

Here, we want to test whether a sample of experimental data conforms to a specific probability distribution. Suppose that the samples are categorical, leading to the primary data

$$X_1, \ldots, X_m, \tag{7.56}$$

where X_i is the number of samples counted in category i, and the total number of samples is

$$\sum_{i=1}^{m} X_i = n. \tag{7.57}$$

For example, in the Luria and Delbrück (1943) experiment described in section 6.1.1, X_i is the number of cultures that contained exactly i mutants. We have a hypothesis H_0 that specifies the probabilities for the different categories of events:

$$p_i = \text{probability that a sample falls in category } i, \tag{7.58}$$

where $\sum_i p_i = 1$. So H_0 states that the set of counts $\{X_1, \ldots, X_m\}$ follows the **multinomial distribution** for n events with event probabilities $\{p_i\}$. For example, in Luria and Delbrück (1943), the hypothesis of virus-induced mutations predicts that p_i is the Poisson distribution.

For an alternative hypothesis H_1, one supposes that the data came from a multinomial distribution whose event probabilities are exactly the frequencies observed in the data. Namely,

$$H_1 : q_i = \frac{X_i}{n}, \tag{7.59}$$

Among all possible multinomial distributions, this choice of event probabilities maximizes the likelihood of the data. Now we compare the likelihood under H_0 to that under H_1. For large n, this likelihood ratio test recommends computing a quantity called **Pearson's chi-square statistic**:

$$X^2 = \sum_{i=1}^{m} \frac{(X_i - E_i)^2}{E_i}, \tag{7.60}$$

where

$$E_i = np_i \tag{7.61}$$

are the expected counts based on hypothesis H_0. Note that X^2 is the sum of the squared deviations between observed and expected counts, normalized by the expected counts. This statistic follows the chi-square distribution with $m - 1$ degrees of freedom:

$$X^2 \sim \chi^2_{m-1}. \tag{7.62}$$

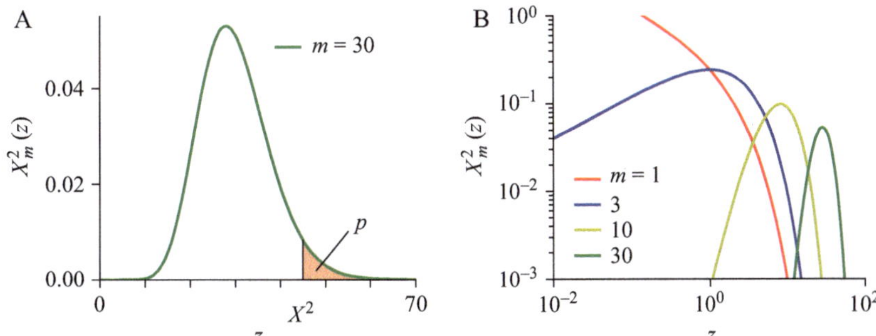

Figure 7.7
The chi-square distribution. A: for $m = 30$ degrees of freedom. Here, a model fit produced a X^2 of 45, and p is the probability of getting that value or greater under the model hypothesis. B: for various degrees of freedom (note the log axes).

7.6.1 The Chi-square Distribution

The **chi-square distribution with m degrees of freedom** is defined as the probability distribution of the sum of squares of m independent, normally distributed variables. Specifically, if

$$X_i \sim \mathcal{N}(0, 1) \tag{7.63}$$

and

$$z = \sum_{i=1}^{m} X_i^2, \tag{7.64}$$

then z follows the distribution

$$f(z) = \chi_m^2(z) = \frac{1}{2^{m/2}\Gamma(m/2)} z^{m/2-1} e^{-z/2}. \tag{7.65}$$

Figure 7.7 shows this distribution for a few values of m. Some notable properties of this distribution include the following:

- The expectation value equals the number of degrees of freedom:

$$E[\chi_m^2] = m. \tag{7.66}$$

So even if the model H_0 is correct, one expects the squared deviation of the data from the model to approximately equal m.

- The width scales with the square root of the degrees of freedom:

$$\text{Var}[\chi_m^2] = 2m. \tag{7.67}$$

- The peak is at $m - 2$:

$$\arg\max_z \chi_m^2(z) = m - 2. \tag{7.68}$$

- In the limit of large m, it tends toward a Gaussian, which is yet another consequence of the central limit theorem:

$$\chi_m^2 \sim \mathcal{N}(m, 2m), \text{ for large } m. \tag{7.69}$$

To assess the quality of a model fit to the measured data, one can locate the measured deviation value of X^2 within the χ^2 distribution predicted by the model (figure 7.7A). If the measured deviation is unusually large, one concludes that the model is a poor fit. This is often summarized by the p-value, which in this case is the probability, under the null hypothesis, of getting a value of X^2 that is equal to or larger than the observed value.

However, note that the χ^2 probability (figure 7.7) is also small for *very low* values of X^2. In that case, the p-value is very close to 1. This indicates a case where the model fit is *too good*, in that the data deviate from the model fit much less than expected based on random sampling. Sometimes this happens when scientists massage their data so they conform to a favorite hypothesis. The statistician Ronald A. Fisher famously reanalyzed the experimental results of Gregor Mendel on garden peas, and he found that the reported outcomes of phenotypes in genetic crosses were suspiciously close to the distribution ratios that Mendel's theory predicted (Fisher (1936)). By pooling all of Mendel's results into one study with 84 degrees of freedom, Fisher obtained a chi-square value of $\chi_{84}^2 = 41.6058$, with an associated p-value of 0.99997.[6] Getting such a precise fit by chance is extremely implausible, of course. The reasons why the data ended up conforming so closely to expectation, even in cases where that expectation is now known to be wrong, remain a subject of debate (Edwards (1986), Franklin et al. (2008)).[7]

7.6.2   Fit to a Parametric Distribution

Suppose that hypothesis H_0 is parametrized, in that the predicted probabilities p_i contain one or more parameters that can be adjusted to the observed data:

$$H_0 : p(X_i) = p_i(\theta_1, \ldots, \theta_k). \tag{7.70}$$

The MLEs of the parameters are those that minimize the normalized sum of squared deviations:[8]

$$X^2(\boldsymbol{\theta}) = \sum_{i=1}^{m} \frac{(X_i - p_i(\boldsymbol{\theta}))^2}{p_i(\boldsymbol{\theta})} \tag{7.71}$$

such that

$$\hat{\boldsymbol{\theta}} = \arg\min X^2(\boldsymbol{\theta}). \tag{7.72}$$

---

6. Fisher slightly underestimated this, giving a value of 0.99993.

7. Mendel had to publish his work before the chi-square test was invented (Pearson 1900). Today, however, if you are in the business of massaging data for a scientific report, you really need to know about the chi-square distribution. The professional way to fake data that conform to your favorite hypothesis is to make sure that your p-values scatter uniformly across the [0,1] interval. Not *too* uniformly of course, or you'll get in trouble again.

8. Strictly speaking, this is correct in the limit of large n. For small n, the likelihood function may differ from X^2.

Then the remaining squared deviation $X^2(\hat{\boldsymbol{\theta}})$ follows the chi-squared distribution with $m-1-k$ degrees of freedom:

$$X^2(\hat{\boldsymbol{\theta}}) \sim \chi^2_{m-1-k}. \tag{7.73}$$

The **degrees of freedom** are the number of dimensions along which the data can vary from the model's predictions. Here, the data are the multinomial frequencies $q_i = X_i/n$. They span a space of $m-1$ dimensions because they are constrained to sum to 1, $\sum_{i=1}^{m} q_i = 1$. Then the model itself has k dimensions θ_j, which get adjusted to the data. Thus the number of directions along which the data can deviate from the model is $m-1-k$.

Here, the chi-square test was presented in the context of a discrete probability distribution, specifically to assess the fit of a model to the observed distribution. This applies with great generality, but for some distributions, there may be a more sensitive measure of fit available. At the same time, we will see that the approach of summing squared deviations and comparing to the chi-square distribution will serve in many other statistical contexts.

7.6.3   Fit to a Contingency Table

Consider the case when each measurement involves two discrete variables, X and Y, with m and n possible values, respectively. The joint probability distribution is then specified by mn probabilities p_{ij}, where $i = 1, \ldots, m$ and $j = 1, \ldots, n$. The **contingency table** is a matrix of counts M_{ij}, where M_{ij} is the number of times that the joint outcome (X_i, Y_j) was observed.

The null hypothesis H_0 states that the joint probabilities are independent, so

$$p_{ij} = p_i q_j, \tag{7.74}$$

where p_i and q_j are the marginal probabilities of X and Y, respectively

$$p_i = \sum_{j=1}^{n} p_{ij}, \qquad q_j = \sum_{i=1}^{m} p_{ij}. \tag{7.75}$$

Fitting this null model reduces the number of degrees of freedom by $m + n - 2$.[9] Thus, the number d of degrees of freedom in the contingency table is

$$d = (mn - 1) - (m + n - 2) = (m-1)(n-1). \tag{7.76}$$

Given the null hypothesis, the expected number of observations in each cell of the contingency table is $N_{ij} = M p_i q_j$, where M is the total number of observations and the marginal probabilities are given in equation (7.75). The **chi-square statistic** is then

$$X^2 = \sum_{i=1}^{m} \sum_{j=1}^{n} \frac{(M_{ij} - N_{ij})^2}{N_{ij}}. \tag{7.77}$$

---

9. The m marginal probabilities p_i, and the n marginal probabilities q_j are each constrained to sum to 1, so there are $m + n - 2$ degrees of freedom in the marginal probabilities.

If H_0 is true, then X^2 follows the chi-square distribution with $d = (m-1)(n-1)$ degrees of freedom:

$$X^2 \sim \chi^2_{(m-1)(n-1)}. \tag{7.78}$$

Just as in section 7.6.2, the null hypothesis can be rejected if X^2 is too large, but X^2 could also be suspiciously low.

7.7 Nonparametric Tests

Many of the estimators and significance tests discussed so far work on the assumption that the data $\{X_i\}$ are drawn from a certain type of distribution, often a normal distribution. However, frequently there is no good a priori argument for such an assumption, and the distribution of the samples may look distinctly nonGaussian. So-called **nonparametric tests** allow us to test hypotheses without specifying the functional form of an underlying probability distribution.

For example, recall the two-sample t-test used to decide whether one sample of points $\{X_i\}$ has a mean different from another sample $\{Y_j\}$. It assumes that both samples are drawn from Gaussian distributions with the same width. A nonparametric test with a similar goal is the **Mann-Whitney U-test**, which compares each X_i to each Y_j and counts how often $Y_j > X_i$. If the two samples were drawn from the same distribution, then this should happen about half the time. If it happens much more frequently, then the distribution of Y lies above that of X.

In practice, one computes the **Mann-Whitney U statistic**:

$$U = \sum_{i=1}^{m} \sum_{j=1}^{n} S(X_i, Y_j), \tag{7.79}$$

where

$$S(X, Y) = \begin{cases} 0, & \text{if } X < Y \\ \frac{1}{2}, & \text{if } X = Y \\ 1, & \text{if } X > Y. \end{cases} \tag{7.80}$$

If X and Y are drawn from the same distribution, then U should be centered around $mn/2$. The exact shape of the null distribution depends on the sample sizes m and n. For large samples (>20), the distribution is again well approximated by a Gaussian. As for many other statistical tests, the standard scientific computing packages offer the U-distribution, its cumulative, and p-values for any given observation of U. Again, the test can be evaluated in a one-tailed mode if the alternative hypothesis states that Y is on average larger than X, or in a two-tailed mode if Y could be either larger or smaller.

7.8   Other Statistical Tests

There is a large smorgasbord of statistical tests adapted to one situation or another. The general framework is as follows:

1. Identify a null hypothesis H_0 about the data and an alternative hypothesis H_1.
2. Choose an appropriate statistic X (namely, a function of the data that will take on different values under H_0 and H_1). Often that statistic can be derived from the likelihood ratio of H_0 and H_1.
3. Derive the null distribution $P_0(X)$ (namely the probability distribution of X if H_0 is true).
4. Interpret the observed value of X in that distribution. Typically, one rejects H_0 if X is at an extreme value.
5. Compute a p-value as the integral under $P_0(X)$ at more extreme values than the observed X.
6. If H_1 contains a measure of effect size, quote the effect size with associated confidence limits.

For more instances of these recipes, see, for example, Kanji (2006). The theory behind this framework is elaborated well in Rice (2006).

7.9   Linear Regression

Suppose that we have performed an experiment where we manipulated a variable X and measured the resulting outcome Y. The data set consists of pairs $(X_1, Y_1), \ldots, (X_m, Y_m)$; see the example in figure 7.8.

We want to check for any systematic relationship between X and Y, and the simplest such relationship is linearity: when X changes, then Y changes by some proportional amount. Formally, the model for the data is as follows:

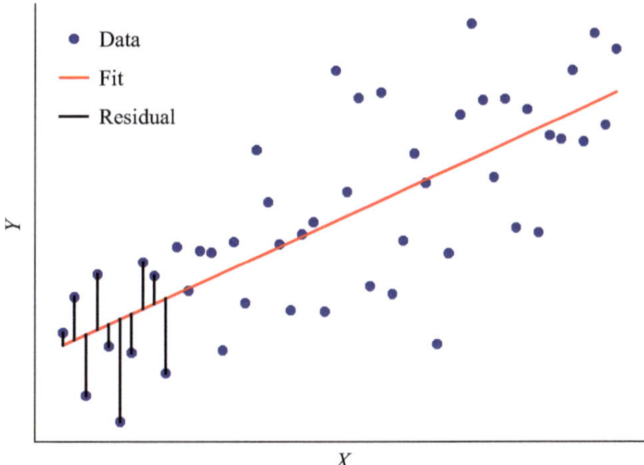

Figure 7.8
A simulated experiment with independent variable X and measurements Y. Linear regression seeks to find a linear fit (red) that minimizes the sum of squares of all the vertical residuals (black).

- For any value of X, Y is distributed according to a normal distribution around a mean of $\mu(X)$ with variance σ^2:

$$f(Y|X, \mu(X), \sigma) = \frac{1}{\sqrt{2\pi}\sigma} \exp\left(-\frac{(Y - \mu(X))^2}{2\sigma^2}\right). \tag{7.81}$$

Here, $\mu(X)$ represents the "true" value of Y for that particular X, and σ is the experimental uncertainty in the measurement.[10]

- The true value $\mu(X)$ depends linearly on x:

$$\mu(X) = a + bX. \tag{7.82}$$

The model has three parameters: a, b, and σ. Linear regression simply finds the maximum-likelihood estimators (MLEs) for all three, along with their confidence limits. As before, we write the likelihood function for the data:

$$L(a, b, \sigma | \{(X_i, Y_i)\}) = \prod_{i=1}^{m} f(Y_i | X_i, a, b, \sigma) \tag{7.83}$$

and the log-likelihood:

$$\log L(a, b, \sigma | \{(X_i, Y_i)\}) = -\frac{1}{2\sigma^2} \sum_{i=1}^{m} (Y_i - a - bX_i)^2 - m\log\sigma - m\log\sqrt{2\pi}. \tag{7.84}$$

From this expression, it is already apparent that the offset a and slope b of the fit line will be chosen so as to minimize the **residual sum of squares**:

$$\mathrm{RSS} = \sum_{i=1}^{m} (Y_i - \hat{a} - \hat{b}X_i)^2. \tag{7.85}$$

This is the sum of all the squared vertical distances of data points from the fit line (figure 7.8). For this reason, linear regression is also called **least squares regression**. In general, one can see that fitting methods that minimize the squared residual of the data are based on an implicit assumption of normally distributed errors.

By maximizing equation (7.84) with respect to all parameters, one finds the maximum likelihood estimators:

$$\hat{b} = \frac{S_{XY}}{S_{XX}}$$

$$\hat{a} = \overline{Y} - \hat{b}\overline{X} \tag{7.86}$$

$$\hat{\sigma}^2 = \frac{\mathrm{RSS}}{m},$$

---

10. Here, we assume that X has negligible uncertainty, for example because it is easily controlled by the experimenter. In other situations, both X and Y are uncertain, which requires a slightly different approach. One option is principal component analysis, which will be presented in section 8.5.1.

where the **sample means** are

$$\overline{X} = \sum_{i=1}^{m} X_i$$
$$\overline{Y} = \sum_{i=1}^{m} Y_i \tag{7.87}$$

and one defines various **sums of squares** as

$$S_{XX} = \sum_{i=1}^{m} (X_i - \overline{X})^2$$
$$S_{YY} = \sum_{i=1}^{m} (Y_i - \overline{Y})^2 \tag{7.88}$$
$$S_{XY} = \sum_{i=1}^{m} (X_i - \overline{X})(Y_i - \overline{Y}).$$

The residual sum of squares given in equation (7.85) now becomes

$$\text{RSS} = S_{YY}\left(1 - \frac{S_{XY}^2}{S_{XX}S_{YY}}\right). \tag{7.89}$$

For small m, it is preferable to replace the MLE for σ^2 with a so-called **unbiased estimate**; namely,

$$s^2 = \frac{\text{RSS}}{m-2}. \tag{7.90}$$

This compensates for the fact that the fit line has been adjusted to the data with the two parameters $\hat{a}$ and $\hat{b}$, removing 2 degrees of freedom from the variation of the data.

Following the usual methods for parameter estimation (section 7.1.1), one can also compute the standard errors for the parameters, in particular

$$s_a{}^2 \equiv \text{Var}[\hat{a}] = \frac{1}{m}\frac{s^2\overline{X^2}}{S_{XX}}$$
$$s_b{}^2 \equiv \text{Var}[\hat{b}] = \frac{1}{m}\frac{s^2}{S_{XX}}. \tag{7.91}$$

Under the model assumptions, the MLE deviates from the true parameter according to the t-distribution with $m-2$ degrees of freedom; namely,

$$\frac{\hat{a}-a}{s_a} \sim t_{m-2}$$
$$\frac{\hat{b}-b}{s_b} \sim t_{m-2}. \tag{7.92}$$

A very common question is **whether there is any significant dependence of Y on X**. One answer is to compute the p-value for the null hypothesis that the slope $b = 0$. This amounts to a t-test on the statistic $\hat{b}/s_b$ with $m - 2$ degrees of freedom. A more informative answer is to quote confidence limits on the slope parameter. For example,

$$b = \hat{b} \pm s_b \quad \text{(standard error)} \tag{7.93}$$

or

$$b = \hat{b} \pm s_b \quad C(t_{m-2}, 0.025) \, (95\% \text{ confidence limit}), \tag{7.94}$$

where $C(t_{m-2}, 0.025)$ is the inverse cumulative function defined in equation (7.48). If the confidence limits include the value of 0, then the slope is not significant.

Another common measure for the strength of the dependence is the **correlation coefficient**:

$$r = \frac{S_{XY}}{\sqrt{S_{XX}S_{YY}}} = \hat{b}\sqrt{\frac{S_{XX}}{S_{YY}}}. \tag{7.95}$$

The correlation coefficient varies from +1 for a perfect positive correlation between X and Y to -1 for perfect anti-correlation. To evaluate significance values or confidence limits on r, one can use the fact that

$$\sqrt{m-2}\,\frac{r}{\sqrt{1-r^2}} \sim t_{m-2} \tag{7.96}$$

follows a t-distribution with $m - 2$ degrees of freedom.

In applications involving linear regression, another coefficient that is often quoted is the **coefficient of determination**, or "R-squared," which is defined as the square of the correlation coefficient:

$$r^2 = \frac{S_{XY}^2}{S_{XX}S_{YY}}. \tag{7.97}$$

From equation (7.89), one can see that r^2 is the fraction of the total variance in Y that is explained by the linear dependence on X.

A good statistical analysis package will compute these and many other desired metrics of the fit, including confidence intervals for the fitted relationship; see figure 7.9.

7.9.1 Multiple Regression

These concepts generalize easily to a model where the measurement Y depends on several independent variables $X_j, j = 1, \ldots, n$. We suppose that the relationship is additive and perturbed by an additive Gaussian error:

$$Y = \sum_{j=1}^{n} b_j X_j + E, \tag{7.98}$$

where

$$E \sim N(0, \sigma^2). \tag{7.99}$$

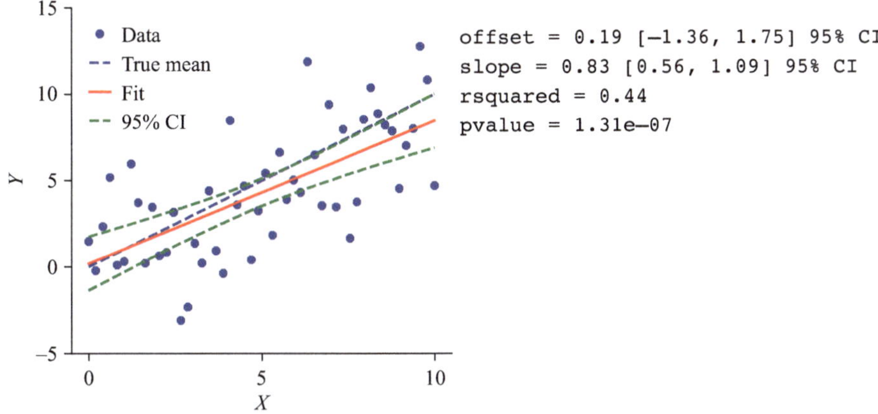

Figure 7.9
A linear fit to the data of figure 7.8. The results include maximum likelihood values for offset and slope, along with 95% CIs on both parameters. These data were simulated by adding Gaussian noise to the identity line, with offset 0 and slope 1. Note that this true mean (blue dashes) falls inside the 95% confidence limits of the fit (green dashes).

The data consist of m measurements $(Y_i; X_{i1}, \ldots X_{in}), i = 1, \ldots, m$. We want to find the optimal values of the weights $b_j, j = 1, \ldots, n$. The log likelihood function is

$$\log L(\mathbf{b}, \sigma) = -\frac{1}{2\sigma^2} \sum_{i=1}^{m} \left(Y_i - \sum_{j=1}^{n} b_j X_{ij} \right)^2 - m \log \sigma - m \log \sqrt{2\pi}. \tag{7.100}$$

As in equation (7.83), the only term in the log likelihood that depends on the parameters b_j is the residual sum of squares:

$$\text{RSS} = \sum_{i=1}^{m} \left(Y_i - \sum_{j=1}^{n} b_j X_{ij} \right)^2 \tag{7.101}$$

$$= (\mathbf{y} - \mathbf{Xb})^\top (\mathbf{y} - \mathbf{Xb}),$$

where the vectors are $\mathbf{y} = [Y_i]$, $\mathbf{b} = [b_j]$ and the matrix $\mathbf{X} = [X_{ij}]$. Differentiating with respect to $\mathbf{b}$ leads to a condition for the minimum:[11]

$$0 = \nabla_{\mathbf{b}} \text{RSS} = -2\mathbf{X}^\top (\mathbf{y} - \mathbf{Xb}) \tag{7.102}$$

so the MLE parameters are

$$\hat{\mathbf{b}} = \left(\mathbf{X}^\top \mathbf{X} \right)^{-1} \mathbf{X}^\top \mathbf{y}. \tag{7.103}$$

---

11. This is a minimum rather than some other extremum, as long as the matrix $\mathbf{X}^\top \mathbf{X}$ is positive definite. In turn, this is guaranteed as long as all the variables X_j are linearly independent. If, however, two of the variables are proportional, $X_{ij} = X_{ik}$ for all i, then the coefficients b_j and b_k are redundant and there is no unique solution.

Similar to the case of linear regression in equation (7.90), an unbiased estimate of the error variance σ^2 is

$$s^2 = \frac{\text{RSS}}{m-n}, \tag{7.104}$$

where $m-n$ is recognized as the number of degrees of freedom in the data (m data points and n fit parameters). Based on this, one can derive the uncertainty of the MLE parameters and the standard errors are

$$s_{b_j} = s\sqrt{c_{jj}}, \tag{7.105}$$

where

$$\mathbf{C} = [c_{ik}] = \left(\mathbf{X}^\top \mathbf{X}\right)^{-1}. \tag{7.106}$$

Again, the deviation from the true parameter follows a t-distribution:

$$\frac{b_j - \hat{b}_j}{s_{b_j}} \sim t_{m-n}. \tag{7.107}$$

If the variables X_j are correlated with each other, then the uncertainty in the corresponding parameters b_j will be correlated as well. For example, suppose that two of the X_j are almost identical, then one can trade off the corresponding b_js without affecting the fit. In detail, one finds that the covariance among the fit parameters is

$$\text{Cov}[b_j, b_k] = \langle (b_j - \hat{b}_j)(b_k - \hat{b}_k)\rangle = s^2\,\mathbf{C}. \tag{7.108}$$

Note that all the results presented above for standard linear regression in equation (7.86) and the subsequent ones are a special case of the formalism derived here, where there are just two independent variables X_1 and X_2. The first is set equal to 1 throughout: $X_1 = 1$. The second is set to the independent variable: $X_2 = X$. Then the linear regression line is defined by the offset $a = b_1$ and the slope $b = b_2$.

One can therefore extend the common linear regression between two variables X, Y to a more general functional fit. For example, for a polynomial fit

$$Y = \sum_{j=0}^{n-1} b_j X^j + E, \tag{7.109}$$

one simply defines n variables $X_j = X^j$ and then finds the optimal b_j from multiple regression as before. Any other functional fit that is linear in the coefficients can be optimized the same way; see figure 7.10.

7.9.2 Model Fitting

Stepping back for a moment, why does multiple regression offer such a simple analytical solution to the process of model fitting? This traces back to the log likelihood function (in equation 7.100), which depends on the fit parameters only through a

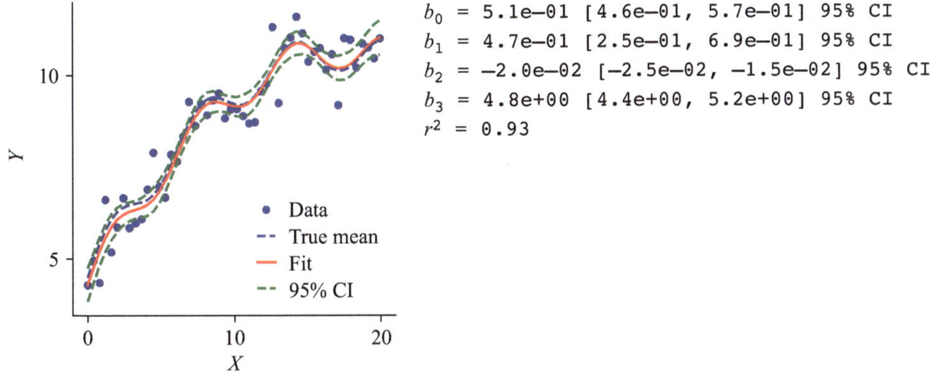

Figure 7.10
A regression fit of the model $Y = b_0 + b_1 X + b_2 (X-5)^2 + b_3 \sin X$. Display as in figure 7.9.

simple quadratic form. The quadratic form is guaranteed positive definite,[12] so it makes a paraboloid in the space of the fit parameters b_j. That parabolic bowl has a single minimum, and it is easy to find.

If the model for Y is not linear in the parameters b_j, then these simple conditions don't apply. However, there are models for which the negative log likelihood still makes a bowl in parameter space, with a single global minimum. Such models are called "convex," and they can be optimized by well-developed routines for finding the minimum. If the likelihood function is not convex, with multiple local minima, then optimization becomes more complicated.

7.10 Bootstrapping

In statistical analysis, **bootstrapping** is the answer to "What do I do if nothing is known about the statistical properties of the quantity I am estimating?" Following the framework that we have presented, the situation is that we have independent measurements $\{X_i\}$ produced by some unknown process that includes random errors. We want to estimate a quantity θ that is a function of the $\{X_i\}$, where θ might be, for example, a parameter in a model that we have for the data. From the existing data, we get just one estimate of that quantity, $\hat{\theta} = \theta(\{X_i\})$. Now we want to express our confidence in that estimate. If we could repeat the whole experiment many times, what would the distribution $P(\hat{\theta})$ be across many such trials? Given such a distribution $P(\hat{\theta})$, we could then quote confidence limits or quote p-values regarding various hypotheses about θ.

Sometimes you cannot repeat the experiment many times because the data are expensive and there is just one data set to work with. In earlier sections, we used assumptions about the distribution $P(\{X_i\})$ from which the data were drawn and a likelihood function for θ to derive the posterior distribution $P(\theta)$. But what if none of that knowledge is available or trustworthy?

Instead of getting more data from new experiments, bootstrapping simply resamples the old data. It assumes that the data sample $\{X_i\}$ is a pretty good representation

---

12. …because $\mathbf{X}^\top \mathbf{X}$ is a positive definite matrix.

of the real distribution of X. You create a new data set $\{X_i^*\}$ by randomly and independently drawing m samples from that set of $\{X_i\}$.[13] Obviously, this resampled set will resemble the original one, but some values of X will be omitted and others appear multiple times. Then you estimate θ from the resampled data:

$$\hat{\theta}^* = \theta(\{X_i^*\}). \tag{7.110}$$

Finally, you repeat resampling many times and accumulate the results in the so-called **bootstrap distribution** $P_B(\hat{\theta}^*)$.

The simplest use of this bootstrap distribution is to treat it as an approximation to the true distribution $P(\theta)$ that we would get from repeating the experiment many times. For example, one may estimate the standard error of θ based on the variance of the bootstrap distribution:

$$s_{\theta^*} = \sqrt{\text{Var}[\hat{\theta}^*]} \tag{7.111}$$

and then report

$$\theta = \hat{\theta} \pm s_{\theta^*} \quad \text{(standard error)} \tag{7.112}$$

Alternatively, one can use quantiles of the bootstrap distribution to quote confidence limits, such as

$$\hat{\theta}^*_{2.5 \text{ percent}} < \theta < \hat{\theta}^*_{97.5 \text{ percent}} \quad (95\% \text{ CI}), \tag{7.113}$$

where $\hat{\theta}^*_{2.5 \text{ percent}}$ and $\hat{\theta}^*_{97.5 \text{ percent}}$ are the 2.5 percent and 97.5 percent quantiles of the bootstrap distribution.

At first, it may seem unreasonable that this reuse of old data should work, like getting something for nothing. Indeed, the name of the method invokes the impossibility of pulling yourself out of the swamp by your own bootstraps. On the other hand, bootstrapping does not promise to improve your estimate of the statistic θ. It will merely tell you how uncertain that estimate is.

This interpretation of the bootstrap distribution sometimes goes by the name "percentile method" because it simply uses quantiles of the distribution for CIs. There are more sophisticated recipes for bootstrapping that get closer to the true distribution $P(\hat{\theta})$ under certain assumptions.[14] One commonly used variant is the **bias-corrected and accelerated bootstrap**, available in many statistical analysis packages.

For illustration, here is an application of bootstrapping to the outcome of multiple linear regression (figure 7.11). The data are a set of $\{(X, Y)_i\}$ values. Using the methods from section 7.9, we have performed a polynomial fit of the type

$$Y_i = \sum_{j=0}^{n-1} b_j X_i^j + E_i \tag{7.114}$$

---

13. Here, "independently" means "drawing with replacements."

14. Note that there is some contradiction even among textbooks with regard to these choices. For example, the recommendation in Rice (2006) conflicts with the inventors of the bootstrap in Efron and Tibshirani (1993).

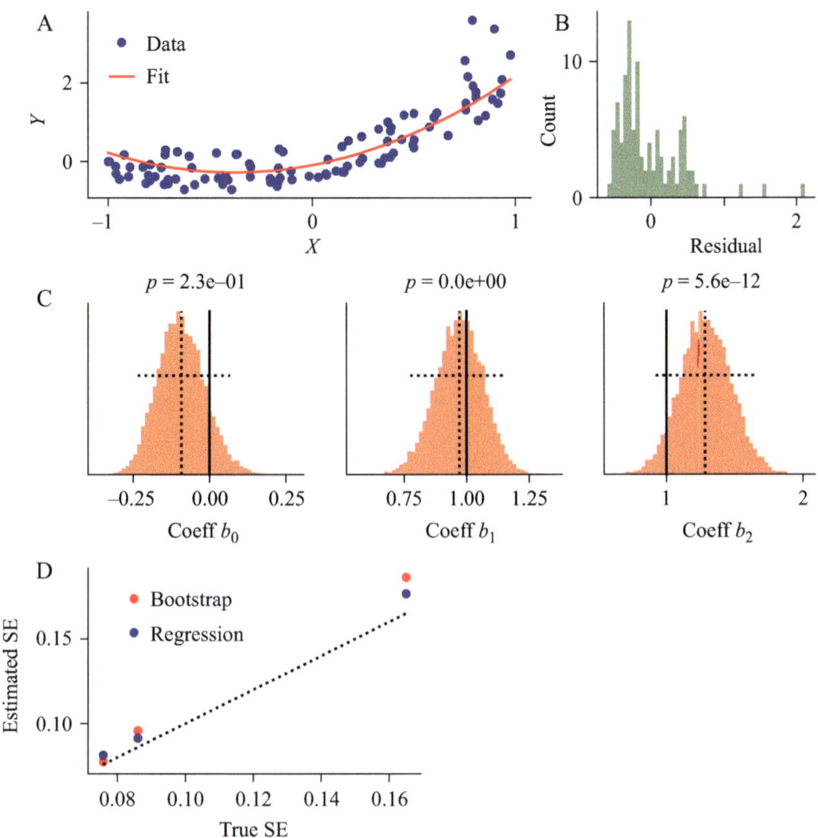

Figure 7.11

A bootstrap analysis of coefficients in a polynomial fit. (A) A regression fit of the model $Y_i = b_0 + b_1 X_i + b_2 X_i^2 + E_i$ to a data set $\{(X, Y)_i\}$. (B) Histogram of the residuals E_i. Note the skewed, nonGaussian shape of this distribution. (C) Bootstrap of the regression fit using $n = 10{,}000$ resamplings of the data. For each coefficient, the histogram shows the bootstrap distribution; solid vertical line = true value of that coefficient in the model that generated the data; dotted line: estimate from the fit to the original data; horizontal line: 95% CI of the bootstrap distribution. p-value tests the hypothesis that the coefficient is zero. (D) Comparison of standard errors derived from repeating the experiment many times (True SE); from a bootstrap of the original data set; and from the formulas for multiple regression.

by minimizing the residual sum of squares $\text{RSS} = \sum_{i=1}^{m} E_i^2$. The residuals from the fit do not look normally distributed, as shown in figure 7.11B, so we cannot formally rely on the linear regression formalism to deliver the correct standard errors for the polynomial coefficients. Bootstrap ping, to the rescue! We resample the original data by picking randomly and independently $m = 100$ data pairs $\{(X, Y)_i^*\}$.[15] Then we perform multiple regression on the resampled data. We do this $n = 10{,}000$ times to get a bootstrap distribution for each of the coefficients, as in figure 7.11C. From that, we derive the bootstrap standard error, s_{θ^*}. In turn, that allows us to quote a p-value for the null hypothesis that the polynomical coefficient is zero, as in figure 7.11C.

---

15. ... of course, keeping each pair (X, Y) together.

Here, we have the benefit of knowing the true model that produced the data: It was in fact a polynomial relationship as in equation (7.114), with an exponentially distributed noise term:

$$b_0 = 0,\ b_1 = 1,\ b_2 = 1,\ E \sim \mathrm{Exp}(\nu),\ \nu = 0.5. \tag{7.115}$$

Note that all the true coefficients are in fact within the 95 percent confidence limits of the estimates; see figure 7.11C.

In this artificial situation, we can easily do more "experiments" by simulating additional data sets produced from the same model. Figure 7.11D compares the standard error derived from bootstrapping to the true standard error derived from doing new experiments. We also compare to the estimate from the multiple regression formulas in equation (7.105), which are based on the inadequate assumption of a normal noise distribution. As it happens, all three estimates are in quite good agreement.

As illustrated here, bootstrapping is a convenient approach to estimating the distribution of a statistic, because computing power is readily available, and because any statistic that can be computed from the sample data can be queried without having a theoretical understanding of its distribution.

8.1   Random Walks and Diffusion

Many processes in biology are driven at their core by random events. On the smallest scales, thermal fluctuations play an essential role: the assembly and disassembly of a protein polymer, the random stepping of a molecular motor, the thermal opening and closing of an ion channel, the random meandering of a signaling molecule through the cytoplasm. On a large scale, the dynamics of a population are governed by births and deaths among its many individuals, events that are sufficiently unpredictable to be treated as random variables.

When a variable executes many independent random steps in sequence it leads to a **random walk**. A canonical example is the position of a small particle, like a calcium ion, buffeted by thermal collisions with molecules of the surrounding fluid, as shown in figure 8.1. If we zoom out from this molecular picture to consider many such random variables collectively, such as the concentration of all calcium ion in a cell, then we observe a process of mass transport called **diffusion**. This chapter will elaborate on the dynamics of random walks and diffusion.

8.1.1   Brownian Motion

The earliest published account of random thermal motion comes from Robert Brown, a biologist interested in the process of pollination (Brown (1828)). While inspecting pollen grains suspended in water using a simple microscope, he "observed many of them very evidently in motion." The motions "arose neither from currents in the fluid, nor from its gradual evaporation, but belonged to the particle itself." Brown at first suspected the particles to be "animated," but soon confirmed that perfectly inorganic substances, when ground into a dust, produced the same type of motion. Physicists largely ignored these phenomena of "Brownian motion" until the early twentieth century, when Einstein showed how they accord with predictions from the broader framework of kinetic theory (Einstein (1905), Brush (1968)).

Suppose that we could track a molecule of oxygen suspended in water and let us just follow its movements along the x-direction. The molecule's kinetic energy is $kT/2$, where T is the temperature in degrees Kelvin and $k = 1.38 \times 10^{-23}$ J/K is Boltzman's constant, so it flies along at about 100 m/s. However, it doesn't get very far: every 10^{-13} s or so, it bangs into a water molecule that changes its speed and direction. In fact, the mean free path during which it flies straight is only 10^{-11} m, about 1/10 the size of a hydrogen atom. All this is to say that the individual step of such a Brownian particle is so small in size and duration that for all practical purposes, we will only ever have to worry about the accumulated effect of many steps.

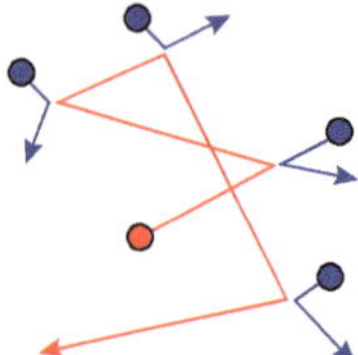

Figure 8.1
A Brownian particle (red) buffeted by collisions with molecules of the surrounding medium (blue).

8.1.2 Random Walk in One Dimension

The simplest mathematical approximation of Brownian motion is a discrete random walk (figure 8.2). Consider a particle moving in one dimension with discrete steps. The particle starts at $x = 0$. At every time step, it moves either right to $x + 1$, with probability p, or left to $x - 1$, with probability $q = 1 - p$.[1] So if

$$x_n = \text{position of the particle after } n \text{ steps,} \tag{8.1}$$

then

$$x_{n+1} = \begin{cases} x_n + 1 & \text{with probability } p \\ x_n - 1 & \text{with probability } q = 1 - p. \end{cases} \tag{8.2}$$

What is the probability distribution $P(x_n)$ for the position of the particle after n time steps? Suppose that the n steps included m steps right and $n - m$ steps left. Then m follows the binomial distribution given in equation (6.20):

$$m \sim \mathrm{Bin}(n, p). \tag{8.3}$$

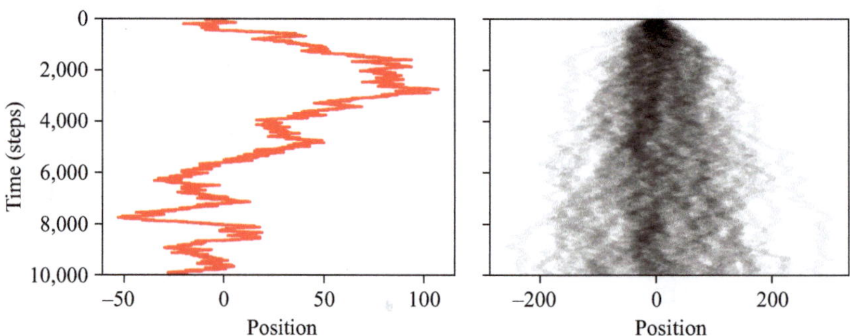

Figure 8.2
Left: Position as a function of time for a particle that performs an unbiased random walk moving right or left at each time step with equal probability. Right: 100 such random walks superposed.

---

1. For a Brownian particle, steps to the left and right are equally probable, but with a little extra effort, we may as well consider this more general case, where $p \neq q$.

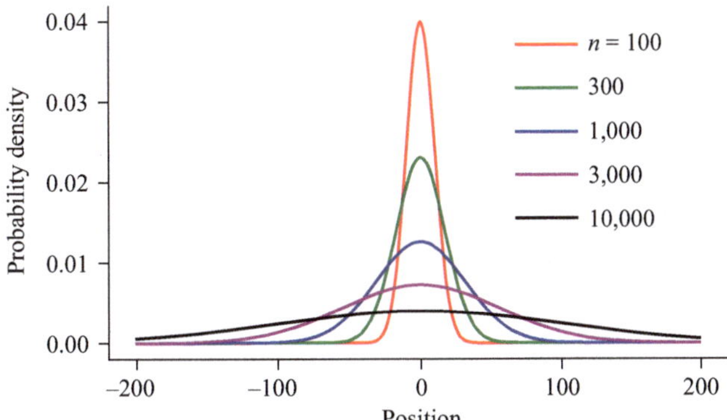

Figure 8.3
Probability density of a random walker with $p = 1/2$ after a large number of n steps.

The corresponding position of the particle is

$$x_n = m - (n - m) = 2m - n. \tag{8.4}$$

So the probability of being at position x after n steps is

$$P(x; n) = \binom{n}{\frac{n+x}{2}} p^{\frac{n+x}{2}} q^{\frac{n-x}{2}}. \tag{8.5}$$

This distribution is shown in figure 8.3. It has a mean μ and variance σ^2, given by

$$\mu = n(p - q), \qquad \sigma^2 = 4npq. \tag{8.6}$$

For large n, we can invoke the central limit theorem to state that

$$P(x; n) = \frac{1}{\sqrt{2\pi\sigma^2}} e^{-\frac{(x-\mu)^2}{2\sigma^2}}. \tag{8.7}$$

So after many steps, the random walker has a bell-shaped probability distribution that follows a Gaussian profile. The width of that Gaussian grows as the square root of the number of steps.

8.1.3   The Diffusion Coefficient
Returning to the real world, what can we conclude about the motion of a Brownian particle? The number of collisions that it undergoes is huge, but strictly proportional to time. After a few nanoseconds, there have been so many collisions that the central limit theorem kicks in. So we can immediately conclude that the particle's position has a Gaussian distribution whose width σ grows proportionally to the square root of time:

$$\sigma = \sqrt{2Dt}. \tag{8.8}$$

The proportionality constant D is called the **diffusion coefficient**. It completely characterizes the particle's behavior under thermal motion.

Table 8.1
Approximate distance versus time for a
small molecule diffusing in water

Time	Distance
1 ms	1 μm
100 ms	10 μm
10 s	100 μm
1,000 s	1 mm
1 day	10 mm

For biological applications, it is useful to remember a couple of order-of-magnitude
numbers:

- For a small molecule (molecular weight up to a few hundred) in water, $D \approx 10^{-5} \mathrm{cm}^2/\mathrm{s}$
- For a protein moving laterally in a cell membrane, $D \approx 10^{-9} \mathrm{cm}^2/\mathrm{s}$

Note the physical dimensions of the diffusion coefficient: distance^2/time. Clearly, this
is not a velocity! The typical distance traveled via diffusion is proportional *not* to time,
but to the square root of time. Table 8.1 lists those distances for a small molecule in
water.

So a small signaling molecule can equilibrate across a typical cell body in 0.1 s,
but if it needs to get 1 cm down the axon of a neuron that would take forever. Clearly,
thermal transport is not sufficient there.

8.1.4   Fick's Laws of Diffusion

Let us now imagine a very large number of molecules, all executing Brownian motion
independently of each other. Again, we will model this as a discrete random walk along
the x-axis. Suppose that after j time steps, there are $N_{i,j}$ particles located at position i
(figure 8.4):

$$N_{i,j} = \text{number of particles at position } i \text{ after step } j. \qquad (8.9)$$

In the next step, half the particles at location i step to the right and the other half to
the left. So the net number of particles moving across the border from i to $i+1$ is:

$M_{i,j} = \text{number of particles moving from position } i \text{ to } (i+1) \text{ during step } (j+1)$

$$= \frac{1}{2}\left(N_{i,j} - N_{i+1,j}\right). \qquad (8.10)$$

So the new number of particles becomes

$$N_{i,j+1} = N_{i,j} + M_{i-1,j} - M_{i,j}. \qquad (8.11)$$

To connect to real-world units, we define

$$\Delta x = \text{size of a step along the } x\text{-axis}$$

$$\Delta t = \text{duration of a step}. \qquad (8.12)$$

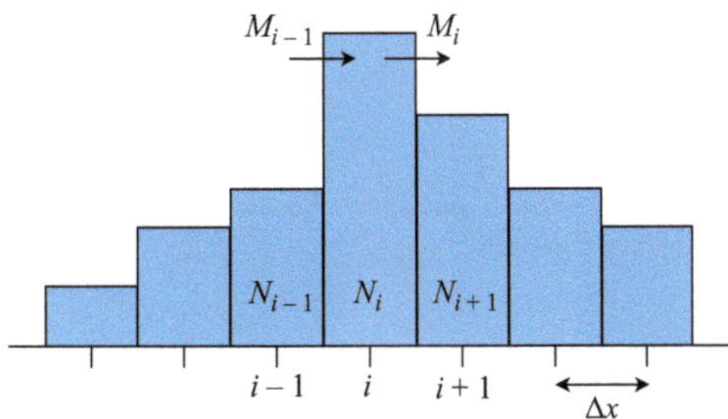

Figure 8.4
A distribution of particles undergoing independent random walks.

Then we take the continuum limit by allowing $\Delta x \to 0$ and $\Delta t \to 0$ while keeping the diffusion coefficient constant $\frac{1}{2}\frac{(\Delta x)^2}{\Delta t} = D$. In that limit,

$$\frac{N_{i,j}}{\Delta x} \to C(x,t) = \text{concentration of particles at position } x = i\Delta x \text{ and time } t = j\Delta t \quad (8.13)$$

and

$$\frac{M_{i,j}}{\Delta t} \to J(x,t) = \text{flux of particles at position } x = i\Delta x \text{ and time } t = j\Delta t. \quad (8.14)$$

Equation (8.10), after dividing by Δt, is

$$\frac{M_{i,j}}{\Delta t} = \frac{1}{2}\frac{(\Delta x)^2}{\Delta t}\frac{\left(N_{i,j} - N_{i+1,j}\right)}{(\Delta x)^2}, \quad (8.15)$$

which becomes, in the continuum limit

$$J(x,t) = -D\frac{\partial C(x,t)}{\partial x}. \quad (8.16)$$

Similarly, equation (8.11), after dividing by Δt and Δx, becomes in the continuum limit

$$\frac{\partial C(x,t)}{\partial t} = -\frac{\partial J(x,t)}{\partial x}. \quad (8.17)$$

Equations (8.16) and (8.17) are called **Fick's laws of diffusion**. They relate the local flux of particles to the concentration. These two partial differential equations can be combined to produce the **diffusion equation**:

$$\frac{\partial}{\partial t}C(x,t) = D\frac{\partial^2}{\partial x^2}C(x,t). \quad (8.18)$$

8.1.5 Qualitative Behavior of the Diffusion Equation

Qualitatively, diffusion acts so as to "smooth" the concentration profile over time, as illustrated in figure 8.5. Around a peak in the profile of $C(x,t)$, the second spatial derivative is negative, so according to equation (8.18), the concentration here will decrease. The simple reason is that this region is flanked on both sides by an outward-sloping

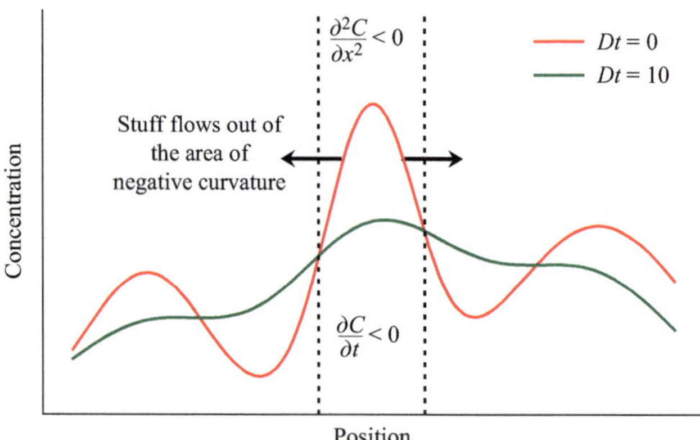

Figure 8.5
The evolution of a concentration profile under diffusion.

concentration gradient, which leads to particles flowing out of that region. On the other hand, at a local minimum, the second spatial derivative is positive, there is an inward sloping gradient on both sides, so this concentration will increase. The net effect is that **peaks of concentration get flattened and valleys get filled in.** The final state at long times tends to have no peaks or valleys, unless some special boundary conditions apply.

8.1.6 Random Walks and Diffusion in Higher Dimensions
Many biological motions take place in two or three dimensions. One can model three-dimensional (3D) Brownian motion as a random walk on a 3D coordinate grid, with the particle taking steps to a neighboring grid point simultaneously in all three directions. So during one step, x, y, and z all change by ± 1. Because the three random walks take place independently, we can consider each coordinate on its own, each of which behaves just like the one-dimensional (1D) case discussed in section 8.1.4.

Back in the real world, if a particle with diffusion coefficient D starts at the origin, then after time t, all three position variables will be distributed like a Gaussian with variance $2Dt$:

$$P_x(x) = \frac{1}{\sqrt{4\pi Dt}} e^{-\frac{x^2}{4Dt}}$$

$$P_y(y) = \frac{1}{\sqrt{4\pi Dt}} e^{-\frac{y^2}{4Dt}} \qquad (8.19)$$

$$P_z(z) = \frac{1}{\sqrt{4\pi Dt}} e^{-\frac{z^2}{4Dt}}.$$

The Euclidean distance from the starting point is $r = \sqrt{x^2 + y^2 + z^2}$ and its variance is

$$\langle r^2 \rangle = \langle x^2 + y^2 + z^2 \rangle = 6Dt. \qquad (8.20)$$

Fick's laws relate the flux of particles to the concentration. The 3D versions are

$$\mathbf{J} = -D\nabla C(\mathbf{r}, t)$$

$$\frac{\partial}{\partial t} C(\mathbf{r}, t) = -\nabla \cdot \mathbf{J}. \qquad (8.21)$$

Here, $\mathbf{r} = (x, y, z)$ refers to the position in three dimensions; C is the concentration and $\mathbf{J}$ is the flux of particles. The term

$$\nabla C = \left(\frac{\partial C}{\partial x}, \frac{\partial C}{\partial y}, \frac{\partial C}{\partial z} \right) \tag{8.22}$$

is the gradient (i.e. multidimensional derivative) of the concentration and

$$\nabla \cdot \mathbf{J} = \frac{\partial J}{\partial x} + \frac{\partial J}{\partial y} + \frac{\partial J}{\partial z} \tag{8.23}$$

is the divergence of the flux field (note that this is *not* the same as the gradient; it is the dot product of the gradient operator with the flux $\mathbf{J}$).

Again, one can combine the two laws into one diffusion equation:

$$\frac{\partial}{\partial t} C(\mathbf{r}, t) = D\nabla \cdot \nabla C(\mathbf{r}, t)$$
$$= D\nabla^2 C(\mathbf{r}, t), \tag{8.24}$$

where $\nabla \cdot \nabla \equiv \nabla^2 \equiv \Delta$ is the **Laplacian operator**. In 3D Cartesian coordinates, this is simply

$$\nabla^2 = \frac{\partial^2}{\partial x^2} + \frac{\partial^2}{\partial y^2} + \frac{\partial^2}{\partial z^2}. \tag{8.25}$$

Depending on the spatial symmetries of the problem at hand, it can be more convenient to work in a different coordinate system, and some caution is required around differential operators. For example, in the 3D spherical coordinate system (section 1.5.1) with coordinates (r, θ, ϕ), the Laplacian is

$$\nabla^2 \equiv \frac{1}{r^2} \frac{\partial}{\partial r} \left(r^2 \frac{\partial}{\partial r} \right) + \frac{1}{r^2 \sin \theta} \frac{\partial}{\partial \theta} \left(\sin \theta \frac{\partial}{\partial \theta} \right) + \frac{1}{r^2 \sin^2 \theta} \frac{\partial^2}{\partial \phi^2}. \tag{8.26}$$

8.1.7 Solving the Diffusion Equation

In addition to the transport of particles by Brownian motion, the diffusion equation covers other phenomena of mass transport, like the conduction of heat, or the movement of electric charge in an electrolyte. In a typical problem, one is given an **initial condition** of the profile $C(x, t = 0)$ and wants to know the future concentration profile $C(x, t)$. Beside the initial condition, one has to also deal with **boundary conditions** that specify what happens at the edges of the space or other special locations. In some cases, one is mostly interested in the final **steady-state solution** at very long times $C(x, t = \infty)$. Here, we touch on some of these methods for solving the diffusion equation. These may help you devise at least an approximate solution to your problem. For tough problems, one can always resort to lookup via the Google search bar. Back when people read books, a classic collection of solutions could be found in Carslaw and Jaeger (1986).

8.1.7.1 Superposition

The diffusion equation falls in the class of **linear partial differential equations**. This simply means that the function of interest $C(x, t)$ and its derivatives appear only with a power of 1. As a consequence, the solutions of the differential equation obey the **superposition principle**: If two functions $C_1(x, t)$ and

$C_2(x, t)$ are both solutions to the diffusion equation, then any linear combination $\lambda_1 C_1(x, t) + \lambda_2 C_2(x, t)$ is also a solution.

We encountered this superposition idea before in the more general treatment of linear systems in section 3.1.2. Here again, we will see that it has powerful consequences.

A simple way to understand superposition is to remember that at time $t = 0$, the initial concentration profile $C(x, 0)$ is made of many independent particles. Each of these executes a random walk, independent of the others. If we arbitrarily divide those particles into a red group $C_1(x, 0)$ and a blue group $C_2(x, 0)$, they will still produce the exact same profile $C(x, t)$ later on. Note that this argument relies on there being no interaction between the particles. If they do interfere with each other, the differential equation will not be linear, and superposition no longer applies.

8.1.7.2　Green's function　This argument leads to another conclusion: We can solve for the future profile $C(x, t)$ if we simply know what happens to the probability density of each individual particle over time. After all, the full profile is simply the sum of the individual particle densities.

So let us suppose that the probability density of a particle that starts at location x' develops according to

$$G(x; x', t) = \text{probability that a particle is at location } x \text{ at time } t$$
$$\text{if it started from } x' \text{ at time 0.} \tag{8.27}$$

Technically, this is called the **Green's function** of the diffusion problem. Obviously, at time $t = 0$, the particle is certain to be at $x = x'$, so

$$G(x; x', t = 0) = \delta(x - x'), \tag{8.28}$$

where $\delta(x)$ is the delta function. We can write the initial profile as a sum over these delta functions:

$$C(x, 0) = \int_{x'} C(x', 0)\, G(x; x', 0)\, dx'. \tag{8.29}$$

Then we allow each of the particles to evolve according to its Green's function and sum again to get the solution:

$$C(x, t) = \int_{x'} C(x', 0)\, G(x; x', t)\, dx'. \tag{8.30}$$

A simple example is diffusion in free space. Suppose that there are no boundaries anywhere, so the entire x-axis is available. Then we already know the Green's function: A particle will diffuse according to the spreading Gaussian of equation (8.7):

$$G(x; x', t) = \frac{1}{\sqrt{4\pi D t}} e^{-\frac{(x-x')^2}{4Dt}}. \tag{8.31}$$

All places along the x-axis behave the same way, so this same Green's function applies no matter where the particle starts. Therefore, the solution with initial condition $C(x, 0)$

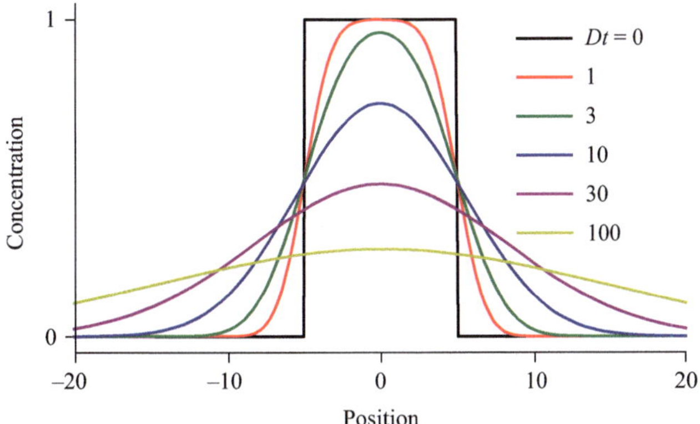

Figure 8.6
Diffusion from an initial square bolus of particles.

is simply

$$C(x,t) = \int_{x'} C(x',0)\, \frac{1}{\sqrt{4\pi Dt}} e^{-\frac{(x-x')^2}{4Dt}}\, dx'. \tag{8.32}$$

Figure 8.6 shows the time-dependent solution when the initial condition is a bolus of particles with a square concentration profile.

8.1.7.3 Boundary conditions   At boundaries in the space, one generally considers two kinds of conditions:

- Reflecting boundary: Particles bounce off this surface. That means that there can be no flux of particles into or out of the surface. So the boundary condition is that everywhere on the surface,

$$\mathbf{J}(\mathbf{r},t)\cdot\mathbf{n}=0 \tag{8.33}$$

  where $\mathbf{n}$ is the normal vector to the surface.
- Absorbing boundary: Particles get swallowed by this surface, never to appear again. That means the concentration of particles is zero everywhere on the surface:

$$C(\mathbf{r},t)=0. \tag{8.34}$$

Some simple boundary problems can be solved with so-called **mirror sources**. For example, suppose that a particle diffuses in one dimension, starting at $x=a$, but there is a reflective boundary at $x=0$, so its motion is constrained to the right half of the x-axis only. We can imagine instead that there is no boundary at all, but a second particle starts out at $x=-a$, in the mirror-reflected position of the true particle (figure 8.7). For every time that the true particle random-walks through the boundary to $x<0$, the mirror particle random-walks out of the boundary in the opposite direction.

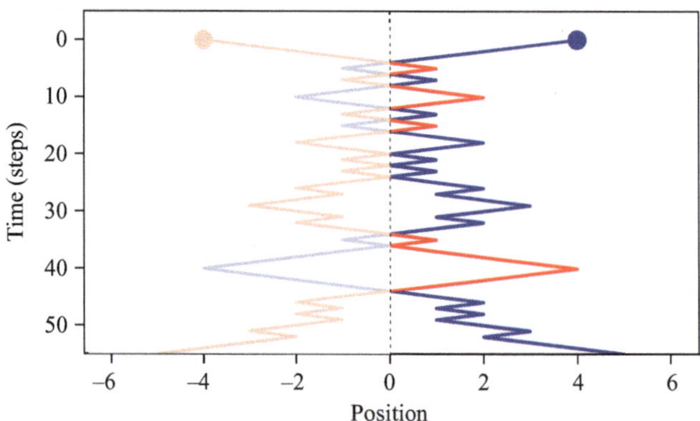

Figure 8.7
Mirror sources: To simulate the random walk of a particle with a reflecting boundary at $x = 0$, we imagine two mirror particles (blue and red) executing mirror-symmetric walks, but without a barrier. Whenever the red particle crosses into the right half, it looks like the blue particle bounced off the barrier.

So in the right half of the space, we can simply add the density of the true and the virtual particles to get the solution:

$$C_{\text{ref}}(x, t) = \frac{1}{\sqrt{4\pi Dt}} \left(e^{-\frac{(x-a)^2}{4Dt}} + e^{-\frac{(x+a)^2}{4Dt}} \right). \tag{8.35}$$

This perfectly emulates the reflection of a particle at a reflecting boundary.

Similarly, for an absorbing boundary, we add an antiparticle in the mirror position (i.e., one with "negative probability"). At the surface, the densities of the two particles precisely cancel each other out, thus enforcing the condition $C(x, t) = 0$:

$$C_{\text{abs}}(x, t) = \frac{1}{\sqrt{4\pi Dt}} \left(e^{-\frac{(x-a)^2}{4Dt}} - e^{-\frac{(x+a)^2}{4Dt}} \right). \tag{8.36}$$

8.1.8 Steady-State Solutions

After a long time t, diffusion systems typically settle into a steady state where nothing changes anymore. Based on equation (8.24), that means

$$\nabla^2 C(\mathbf{r}) = 0. \tag{8.37}$$

The solutions to this equation[2] depend entirely on the boundary conditions. Here are some examples.

Example 8.1 (Diffusion in a box) Suppose that a volume is entirely enclosed by a reflecting boundary. Then the concentration within the volume will eventually settle down to a constant value everywhere:

$$C(\mathbf{r}) = c. \tag{8.38}$$

---

2. This is called **Laplace's equation** and also appears in electrostatics, where it governs the electric potential in charge-free space. Sometimes you can crib a diffusion solution from an electrostatics book.

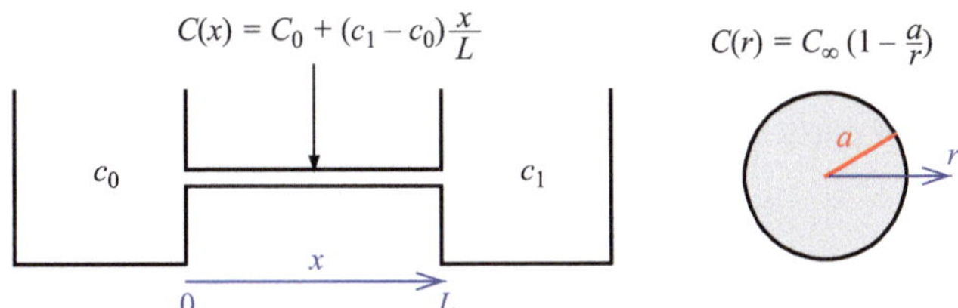

Figure 8.8
Geometry of diffusion examples. Left: Pipe between two stirred tanks. Right: An absorbing sphere in an infinite tank.

Clearly, this satisfies $\nabla^2 C(\mathbf{r}) = 0$. Also, the flux is zero everywhere: $\mathbf{J}(\mathbf{r}) = -D\nabla C(\mathbf{r}) = 0$, which satisfies the reflecting condition at the boundary of the space. $\square$

Example 8.2 (Diffusion between two stirred compartments) Imagine a thin pipe between two water tanks (figure 8.8). Each tank is kept at a constant concentration of the solute. If $x \in [0, L]$ is the position along the pipe, then the boundary conditions for $C(x)$ are

$$C(0) = c_0, \, C(L) = c_1. \tag{8.39}$$

Along the pipe, the concentration is

$$C(x) = c_0 + (c_1 - c_0)\frac{x}{L}. \tag{8.40}$$

The gradient $\frac{\partial C}{\partial x} = \frac{c_1 - c_0}{L}$ is constant along the pipe, so $\frac{\partial^2 C}{\partial x^2} = 0$ satisfies equation (8.37). Also, there is a constant flux of solute along the pipe of strength $J = -D\frac{\partial C}{\partial x} = -D\frac{c_1 - c_0}{L}$ from the high-concentration to the low-concentration tank. $\square$

Example 8.3 (Diffusion to an absorbing sphere) Imagine a sphere of radius a immersed in an infinite tank (figure 8.8). The sphere absorbs all the solute particles that hit its surface. Because of spherical symmetry, the concentration $C(r)$ will depend only on the distance r from the center of the sphere. At the surface of the sphere, $C(a) = 0$. Far from the sphere, the concentration is maintained at $C(\infty) = C_\infty$. In between, the solution to equation (8.37) is

$$C(r) = C_\infty \left(1 - \frac{a}{r}\right). \tag{8.41}$$

To verify this, recall the form of the Laplacian differential operator in spherical coordinates in equation (8.26). $\square$

Example 8.4 (What is the "diffusion-limited reaction rate"?) For two molecules A and B to react, they must diffuse to within molecular dimensions of each other. Suppose that molecule A is held fixed at the origin and molecules of type B are present at an average concentration C_∞. We want to know at what rate per unit time molecules of type B get to within the reaction radius a of the origin. So imagine an absorbing sphere of radius a that destroys all the molecules that touch its surface. Given the result in equation (8.41), the steady-state concentration profile is

$$C(r) = C_\infty \left(1 - \frac{a}{r}\right). \tag{8.42}$$

The resulting flux of particles at the surface of the sphere is

$$J(a) = -D\frac{\partial}{\partial r}C(a) = \frac{D}{a^2}C_\infty. \tag{8.43}$$

and the rate at which particles hit the surface is

$$R = J(a)4\pi a^2 = 4\pi aDC_\infty. \tag{8.44}$$

This rate is proportional to the concentration of molecules C_∞ and the proportionality constant is called the **diffusion-limited reaction rate**, k_D. If we choose a to be a typical molecular dimension of 0.1 nm, and D a typical diffusion coefficient of 10^{-5} cm^2/s, then

$$k_D = R/C_\infty = 4\pi aD \approx 10^9 \text{M}^{-1}\text{s}^{-1}. \tag{8.45}$$

$\square$

8.2   Random Time Series

Experimental measurements often involve recording a quantity over time and trying to infer some structure from these measurements. Typically, the measurements are taken at discrete times t_i, yielding values y_i. Such a sequence of measurements $\{(y_1, t_1), (y_2, t_2), \ldots, (y_n, t_n)\}$ is called a "time series." A **random time series** is a function $\{(y_i, t_i)\}$ whose evolution is stochastic and not uniquely determined by the initial conditions. Examples include the position of a particle following Brownian motion, the number of mutations on a chromosome over time, the number of bacteria in a growing population, the electric current flowing across a cell membrane, and the fluorescence intensity of a chromophore, just to name a few.

A random time series can be characterized completely by specifying the joint probability distribution for its values at the various times:

$$
\begin{aligned}
P_n\left(y_1, t_1; \ldots; y_n, t_n\right) \mathrm{d}y_1 \ldots \mathrm{d}y_n = \\
= \text{Prob}\left(y(t_1) \in [y_1, y_1 + \mathrm{d}y_1], \ldots, y(t_n) \in [y_n, y_n + \mathrm{d}y_n]\right).
\end{aligned}
\tag{8.46}
$$

Of course, this is a huge object with almost infinitely many parameters. Fortunately, there are special conditions under which the probability distribution simplifies, to the point where one can capture its essence in a finite experiment and use it to make practical predictions.

8.2.1   Stationary Process

A **stationary process** is one whose rules don't change over time. This means that any given sequence of measurements $\{(y_1, t_1), (y_2, t_2), \ldots, (y_n, t_n)\}$ is as equally probable now as it was some time ago. The joint probability distribution depends only on time differences, not on the absolute time:

$$P_n\left(y_1, t_1; y_2, t_2; \ldots; y_n, t_n\right) = P_n\left(y_1, 0; y_2, t_2 - t_1; \ldots; y_n, t_n - t_{n-1}\right). \tag{8.47}$$

Often, one can argue from first principles that a process should be stationary, for example because none of the external constraints have changed in a long time, and the system has somehow found equilibrium.

8.2.2 Markov Process

A **Markov process** is a special type of stationary process whose future evolution is completely determined by the most recent value. How the system arrived at that value is not important: the history of the random variable plays no role in its future. This applies to many important processes, like random walks, protein state transitions, and US foreign policy (to good approximation):

$$P_n\left(y_n, t_n | y_1, t_1; y_2, t_2; \ldots; y_{n-1}, t_{n-1}\right) = P_2\left(y_n, t_n | y_{n-1}, t_{n-1}\right). \tag{8.48}$$

This is a very powerful simplification, as we only need to consider transitions from the current time point to the next. A Markov process is completely determined by the **instantaneous distribution** $P_1(y_1)$, where

$$P_1(y_1)\mathrm{d}y_1 = \mathrm{Prob}\left(y \in [y_1, y_1 + \mathrm{d}y_1]\right), \tag{8.49}$$

and the **transition probability**

$$P_2\left(y_2, t | y_1\right) = \frac{P_2\left(y_1, 0; y_2, t\right)}{P_1(y_1)}, \tag{8.50}$$

where

$$P_2\left(y_2, t | y_1\right) \mathrm{d}y_2 = \mathrm{Prob}\left(y \in [y_2, y_2 + \mathrm{d}y_2] \text{ at time } t, \text{ given that } y = y_1 \text{ at time } 0\right). \tag{8.51}$$

Example 8.5 (Random telegraph signal) This is a simple random process that nonetheless serves as a useful model in many situations of practical importance—namely, any time a system flips back and forth between two states in a historyless fashion (figure 8.10). Examples are chemical binding sites flipping between bound and empty, an enzyme flickering on and off, or an ion channel switching between open and closed. Here, the variable $y(t)$ is binary:

$$y(t) \in \{0, 1\}, \tag{8.52}$$

and it performs transitions from one value to the other at a constant rate: If $y = 0$, then in the next short interval $\mathrm{d}t$, it will switch to 1 with probability $\alpha_{01}\mathrm{d}t$ as shown in figure 8.9. Similarly, if $y = 1$, then it will switch to 0 with probability $\alpha_{10}\mathrm{d}t$. Note that this process is both stationary and Markov: The switching rates α_{01}, α_{10} are constant in time and transitions depend only on the current state of the variable, not on its history.

Based on this definition of the process, we compute the transition probability as follows: Call $P_2\left(1, t | 0\right)$ the probability that $y = 1$ at time t, given that $y = 0$ at time 0. Now let us consider how that probability changes between t and $t + \mathrm{d}t$. One can get $y = 1$ at time $t + \mathrm{d}t$ in two ways: either $y = 0$ at time t, and then the value switches to 1 in the following small interval $[t, t + \mathrm{d}t]$; or $y = 1$ at time t, and there is no switch in the interval $[t, t + \mathrm{d}t]$. The two possibilities are mutually exclusive, so we can write

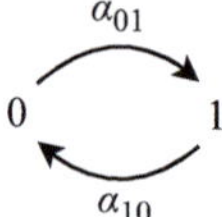

Figure 8.9
A random telegraph signal with transitions between values of 0 and 1.

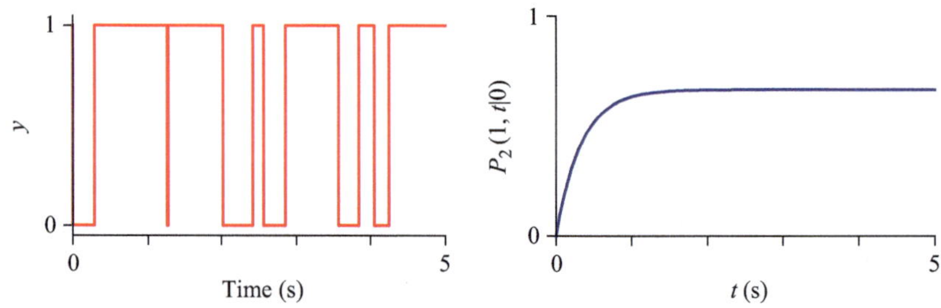

Figure 8.10
Left: Time course of a random telegraph signal with $\alpha_{01} = 2\ \text{s}^{-1}$ and $\alpha_{10} = 1\ \text{s}^{-1}$. Right: Probability that the signal will be 1 at time t, given that it was 0 at time $t = 0$.

$$P_2\left(1, t + \mathrm{d}t \mid 0\right) = P_2\left(0, t \mid 0\right)\alpha_{01}\mathrm{d}t + P_2\left(1, t \mid 0\right)\left(1 - \alpha_{01}\mathrm{d}t\right)$$
$$= (1 - P_2\left(1, t \mid 0\right))\alpha_{01}\mathrm{d}t + P_2\left(1, t \mid 0\right)\left(1 - \alpha_{01}\mathrm{d}t\right). \tag{8.53}$$

So

$$\frac{\mathrm{d}}{\mathrm{d}t}P_2\left(1, t \mid 0\right) = \alpha_{01} - (\alpha_{01} + \alpha_{10})P_2\left(1, t \mid 0\right), \tag{8.54}$$

with the solution

$$P_2\left(1, t \mid 0\right) = \frac{\alpha_{01}}{\alpha_{01} + \alpha_{10}}\left(1 - e^{-(\alpha_{01} + \alpha_{10})t}\right). \tag{8.55}$$

From symmetry, one gets the other transition probabilities:

$$P_2\left(0, t \mid 0\right) = 1 - P_2\left(1, t \mid 0\right)$$
$$P_2\left(0, t \mid 1\right) = \frac{\alpha_{10}}{\alpha_{01} + \alpha_{10}}\left(1 - e^{-(\alpha_{01} + \alpha_{10})t}\right) \tag{8.56}$$
$$P_2\left(1, t \mid 1\right) = 1 - P_2\left(0, t \mid 1\right).$$

Finally, the instantaneous probability that $y = 1$ is obtained from the transition probabilities after a very long time is:

$$P_1(1) = P_2\left(1, t = \infty \mid 0\right) = \frac{\alpha_{01}}{\alpha_{01} + \alpha_{10}} \tag{8.57}$$

and obviously, $P_1(0) = 1 - P_1(1)$. Because this is a Markov process, everything about it can be computed from functions P_1 and P_2. $\square$

8.2.3 Moments of a Random Process

The mean of a random process is defined as

$$\text{Mean} = \langle y(t) \rangle \tag{8.58}$$

and the variance as

$$\text{Variance} = \langle (y(t) - \langle y(t) \rangle)^2 \rangle = \langle y^2(t) \rangle - \langle y(t) \rangle^2, \tag{8.59}$$

where the angle brackets denote the **ensemble average** over different instantiations of the random process that start from the same initial conditions. **If the process is stationary**, then the ensemble average is equal to the time average and no longer depends on absolute time:

$$\langle y(t) \rangle = \bar{y} = \lim_{T \to \infty} \frac{1}{T} \int_0^T y(t) \mathrm{d}t. \tag{8.60}$$

8.2.4 Correlation Function and Power Spectrum

Another second moment of a random process is the **correlation function** $C(\tau)$, which relates values over time:

$$C(\tau) = \langle y(t) y(t + \tau) \rangle. \tag{8.61}$$

For a stationary process, one can again compute the averages over time, and the correlation function depends only on the time difference τ, not on absolute time t:

$$C(\tau) = \lim_{T \to \infty} \frac{1}{T} \int_0^T y(t) \cdot y(t + \tau) \, \mathrm{d}t. \tag{8.62}$$

An important result relates the correlation function of a random process to its power spectrum. Extending the definition in equation (3.28), the power spectrum of a random process is the expectation value of the square modulus of the Fourier transform:

$$P(\omega) = \left\langle |\hat{y}(\omega)|^2 \right\rangle, \tag{8.63}$$

where the expectation is over many instances of the same process.

For a stationary process, the **Wiener-Khintchin theorem** states that

$$P(\omega) = \int_{\tau=-\infty}^{+\infty} C(\tau) e^{-i\omega\tau} \, \mathrm{d}\tau. \tag{8.64}$$

Stated in words: The power spectrum is the Fourier transform of the correlation function.

Example 8.6 (Random telegraph signal) To illustrate these concepts, let us return to the random telegraph process of section 8.5. Recall that this random variable y switches between the values of 0 and 1. Transitions from 0 to 1 happen at constant probability per unit time α_{01} and from 1 to 0 at a rate α_{10}. What is the correlation function for this system?

We need to calculate

$$C(\tau) = \langle y(0)y(\tau)\rangle. \tag{8.65}$$

As the product vanishes when either $y(0)$ or $y(\tau)$ are 0, the only remaining contribution is

$$C(\tau) = \text{Prob}[y(0) = 1 \text{ and } y(\tau) = 1]. \tag{8.66}$$

Using conditional probability, we see that

$$\begin{aligned}
\text{Prob}[y(0) = 1 \text{ and } y(\tau) = 1] &= \text{Prob}[y(0) = 1] \cdot \text{Prob}[y(\tau) = 1 | y(0) = 1)] \\
&= P_1(1) \cdot P_2(1, \tau | 1).
\end{aligned} \tag{8.67}$$

From the results in section 8.5, one finds

$$C(\tau) = \frac{\alpha_{01}}{\alpha_{01} + \alpha_{10}} \cdot \left(\frac{\alpha_{01}}{\alpha_{01} + \alpha_{10}} + \frac{\alpha_{10}}{\alpha_{01} + \alpha_{10}} e^{-(\alpha_{01}+\alpha_{10})\tau} \right). \tag{8.68}$$

The correlation function consists of a decaying exponential. The power spectrum is the Fourier transform of that function. Note that we encountered the power spectrum of a decaying exponential previously in equation (3.30). Here,

$$\begin{aligned}
P(\omega) &= \int_{-\infty}^{+\infty} C(\tau)e^{-i\omega\tau}\,d\tau \\
&= 2\,\text{Re}\left[\int_0^{+\infty} C(\tau)e^{-i\omega\tau}\,d\tau \right], \\
&= 2b(1-b)\frac{\alpha}{\alpha^2 + \omega^2}
\end{aligned} \tag{8.69}$$

where

$$\begin{aligned}
b &= P_1(1) = \frac{\alpha_{01}}{\alpha_{01} + \alpha_{10}} \\
\alpha &= \alpha_{01} + \alpha_{10},
\end{aligned} \tag{8.70}$$

and we have ignored the divergence of the power at $\omega = 0$. Figure 8.11 illustrates the correlation function and power spectrum of this process. $\square$

8.2.5 Discrete Markov Process

Frequently, one approximates a system as taking on a discrete set of states. For example, a protein might exist in one of several discrete conformations, or an animal may be in one of a few behavioral states. If these discrete states are long-lived compared to the transitions between them, then we can take the transitions to be instantaneous. This defines a **discrete stochastic process.**

If, in addition, the process $X(t)$ is stationary and Markovian, then it is called a **discrete Markov process.** This means that transitions from state $X = i$ to state $X = j$ happen at constant probability per unit time α_{ij}, and that transition rate is independent of the prior history of the process (figure 8.12). Note that this is a generalization of the random telegraph described in example 8.5, which exists in only two states: $X \in \{0, 1\}$.

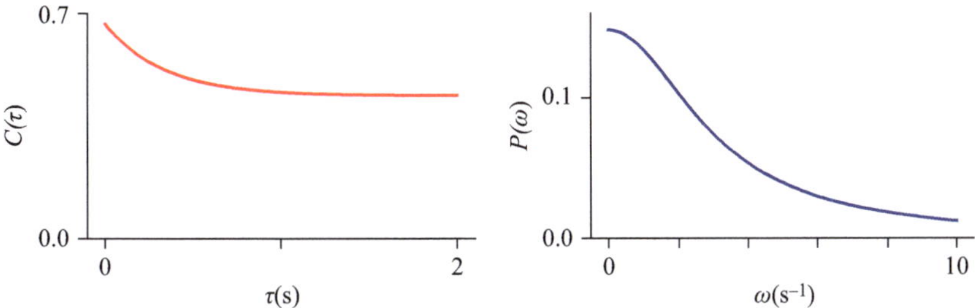

Figure 8.11
Left: Correlation function of a random telegraph signal with $\alpha_{01} = 2$ s^{-1} and $\alpha_{10} = 1$ s^{-1}.
Right: Power spectrum of that same random process.

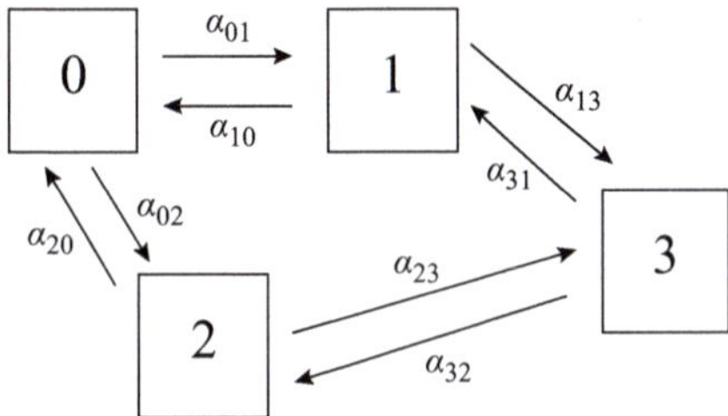

Figure 8.12
A discrete Markov process takes on one of a set of states i, with first-order transitions
happening at rates α_{ij}. In this example, four states are possible.

To understand the evolution of $X(t)$ from any given starting state, one can follow the same approach as for the random telegraph process. First, define a **transition
probability**:

$$P_{ij}(t) = \text{Probability that } X = j \text{ at time } t, \text{ given that } X = i \text{ at time 0.} \tag{8.71}$$

Again, by considering what happens in the last short time interval dt, one finds that

$$\frac{\text{d}}{\text{d}t} P_{ij}(t) = \sum_k P_{ik}(t)\alpha_{kj} - P_{ij}(t) \sum_l \alpha_{jl}. \tag{8.72}$$

Here, the first term includes all the transitions from other states into state j and the
second term are transitions away from state j. Equation (8.72) is called the **master
equation** of the process.

To solve the master equation, note that it can be written in matrix form as

$$\frac{\text{d}}{\text{d}t} \mathbf{P}(t) = \mathbf{P}(t) \cdot \mathbf{Q}, \tag{8.73}$$

where $\mathbf{P}$ is the matrix of all transition probabilities P_{ij}, and $\mathbf{Q}$ is given by

$$\mathbf{Q}_{ij} = \alpha_{ij} - \delta_{ij} \sum_l \alpha_{jl}. \tag{8.74}$$

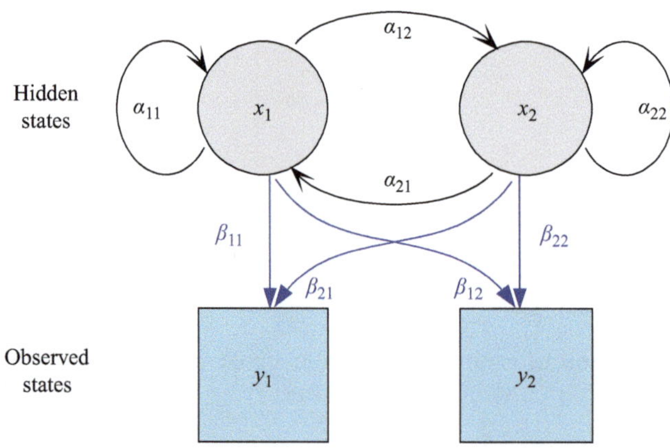

Figure 8.13
An HMM with two hidden states and two observable outcomes.

This is simply the matrix version of the rabbit equation (1.34), so the general solution is

$$\mathbf{P}(t) = \mathbf{P}(0) \cdot e^{\mathbf{Q}t} = e^{\mathbf{Q}t} \tag{8.75}$$

because $\mathbf{P}(0)$ is the identity. As usual, this is most easily evaluated in the eigenbasis of the matrix $\mathbf{Q}$; see section 2.11.2. Using the diagonalizing transform $\mathbf{S}$,

$$\mathbf{P}(t) = \mathbf{S} \cdot \begin{bmatrix} e^{\lambda_1 t} & \cdots & 0 \\ \vdots & \ddots & \vdots \\ 0 & \cdots & e^{\lambda_n t} \end{bmatrix} \cdot \mathbf{S}^{-1}, \tag{8.76}$$

where λ_k are the eigenvalues of the matrix $\mathbf{Q}$. So the transition probabilities take the general form of a sum of exponentials:

$$P_{ij}(t) = \sum_k c_k e^{\lambda_k t} \tag{8.77}$$

where the c_k can be computed from equation (8.76). This fully describes the stochastic evolution of the system from any initial distribution of states.

With these transition probabilities, one can further compute the correlation function or power spectra of the process, following the approach elaborated here for the random telegraph signal (see example 8.6).

8.3   Hidden Markov Models

In section 8.2.5, we considered a system that exists in discrete states X_i, with transition rates α_{ij} among these states constant in time. Sometimes we cannot observe the system's state directly, but rather have to guess it based on some observable Y that is produced in a way that depends on the state X. A formalization of this concept is the **hidden Markov model (HMM)**.

Figure 8.13 shows an example of an HMM with two states $X \in \{x_1, x_2\}$ and an observable that takes one of two values $Y \in \{y_1, y_2\}$. We will treat this as a discrete-time process, where time proceeds in discrete steps. If the system is in state $X = x_i$, then in the next time step, it transitions to state x_j with probability α_{ij}. Also, the system emits an observable $Y = y_k$ with probability β_{ik}. The probabilities are normalized, such that

$$\sum_j \alpha_{ij} = 1 \quad \text{for all } i$$
$$\sum_k \beta_{ik} = 1 \quad \text{for all } i. \tag{8.78}$$

A sample sequence generated by the HMM in the figure would be

$$\begin{aligned}
\text{Hidden state sequence } X(t): \quad & x_1 \quad x_1 \quad x_2 \quad x_2 \quad x_2 \quad x_1 \quad x_2 \quad x_1 \quad x_1 \quad x_2. \\
\text{Observable sequence } Y(t): \quad & y_2 \quad y_2 \quad y_2 \quad y_1 \quad y_1 \quad y_2 \quad y_2 \quad y_2 \quad y_1 \quad y_1.
\end{aligned} \tag{8.79}$$

An HMM is fully determined by the set of states X, the set of observables Y, and the probabilities for transitions α_{ij} and those for emissions β_{ik}:

$$[X, Y, \{\alpha_{ij}, \beta_{ik}\}]. \tag{8.80}$$

There are three types of problems that one wants to solve in the context of HMMs:

- Given a model $[X, Y, \{\alpha_{ij}, \beta_{ik}\}]$, what is the probability of a particular sequence of observations $Y(t)$? Note that several different hidden state sequences may generate the same sequence of observations, and one must take all of them into account. This is done using the **forward algorithm.**
- Given a model $[X, Y, \{\alpha_{ij}, \beta_{ik}\}]$ and a sequence of observations $Y(t)$, what is the most likely sequence of states $X(t)$ that generated it? This is done by the **Viterbi algorithm**, and this sequence of hidden states is called the **Viterbi path.**
- Given an HMM architecture $[X, Y]$ and a sequence of observations $Y(t)$, what are the most likely transition and emission probabilities $[\{\alpha_{ij}, \beta_{ik}\}]$? This is typically done using the **Baum-Welch algorithm**, which in turn uses the **forward-backward algorithm.**

In all these cases, the challenge is that the number of possibilities grows exponentially with the sequence length. These algorithms have been developed to make the calculations computationally feasible. Although a detailed explanation of these algorithms is beyond the scope of this text, we illustrate the ideas with some examples here.

8.3.1 Finding Genes in DNA Sequence

The central dogma in biology states that deoxyribonucleic acid (DNA) is transcribed into messenger ribonucleic acid (mRNA), which is translated into protein. However, during preparation of mRNA from DNA, some sections of sequence are removed by splicing. These sections are called "introns," whereas the remaining segments that ultimately encode the protein are known as "exons." Given a DNA sequence such as

$$\ldots \text{ATGCGACTGCATAGCACTT} \ldots, \tag{8.81}$$

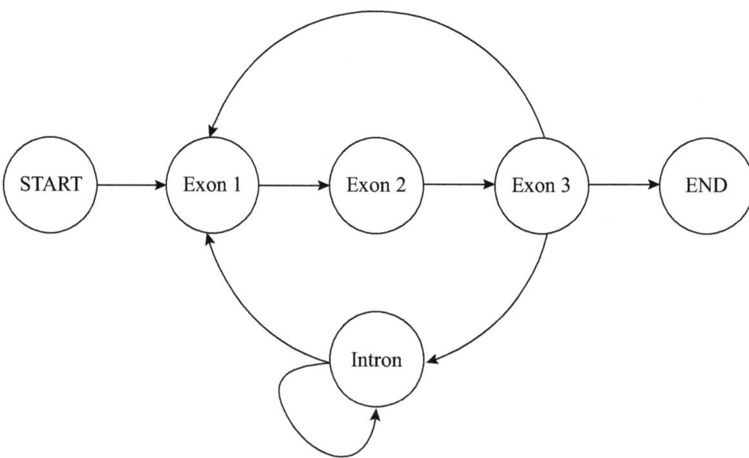

Figure 8.14
An HMM for modeling introns and exons within genes.

it is a challenge to determine which sections are exons and which are introns. Some clues are available because the sequence in exons has to meet certain constraints. Each triplet of nucleotides encodes an amino acid, and thus the frequency of triplets inside exons follows certain statistical regularities. Within introns, the frequencies are different. Effectively, the DNA sequence has a hidden state—exon or intron—at each location. We cannot observe the state directly, but we can see the nucleotides that were "emitted," and their probabilities depend on the state.

One can formalize this argument by imagining that the DNA sequence was produced by a machine that has the mechanics of an HMM (figure 8.14). The machine has three exon states and one intron state. With every step along the DNA, the machine makes a state transition. In state "Exon 1," it emits the first nucleotide of a codon. Then it transitions deterministically to state "Exon 2" and emits the second nucleotide. Next, it moves to state "Exon 3" to emit the last nucleotide of that codon. At the next step, the machine may return to "Exon 1" to deliver another codon of three nucleotides. Or it may switch to state "Intron" and deliver one or more intron nucleotides before returning to "Exon 1."

We would like to determine **what is the most likely sequence $X(t)$ of intron/exon states, given the observed sequence $Y(t)$ of nucleotides.** This requires that we know the transition probabilities α_{ij} among the states of the model and the emission probabilities β_{ik} with which each state generates nucleotides. Both have been tabulated from a large number of genes where the ground truth about introns and exons is known. Then one can use these probabilities to infer the intron/exon state on a novel sequence. This can be accomplished by the Viterbi algorithm.

8.3.2 Sequence Alignment

In the analysis of genetic sequences, either protein or DNA, one often wants to evaluate the similarity between two sequences. For example, we may be interested in understanding how the protein sequence from several species diverged from that of a common ancestor, so as to place those species on a philogenetic tree. In measuring the similarity of two sequences, the first issue is the problem of alignment: Which

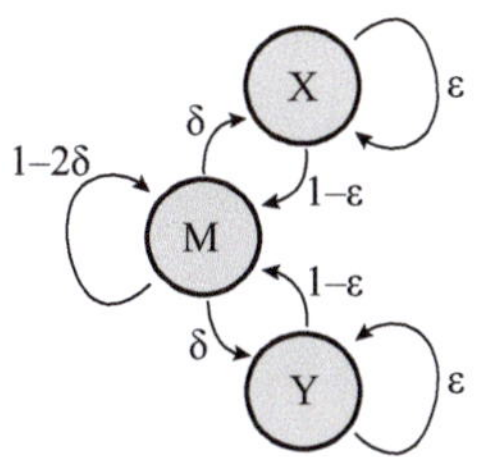

Figure 8.15
An HMM for sequence alignment.

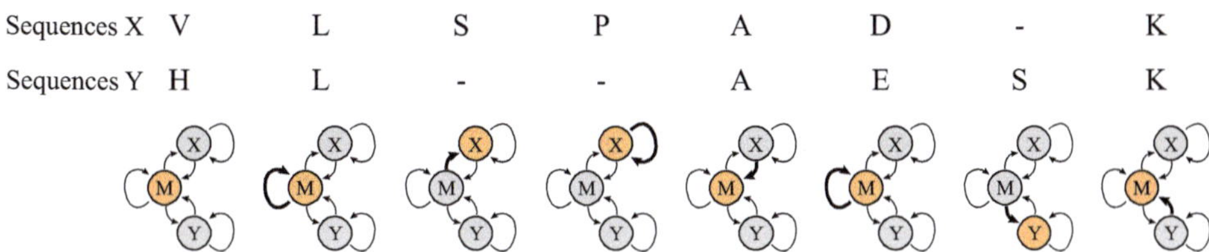

Figure 8.16
A sequence alignment as a path through the HMM.

amino acid in one sequence corresponds to which in the other? Two sequences may differ for three reasons: (1) an amino acid has been mutated to a different one; (2) an extra amino acid has been inserted; or (3) an amino acid has been deleted.

As shown in section (8.3.1), we imagine that the two amino acid sequences $X(t)$ and $Y(t)$ were produced by a machine that acts like an HMM (figure 8.15). The model has three states, M, X and Y. In state M, the model generates a pair of symbols (amino acids) and adds one to sequence $X(t)$ and the other to sequence $Y(t)$. If there is no mutation, the amino acids are the same, and this is the most likely scenario. But sometimes a mutation occurs, so the machine may add, say, alanine to one sequence and glycine to the other. The probability of emitting a particular pair of amino acids (x, y) is $P(x, y)$.

Sometimes, though, the model will switch to one of the other states, X or Y. In this state, the model generates an amino acid for one of the sequences, but not for the other. The rate at which this switch between states happens is controlled by the transition probabilities ϵ and δ, and once the system is in one of these states, it will generate an amino acid with probability $Q(x)$ or $Q(y)$. This allows the model to account for amino acid insertions and deletions.

For a given pair of sequences, the particular path $S(t) \in \{M, X, Y\}$ of the state of the model proposes an alignment between the two sequences—namely, the location of mutations, insertions, and deletions (figure 8.16). Different paths will generate the observed sequences with different likelihoods, and the maximum likelihood path represents the optimal alignment.

The model parameters are the transition probabilities ϵ and δ and the emission probabilities $P(x_i, y_j)$ and $Q(x_i)$. These parameters are determined from experience, following many successful protein alignments. For example, the emission probabilities $P(x, y)$ will depend on factors such as the evolutionary distance between the two species or the specific protein in question. Tables have been compiled empirically that capture these factors, and one such table, called *BLOSUM 50*, is shown in figure 8.17.

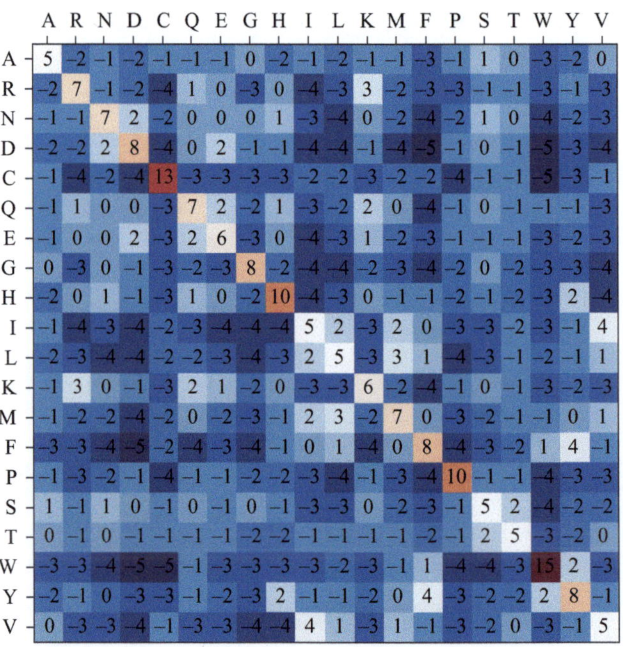

Figure 8.17
The BLOSUM 50 matrix used for protein sequence alignment. The entry (i, j) gives the log-likelihood of adding amino acid i and amino acid j to the sequences X and Y, respectively.

The simple three-state HMM presented here is only a first approximation to the myriad of complex models and refinements that have been developed in the field of sequence alignment. Proteins have different domains, and it is natural to assume that different models will apply to different domains. This can also be incorporated into the analysis.

8.3.3 Further Reading
See Henderson et al. (1997) for an early application of HMMs to gene finding. The general application to sequence analysis is reviewed in Durbin et al. (1998) and Eddy (2004). An application to problems of learning in animals is found in Smith et al. (2004). More theory and applications can be found in Dymarski (2011).

8.4 Point Processes

A **point process** is a series of identical events that are point like in time. A sample from a point process is completely specified by listing the event times $\{t_1, t_2, \ldots, t_n\}$. Some examples encountered in biological research include the following:

- The arrival times of photons at a camera or at a photoreceptor cell
- Times of collision between a ligand and a binding site
- Times of opening of an ion channel
- Times of random mutations occurring in a genome
- Times of action potentials fired by a neuron

Point processes can also be defined in spatial dimensions, such as the locations of all trees of a given species in a field, or the locations of all nerve cells of a given type on the surface of the retina. The following sections will focus on temporal processes, but the treatment translates easily to space.

8.4.1 Intensity Function

A random point process is fully specified by the **conditional intensity function**, which spells out the probability of getting an event in the next small time interval. In general, that probability depends on time, but it also depends on the entire preceding history of the process:

$$P(t|\ldots,t_{-2},t_{-1})\mathrm{d}t = \text{probability of getting an event in } [t,t+\mathrm{d}t]$$

$$\text{as a function of } t \text{ and the entire history} \tag{8.82}$$

$$\{\ldots,t_{-2},t_{-1}\} \text{ of event times prior to } t.$$

8.4.2 Stationary Point Process

As introduced in section 8.2.1, a stationary process does not depend on absolute time, but only on time differences. A point process is **stationary** if a time shift by τ leaves the conditional intensity unchanged; namely,

$$P(t+\tau|\ldots,t_{-2}+\tau,t_{-1}+\tau) = P(t|\ldots,t_{-2},t_{-1}). \tag{8.83}$$

8.4.3 Poisson Process

This is the simplest case of a point process, in which the events happen with constant probability per unit time and independent of history:

$$P(t|\ldots,t_{-2},t_{-1}) = \lambda. \tag{8.84}$$

Classic examples are the arrival of photons from a constant light source and the clicks in a Geiger counter from a radioactive sample. Some characteristics of this process have been elaborated earlier in this book, in sections 6.3.7 and 6.4.6.

The **number of events N observed in a time interval** of length T is a discrete random variable, which follows the Poisson distribution

$$P(N, T) = e^{-\mu}\frac{\mu^N}{N!}, \tag{8.85}$$

where

$$\mu = \langle N \rangle = \lambda T \tag{8.86}$$

is the expectation value of N.

The **time interval τ between successive events** is a continuous random variable that follows the exponential distribution

$$P(\tau) = \lambda e^{-\lambda\tau}. \tag{8.87}$$

Successive time intervals are statistically independent, so the time τ_n to the nth event follows a gamma distribution:

$$\tau_n \sim \text{Gamma}(n, \lambda)$$

$$P(\tau_n) = \frac{\tau_n^{n-1} e^{-\beta \tau_n} \beta^n}{\Gamma(n)}. \tag{8.88}$$

8.4.4 Inhomogeneous Poisson Process

Here, the intensity λ varies with time, but independent of the history of the process:

$$P(t|\ldots, t_{-2}, t_{-1}) = \lambda(t). \tag{8.89}$$

In a simple example, consider a lamp whose intensity $I(t)$ is getting modulated up and down with a dimmer knob. The photon stream from that light source follows an inhomogeneous Poisson process with $\lambda(t) \propto I(t)$.

One can obtain such an inhomogeneous Poisson process from a homogeneous one with $\lambda = 1$ by warping the time axis. Imagine that warp time τ flows faster and slower relative to t according to the intensity $\lambda(t)$:

$$\frac{\mathrm{d}\tau}{\mathrm{d}t} = \lambda(t). \tag{8.90}$$

Then events that are at a constant density of 1 per unit time on the τ-axis get compressed or expanded on the t-axis, exactly so as to achieve the density $\lambda(t)$.

From this argument, it follows that the number of events in any given time interval $[t_a, t_b]$ is again Poisson distributed. The mean number is

$$\mu = \int_{t_a}^{t_b} \lambda(t)\mathrm{d}t, \tag{8.91}$$

and the distribution is

$$N \sim \text{Poiss}(\mu). \tag{8.92}$$

8.4.5 Spectral Analysis of a Point Process

To use the range of frequency-analysis tools developed in section 3.2, it helps to convert the point process into a continuous function of time. For that purpose, each event in the process $\{t_k\}$ contributes a dirac delta function:

$$R(t) = \sum_k \delta(t - t_k). \tag{8.93}$$

This function can be seen as the rate of occurrence of events because its integral delivers the cumulative number of events:

$$\int^t R(t')\mathrm{d}t' = N(t) = \text{number of events before time } t. \tag{8.94}$$

The Fourier transform at frequency ω becomes

$$\hat{R}(\omega) = \int R(t)e^{-i\omega t}\mathrm{d}t = \sum_k e^{-i\omega t_k}. \tag{8.95}$$

This expression has a useful geometric interpretation: $e^{-i\omega t_k}$ is a phasor of unit length in the complex plane, oriented at phase ωt_k. If the t_k happen with a periodicity of $2\pi/\omega$, then all the phasors point in the same direction and their sum will be very large. If, on the other hand, the t_k happen at random times, then the phasors are oriented randomly and their sum will be small. In this way, the Fourier coefficient $\hat{R}(\omega)$ reflects the degree of periodicity of the point process at frequency ω.

8.4.6 Power Spectrum of a Point Process

By applying the definition in equation (8.63) for the power spectrum to the rate function $R(t)$, one finds the power spectrum of a random point process to be

$$P(\omega) = \left\langle \left| \hat{R}(\omega) \right|^2 \right\rangle = \left\langle \left| \sum_k e^{-i\omega t_k} \right|^2 \right\rangle \tag{8.96}$$

where the expectation is over instantiations of the process.[3]

8.4.6.1 Power spectrum of a Poisson process

For example, consider a Poisson point process with intensity λ, extending over the time period $[0, T]$. As discussed in section 3.2.6, we will want to evaluate the power spectrum at frequencies that are multiples of the fundamental $\omega_j = j \cdot 2\pi/T$, $j = 0, 1, \ldots$. Say that N is the number of events observed in any instantiation of the process. We know that N follows the Poisson distribution with mean $\mu = \lambda T$:

$$N \sim \text{Poiss}(\lambda T). \tag{8.97}$$

Suppose now that N is large, so we can ignore the fluctuations of order $\sqrt{N}$. Then the power at zero frequency is simply

$$P(\omega = 0) = \left\langle N^2 \right\rangle \approx (\lambda T)^2. \tag{8.98}$$

To evaluate power at the nonzero frequencies, note that each of the event times t_k is distributed uniformly throughout the interval $[0, T]$. So the phase $\omega_j t_k$ will be uniformly distributed in $[0, 2\pi]$, and thus the phasors $e^{-i\omega_j t_k}$ all point in random and independent directions. The sum of those phasors,

$$z = \sum_k e^{-i\omega t_k}, \tag{8.99}$$

is the sum of N independent random unit vectors in the complex plane. If N is large, one can invoke the central limit theorem to argue that this sum vector has a Gaussian distribution around the origin with a variance that is the sum of the individual variances—namely,

$$\text{Var}[z] = \left\langle |z|^2 \right\rangle = \langle N \rangle. \tag{8.100}$$

---

3. As usual, there are alternative definitions that differ by some normalization factor. If the only goal is to compare power at different frequencies, that doesn't matter.

In conclusion,

$$P(\omega_j) \approx \begin{cases} (\lambda T)^2, & \text{if } j = 0 \\ \lambda T, & j > 0. \end{cases}$$

(8.101)

So the homogeneous Poisson process has equal power at all frequencies (except $\omega = 0$): a **white noise** spectrum.[4]

8.4.7 Shot Noise

Often the discrete events in a random point process are not observed directly, but rather through some signal caused by each event. For example, every photon captured by a photo-detector tube produces a short unitary blip of electric current. We observe the time course of the current and want to infer the occurrence time of the blips. In such a case, the time course of the unitary signal is called the **shot**, and the superposition of all the shots is called **shot noise**.

Note that the shot noise signal $F(t)$ results from a convolution of the point process rate function $R(t)$ and the shape of the individual shot $S(t)$:

$$F(t) = \sum_k S(t - t_k) = \int R(t')S(t - t')dt' = R(t) * S(t).$$

(8.102)

Recall that the Fourier transform of a convolution is equal to the product of the two Fourier transforms (as discussed in section 3.2.4.4). Consequently, the same is true for the power spectrum and

$$P_F(\omega) = P_R(\omega)P_S(\omega).$$

(8.103)

This relationship gets used in both ways: Sometimes we know the shape of the individual shot (e.g. for the photo-detector tube), and this allows us to infer something about the point process that produces the shots. Other times, we are confident about the spectrum of the point process, and thus we can learn something about the shape of the individual shot. In particular, if $R(t)$ is a Poisson point process, then the spectrum $P_R(\omega)$ is white, and therefore the spectrum of the shot $P_S(\omega)$ has the same shape as that of the measured shot noise $P_F(\omega)$. See section 9.4.2 for an example.

8.4.8 Converting a Point Process to a Time Series

For practical calculations, one often converts a point process $\{t_k\}$ into a discrete time series R_i with values of 1 or 0. A common form of conversion is **binning** of the point process: choose a bin width Δt and count the number of events in each bin

$$R_i = \frac{N((i+1)\Delta t) - N(i\Delta t)}{\Delta t}.$$

(8.104)

Clearly, the timing of an event within each bin is lost in the process, so Δt effectively sets the time resolution of any subsequent analysis. Now one can apply the full battery

---

4. If λT is not large, one can follow the same logic to find the exact power spectrum. It will still be white (namely, equal power at all frequencies except $\omega = 0$).

of time-series analysis methods, including spectral analysis and cross-correlation analysis.

However, a drawback of this approach is that it represents the point process very inefficiently: Suppose that there are 100 events in 10 s, and we want to preserve their timing to 1 ms. That produces a time series R_i with 10,000 values, even though $\{t_k\}$ has only 100 values. To compute a correlation of two such processes (by brute force) requires $\sim 10^8$ operations, compared to $\sim 10^4$ if one worked with the event times directly. In the old days when computations were done by hand, no one would have dreamed of making such an inefficient change in representation from point process to time series. These days, when computer speed is hardly a constraint for most scientific computing, the Fast Fourier transform speeds up linear operators, and optimized routines exist that work on sparse arrays, the cost can be negligible. On the other hand, if you are operating in a big data regime that involves lots of events and very high time resolution, you may reconsider your options.

8.4.9 Further Reading
Daley and Vere-Jones (2013) introduce point processes with the full mathematical armamentarium. Brown et al. (2004) discuss additional point process methods in the context of analysis of neural signals.

8.5 Dimensionality Reduction

Several subfields of biology have decidedly entered the area of big data owing to revolutionary new methods for large-scale measurements. Today, one can measure the expression levels of thousands of genes across thousands of different cells; or the activity of thousands of neurons over many thousands of timepoints. To gain any understanding from such high-dimensional data, one must somehow reduce the number of dimensions.

One goal of dimensionality reduction is to find structure in the data. For example, gene expression patterns of 10,000 genes may reduce to a few modules that group together genes with similar dynamics. Another goal is to separate signal from noise: the most dominant patterns in the data should get attention first. Another goal is visualization: we have no way of representing 10,000-dimensional space, but we can draw figures in two dimensions. Finally, one could argue that the whole process of scientific understanding itself is one of dimensionality reduction, such that eventually the meaning of a huge data set can be captured by a few equations interspersed with words of text.

For the purpose of this section, we will assume that the data consist of T data points $\mathbf{x}_j$, $j = 1, \ldots, T$. Each data point consists of N measured variables $\mathbf{x}_j = [x_{1j}, \ldots, x_{Nj}]^\top$. For example, x_{ij} might be the activity of neuron i at time j or the expression of gene i in cell j. Sometimes we will write these data in the form of a single $N \times T$ data matrix:

$$
\mathbf{X} = \begin{pmatrix}
x_{11} & \cdots & \cdots & x_{1T} \\
\vdots & \ddots & & \vdots \\
\vdots & & \ddots & \vdots \\
x_{N1} & \cdots & \cdots & x_{NT}
\end{pmatrix}.
\tag{8.105}
$$

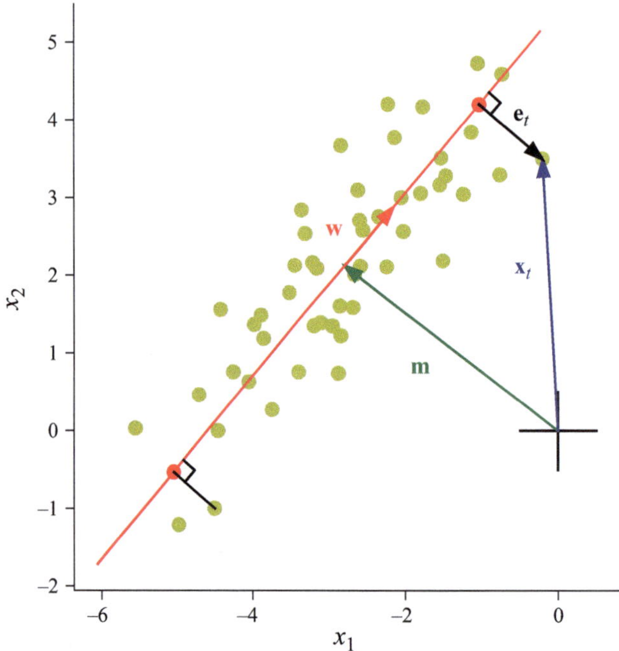

Figure 8.18
A two-dimensional (2D) data set with 50 data points (yellow) and a 1D approximation, illustrating the mean $\mathbf{m}$, the first principal component $\mathbf{w}$, the line corresponding to the 1D approximation (red), two examples of data points projected on that line (red circles), a random data point $\mathbf{x}_t$, and the residual for that data point $\mathbf{e}_t$.

8.5.1   Principal Component Analysis

Consider the $2 \times T$ data set in figure 8.18. Clearly, the two variables x_1 and x_2 vary together, a strong pattern in the data. One is tempted to just draw a line through this data cloud that gets as close as possible to all the data points. That line captures the direction along which the data vary the most, so it serves as a first approximation of the data set. With the proper definitions, discussed next, that line is called the "first principal component" of the data. The direction perpendicular to it is the "second principal component"; this is the direction along which the data vary the least. The algorithm for finding those special directions is called "principal component analysis" (PCA).

To formalize the goal of dimensionality reduction in the language of linear algebra, we want to approximate the data vectors by a linear superposition of just a few basis vectors with the smallest amount of error:

$$\mathbf{x}_j = \mathbf{m} + c_{1,j}\,\mathbf{w}_1 + \cdots + c_{D,j}\,\mathbf{w}_D + \mathbf{e}_j^{(D)}. \tag{8.106}$$

Here, $\mathbf{m}$ is a constant offset vector, $\mathbf{w}_k$ is the kth basis vector or **principal component** of the data cloud, $c_{k,j}$ is the coefficient of component k in data point j, and $\mathbf{e}_j^{(D)}$ is the residual of the approximation with D components for data point j. The goal is to choose $\mathbf{m}$, the $\mathbf{w}_k$, and the $c_{k,j}$ so as to minimize the average squared residual, which represents the **error of the D-dimensional approximation**:

$$E^{(D)} = \frac{1}{T}\sum_{j=1}^{T} \mathbf{e}_j^{(D)\top} \cdot \mathbf{e}_j^{(D)}. \tag{8.107}$$

If D is much smaller than N, and the error $E^{(D)}$ is acceptably small, then one has achieved successful dimensional reduction from N to D dimensions.

How can we find the best choices of the parameters $\mathbf{m}$, $\mathbf{w}_k$, and $c_{k,j}$? We do this by differentiating the error in equation (8.107) with respect to the parameters, as shown in exercise 10.18.

The solution tells us how to perform principal component analysis:

1. Compute the mean of the data cloud and subtract it from each data point

$$\mathbf{m} = \frac{1}{T}\sum_{j=1}^{T} \mathbf{x}_j \tag{8.108}$$

$$\mathbf{y}_j = \mathbf{x}_j - \mathbf{m}.$$

2. Compute the covariance matrix as in equation (6.69) of the data

$$\mathbf{C} = \frac{1}{T}\sum_{j=1}^{T} \mathbf{y}_j \cdot \mathbf{y}_j^{\top}. \tag{8.109}$$

This is an $N \times N$ matrix. It is symmetric (see section 2.11.1) and positive semidefinite, so it is guaranteed to have N eigenvalues, and they are all nonnegative.

3. Find the eigenvalues λ_k of the covariance matrix $\mathbf{C}$ and the associated eigenvectors $\mathbf{w}_k$. Sort them in decreasing order of the eigenvalues: $\lambda_1 > \lambda_2 > \cdots > \lambda_N$. Normalize all the eigenvectors, such that $\mathbf{w}_k^{\top}\mathbf{w}_k = 1$.
4. Then the **principal component representation** of the data is

$$\mathbf{y}_j = \sum_{k=1}^{N} c_{k,j}\,\mathbf{w}_k, \tag{8.110}$$

where

$$c_{k,j} = \mathbf{w}_k^{\top}\mathbf{y}_j. \tag{8.111}$$

Equation (8.110) is an exact representation of the data, and there is no residual error. Every data vector $\mathbf{y}_j$ gets mapped into a coefficient vector $\mathbf{c}_j = [c_{1,j}, \ldots, c_{N,j}]^{\top}$. This coefficient vector also has N dimensions.

To achieve some dimensional reduction, let us cut off the sum in equation (8.110) after the first D terms:

$$\mathbf{y}_j^{(D)} = \sum_{k=1}^{D} c_{k,j} = \mathbf{m} + \sum_{k=1}^{D} \mathbf{w}_k^{\top}\,\mathbf{y}_j\,\mathbf{w}_k. \tag{8.112}$$

Each data point is now mapped onto just D coefficients. The resulting approximation $\mathbf{x}_j^{(D)}$ corresponds to the orthogonal projection of $\mathbf{x}_j$ onto the space spanned by the principal components $\mathbf{w}_1, \ldots, \mathbf{w}_D$ (see the red dots in figure 8.18). **This D-dimensional approximation to the data in equation (8.112) is guaranteed to have the smallest possible residual.** That is the special property of the principal component representation.

How large is the error incurred by going from N to D dimensions? The **total variance of the data set** is

$$V = \frac{1}{T} \sum_{j=1}^{T} \mathbf{y}_j^\top \mathbf{y}_j. \tag{8.113}$$

This variance is equal to the sum of all the eigenvalues λ_k:

$$V = \sum_{k=1}^{N} \lambda_k. \tag{8.114}$$

Furthermore, the error $E^{(D)}$ of the D-dimensional approximation in equation (8.107) is the sum of the "unused" eigenvalues:

$$E^{(D)} = \sum_{k=D+1}^{N} \lambda_k. \tag{8.115}$$

This is also called the **unexplained variance** of the D-dimensional approximation. Vice versa, one says the **explained variance** is

$$V^{(D)} = \sum_{k=1}^{D} \lambda_k. \tag{8.116}$$

How should one choose D? Of course, this depends on the research goals that motivated the PCA. The trade-off is between lower dimensionality (low D) and lower error (high D). One often shows a plot of λ_k versus k, called the **eigenvalue spectrum** or **scree plot** (figure 8.19). If that plot shows a sudden break to lower eigenvalues, that can be a reason to set the cut-off D at the break.

We illustrate these procedures with two example data sets.

8.5.1.1 Example: Spearman's data In 1904, Spearman published "'General Intelligence,' Objectively Determined and Measured." This paper has historical importance as the first notable application of factor analysis, a close relative of PCA. Second, it put the psychological concept of "general intelligence" on a quantitative basis. Figure 8.20 reproduces just one data set from this study. The boys in an English village school were ranked according to their performance in various subjects. Then Spearman measured something seemingly unrelated—namely, their ability to distinguish sounds of different pitches, as well as lights of different intensities and weights of different mass.

Using the notation introduced here, each boy j is represented by a data vector $\mathbf{x}_j$ corresponding to a row in the table that contains the grades in the various subjects

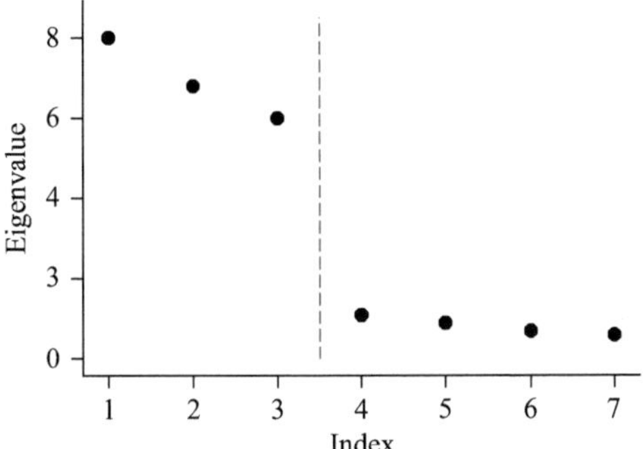

Figure 8.19
Sample scree plot, which suggests keeping only the first three principal components.

EXPERIMENTAL SERIES IV.

High Class Preparatory School for Boys.

A. Original Data.

| Age | | Pitch | Place in School (*before modification to eliminate Age*). | | | | | | | | | | | | Music |
| Years | Months | Discrim. Thres. in ⅓ v. d., October, 1902 | Classics | | | French | | | English | | | Mathem. | | | Ranked by Music Master |
			Xmas, 1902	Easter, 1903	July, 1903	Xmas, 1902	Easter, 1903	July, 1903	Xmas, 1902	Easter, 1903	July, 1903	Xmas, 1902	Easter, 1903	July, 1903	
12	6	2	8	7	4	5	3	3	4	3	3	4	2	3	8
12	4	3	11	12	10	13	13	10	13	13	11	12	13	11	9
9	8	3	19	18	15	21	19	16	23	21	18	21	19	17	6
13	7	4	2	2	1	2	2	1	2	2	1	7	7	7	3
10	4	4	21		19	22		23	22		20	21		24	16
10	7	4	23	23	22	26	23	22	28	25	23	29	25	23	1
13	6	5	3			3			3			3			21
11	10	5	6	4	3	7	6	5	6	6	2	9	8	6	
10	1	5	29	26	24	23	25	21	27	26	22	25	23	19	7
11	1	6	20	20	18	20	21	18	21	20	19	17	16	15	14
13	4	7	1	1		1	1		1	1		1	1		5
10	6	7	26	24	21	27	16	13	26	19	17	22	18	16	11
12	3	7	18	17	16	17	20	19	25	23	21	19	17	14	20
13	1	8	5	5	5	4	4	2	5	8	5	5	4	1	4
11	1	10	22	19	17	19	18	17	20	17	15	23	21	21	18
9	9	10	33	29	27	33	29	27	33	27	27	32	29	27	17
10	4	11	28	25	23	30	27	24	18	18	13	30	27	22	
13	0	11	4	3	2	6	5	4	7	4	4	2	3	4	
10	2	11	7	6	6	12	7	6	8	5	8	11	9	8	
13	0	11	12	11	11	11	11	12	15	16	16	6	5	2	12
12	0	11	17	16		16	15		24	22		24	24		15
12	11	12	9	8	7	8	8	7	9	7	7	14	12	12	
13	1	14	10	9	8	10	9	8	11	10	9	10	10	9	13
10	4	14	27	21	14	24	22	15	17	11	10	26	20	18	2
10	1	15	24	22	20	18	17	14	29	24	24	18	15	13	
12	6	15	14	13	12	15	14	11	10	9	6	8	6	5	10
10	8	15	30	27		29	26		30	29		28	26		
12	8	18	16	15	13	25	24	20	14	14	12	20	21	20	19
9	5	20	32		25	31		25	32		26	33		26	
11	2	24	15	14	9	14	12	9	16	15	14	13	11	10	
10	9	50	25			28			19			15			
10	11	> 60	31	28	26	32	28	26	31	28	25	31	28	25	22
13	7	> 60	13	10		9	10		12	12		16	14		

Figure 8.20
An excerpt from Spearman's 1904 data set.

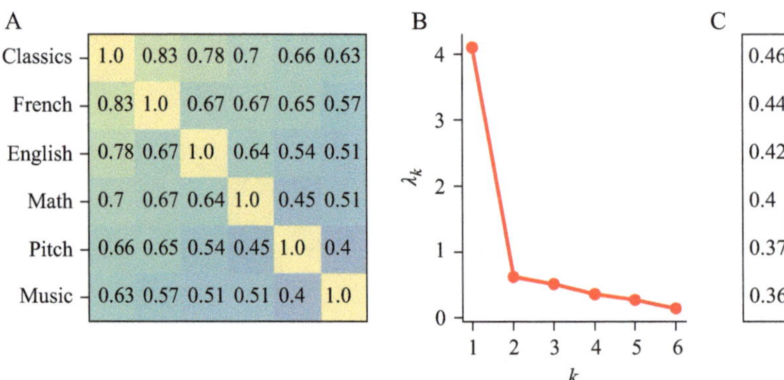

Figure 8.21
Principal component analysis of data in Spearman (1904), "Experimental Series IV." A: Correlation matrix of the scores of $T = 33$ boys in $N = 6$ school subjects including pitch discrimination. B: Scree plot of the eigenvalues λ_k. The first eigenvalue accounts for 68 percent of the variance. C: Coefficients of the first principal component arranged as in panel A.

and sensory tests. Figure 8.21 presents a principal component analysis of Spearman's "Experimental Series IV": The correlation matrix $\mathbf{C}$ (figure 8.21A) shows strong covariance of the scores across all subjects, including pitch discrimination. In other words, the typical boy tended to fare well or poorly in all subjects. This is reflected in the eigenvalue spectrum (figure 8.21B) which shows that a single principal component accounts for 68 percent of the variance. The coefficients of that component (figure 8.21C) are indeed positive along all the subjects. [5]

Spearman concluded that a single factor explains much of the boys' performance in class, but also on seemingly unrelated tests of perceptual discrimination. That factor eventually became known as "general intelligence."

8.5.1.2 Example: Dimensionality reduction of neuronal population activity Consider the example data set shown in figure 8.22A, which corresponds to the normalized activity of 50 neurons[6]. It is an $N \times T$ data matrix $\mathbf{Y}$, with N being the number of neurons and T being the number of time points. The activity is measured using a calcium indicator, a fluorescent probe expressed inside neurons that increases its fluorescence when the calcium concentration increases. Because the intracellular concentration of calcium increases when a neuron is active, the fluorescence can be used as a proxy for neuronal activity.

At first sight, one gets the impression that several neurons share the same pattern of activity. Instead of considering 50 neurons, can we collect then into a smaller number of "neuronal components" that are linear combinations of the original neurons

---

5. The literature on PCA suffers from a good amount of redundant and confusing nomenclature, including terms like "factors," "weights," "loadings," and "scores." In this book, we use the term "principal component" to refer to one of the eigenvectors of the covariance matrix. We use "PC coefficient" for the coefficient of a data point along that principal component.
6. Normalized activity implies that the fluorescence time series for each neuron is mean subtracted and has unit standard deviation—see section (8.5.1.3).

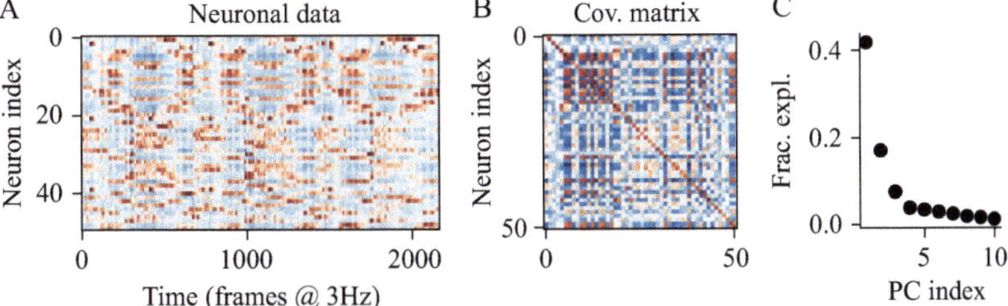

Figure 8.22
A: The activity of 50 neurons reported by a genetically encoded calcium indicator as a function of frame number, where frames were collected at 3 Hz. B: The covariance matrix computed from (A). C: Fraction of the variance explained by each of the first ten principal components.

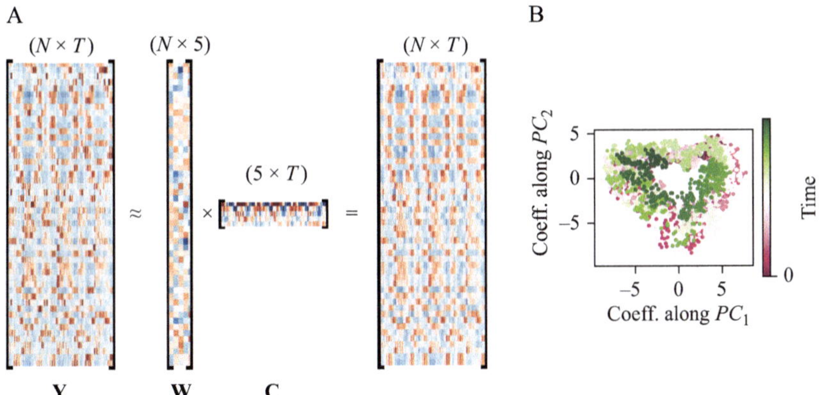

Figure 8.23
Dimensionality reduction of the activity of $N = 50$ neurons using PCA. A: The images illustrate the matrices $\mathbf{Y} \approx \mathbf{W}\mathbf{C}$. The dimensions of the matrices are shown on the top. The first principal component is the first column of $\mathbf{W}$. The coefficient of the data along the first principal component is given by the first row of $\mathbf{C}$. This row has larger coefficients than those in the second row, and so on. This is a manifestation that the first PC explains more of the variance than the second PC, and so on. B: We now keep just the first two neuronal components. The temporal activity of the 50 neurons is reduced to a trajectory in this 2D space. The time course is color-coded from magenta to green.

that explain most of the activity? In order to test this, one might perform principal component analysis.

Following the procedure described in the beginning of this section referring to 8.5.1, we can compute the covariance matrix, shown in figure 8.22B. The eigenvectors of this matrix are the neuronal principal components, and the eigenvalues report the variance explained by each one. We plot these in the scree plot shown in figure 8.22C. The first five principal components together contribute almost 75 percent of the variance in the data set (with the first two contributing almost 59 percent).

Using the first five principal components we can dimensionally reduce our data as shown in figure 8.23A. This shows that although although there are 50 neurons, they are strongly correlated in their time-varying activity. About 75 percent of the variance in

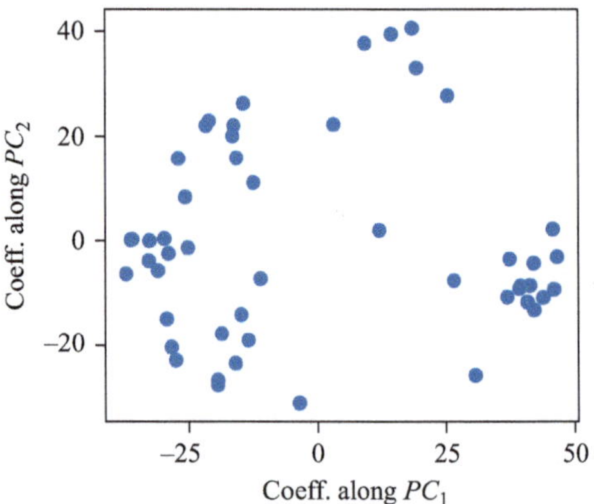

Figure 8.24
Scatterplot of the 50 neurons showing their coefficients along the first two temporal principal components.

their activity occurs in a five-dimensional subspace, and 59 percent in just two dimensions. In figure 8.23B we plot the activity as a trajectory in the space spanned by the first two principal components. Here, we can observe that it circles around the origin with changing angular velocity. [7]

What we have performed here is **neuronal PCA**. We could have also performed **temporal PCA** by computing the covariance matrix of the N-dimensional data vectors $\mathbf{y}_j$ (one for each time point j) and then performing PCA on that matrix. That covariance matrix is a $T \times T$-dimensional matrix so its eigenvectors, the temporal principal components, would be T-dimensional. Temporal PCA would give us the time courses that explain most of the variance across all the neurons. In fact, we will cluster the neurons according to the coefficients of their fluorescence along the first two temporal principal components in section 8.5.3 (see also figure 8.24).

8.5.1.3   Normalization and other preprocessing in PCA   Sometimes, the components x_i that make up the data vector $\mathbf{x} = [x_1, \ldots, x_N]^\top$ represent very different variables. For example, Spearman's measurement of pitch discrimination obviously uses a different scale from the grades of the mathematics teacher. In the example involving neuronal data, the fluorescence of every neuron will depend on its size and how much fluorophore it expresses. In other cases, the measurements may be of entirely different physical quantities, like temperature and precipitation. Obviously, one needs to account for such differences in units before computing the covariance matrix.

One popular method for normalization adjusts the scale on each variable so they all have the same sample variance in the data set. This is known as **z-scoring** the data: subtract the sample mean and divide by the sample standard deviation. This leads to a preprocessed data set:

$$\mathbf{y}_j = [y_{1,j}, \ldots, y_{N,j}]^\top, \tag{8.117}$$

---

7. These data are from Petrucco et al. (2023) and in fact correspond to the heading direction neuronal network of larval zebrafish, which keeps track of the direction the fish is heading toward as it swims.

where

$$y_{i,j} = (x_{i,j} - m_i)/s_i, \tag{8.118}$$

and

$$m_i = \frac{1}{T} \sum_j x_{i,j}, \tag{8.119}$$

is the sample mean of the ith component and

$$s_i = \sqrt{\frac{1}{T} \sum_j (x_{i,j} - m_i)^2} \tag{8.120}$$

is the sample standard deviation of the ith component. Mathematically, this is equivalent to using the correlation matrix rather than the covariance matrix for the eigenvalue analysis. This is the method that we used for Spearman's data.

A different approach comes from considering experimental uncertainties: If component x_i of the data vector is affected by measurement error σ_i, then it may make sense to normalize each component by its uncertainty. That is, preprocess the data to

$$y_{i,j} = (x_{i,j} - m_i)/\sigma_i. \tag{8.121}$$

In that case, the total residual E in equation (8.107) takes on the character of a χ^2 statistic (see section 7.6.1), such that minimizing E is like maximizing the likelihood.

Other preprocessing steps include nonlinear transforms. For example, if x_i has a log-normal distribution in the data set, then a logarithmic transform $y_i = \log x_i$ will produce a more nearly normal variable, and thus a nicer shape to the data cloud.

Whatever preprocessing steps you apply, try to understand why you are doing it and what the consequences are for structures that might appear in the processed data.

8.5.2 Other Dimensionality Reduction Techniques: NNMF and ICA

Both linear regression and PCA can be interpreted as reducing the dimension of an $N \times T$ data matrix $\mathbf{X}$ according to

$$\mathbf{X} \approx \mathbf{WC}, \tag{8.122}$$

where $\mathbf{W}$ is $N \times D$ and $\mathbf{C}$ is $D \times T$. The rows of $\mathbf{C}$ form the basis of the new D-dimensional space and $\mathbf{W}$ are the components of the data in terms of this basis.

Linear regression and PCA impose different constraints on the basis set C. Linear regression minimizes the unexplained variance in the dependent variables, whereas PCA minimizes the total unexplained variance along all dimensions. PCA leads to an orthogonal basis set, consisting of the principal components, which can be computed as the eigenvectors of the correlation matrix.

There are versions of dimensionality reduction that call for different conditions on the basis vectors. For example, **independent component analysis (ICA)** tries to explain the data as the weighted sum of a small number of signals, just like PCA, but in this case, with the requirement that these signals should be statistically independent of each other, as opposed to uncorrelated, which was the condition imposed by PCA. Another version, called **nonnegative matrix factorization (NNMF)**, imposes the constraint that the matrices **W** and **C** contain no negative elements. This arises when the data in question are naturally constrained nonnegative, such as intensities, concentrations, neuronal firing rates, and probabilities. These methods don't come with an analytical closed-form solution, like PCA, but efficient algorithms exist for deriving the components numerically.

8.5.3   Clustering

Dimensionality reduction simplifies the data distribution by allowing us to focus on a subspace of the original measurement space. However, within that subspace, the data are still widely distributed. An even greater simplification could be accomplished if the data form discrete clusters within the subspace. Hence there is broad interest in techniques that identify discrete clusters in a distribution.

Continuing with the 50-neuron data set here, we now replot the neuronal responses in the space of the first two temporal principal components (figure 8.24).[8] Note that these two components already explain 59 percent of the variance in the data set. One does get the vague impression that the points bunch together in certain regions of the space.

A popular method to identify clusters is called **k-means clustering**. This algorithm asks the user what number k of clusters should be found, and given k, it identifies the optimal allocation of data points to k clusters. Its criterion is to minimize the sum of the squared distances between every point and the cluster centroid to which it is assigned.

Choosing the number of clusters for k-means is not a trivial matter. One way is to apply the **elbow method**, similar to the interpretation of scree plots in PCA: repeat the analysis for a range of cluster numbers k, plot the sum of the squared distances s as a function of k, and check whether there is a critical number k beyond which s no longer decreases very much. This transition from a steep decrease in $s(k)$ to a more gradual decrease is called "an elbow." Figure 8.25 shows $s(k)$ for the neuronal data presented here. In this case, the elbow is not obviously apparent, so we show the clustering into three and six clusters, respectively, to serve as a comparison. Figure 8.26 shows that each of the clusters represents a different time course of neural activity.

8.5.3.1   Example: Otsu's image background separation method   Image processing is important in many biological applications, and many experiments rely on the correct quantification of "particles" within these images. An important step in this analysis is separating the background of the image from the "particles" that need to be counted. These "particles" can be fluorescent cells, stained mitochondria, or birds flying in the sky.

Otsu's method is a simple method that uses k-means clustering to do that. It assumes that the pixel intensities cluster into two groups, one corresponding to the background and the other to the foreground. [9] Figure 8.27 shows how this is applied.

---

8. Note that in section (8.5.1.2), we performed neuronal PCA, in which each PC was a linear combination of the 50 neurons. Here, we have performed temporal PCA, in which each PC is a fluorescence time series, and the fluorescence time series of each individual neuron can be expressed as a linear combination of these PCs.

9. Note that this is a nontrivial assumption. A continuum of values can always be clustered into two clusters, although this does not necessarily mean that there are two clusters.

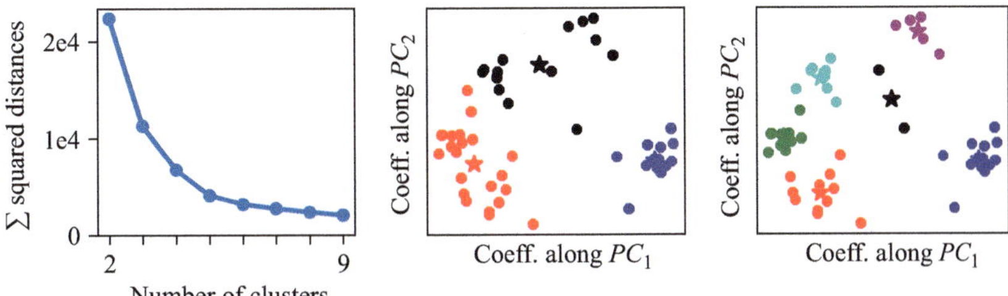

Figure 8.25
K-means clustering of the 50 neurons in the space of the first two PCs. Left: Unexplained variance as a function of the number of clusters. Middle: Three clusters and their centroids (stars). Right: Six clusters and their centroids (stars).

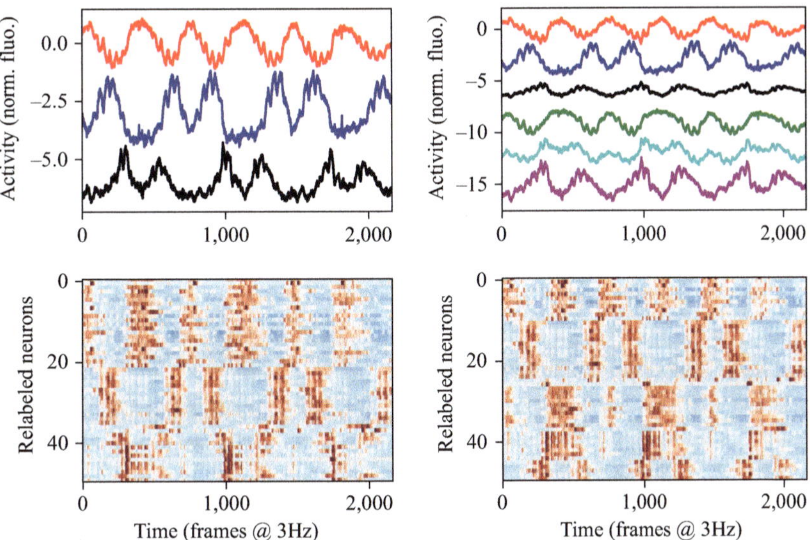

Figure 8.26
Activity traces corresponding to the cluster centroids of the figure 8.25, for both three and six clusters. Activity traces of neurons are reordered in such a way that neurons belonging to the same cluster appear together for the three and six clusters shown here.

There are many extensions of this simple algorithm; the easiest one to understand is a two-Gaussian mixture model (discussed next). Nevertheless, they all rely to some extent on clustering pixel intensities, taking into account assumptions on the distributions of these intensities or the morphology of the particles of interest.

8.5.3.2 Other clustering algorithms The k-means clustering algorithm presented here is by no means the only one. In fact, k-means implicitly assumes certain features that are not always assured, such as that all clusters have a spherical shape and the same variance, and similar numbers of members.

Other clustering methods relax some of these assumptions. Gaussian mixture models (GMMs), for example, do not assume a spherical distribution or equal variance. They return the probability that each sample belongs to each cluster. Nevertheless, GMMs

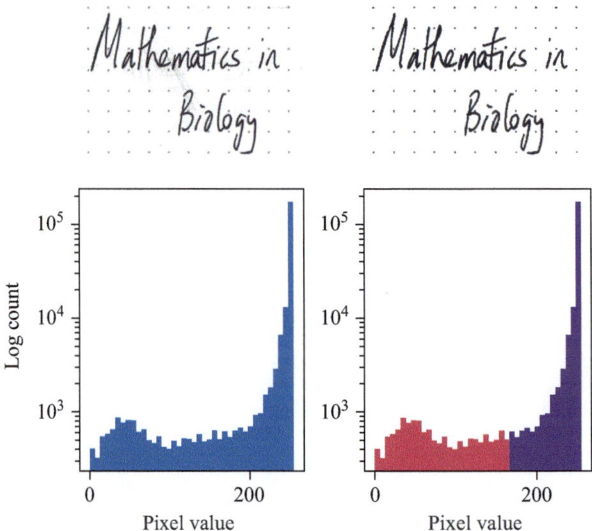

Figure 8.27
Left: Grayscale image (top) and the associated histogram (bottom). Right: Otsu's method assumes that pixel intensities cluster into two values, corresponding to background and signal, respectively. It divides the pixel intensities into these two clusters (shown in magenta and purple) and returns a binarized image (top).

do use the expected cluster number as an input into the algorithm, which needs to be decided a priori or explored empirically, just as in k-means.

Other algorithms, such as hierarchical clustering, return a branching tree where each sample is a leaf. By considering clusters to be small twigs, or small or large branches, the data can be separated into different numbers of clusters in a graded fashion. It is also possible to use more exotic distance metrics, other than the standard Euclidean metric.

On the whole, clustering is a bit of an art form. It is always possible to perform clustering, even when the data are drawn from a continuous distribution. A number of quantitative criteria have been proposed to evaluate the significance of clusters. In practice, it is important to find some visualization of the clusters, so the user can gain intuition for the results and evaluate visually whether the clusters make sense and can be interpreted usefully for the research purpose at hand.

8.6 Information Theory

Life is an interplay of energy, entropy, and information.
    —Eigen (2019)

Public discourse these days is awash with loose talk of "information," often with numbers thrown in the mix, measured in gigabits or terabytes. Typically, this arises when a message needs to be transmitted from one place to another, such as to stream a television show on your monitor or to store a large document in a file. Such signal transmission also is ubiquitous in biological systems: the genome communicates through cellular machinery to specify the cell's proteome; cells signal to each other in the course of development or an immune response; the eye signals to the brain with messages about our visual surroundings; the brain signals to muscles in order to respond. Understanding these processes (and more) benefits from a rigorous quantitative treatment of this substance called **information.**

Fundamentally, information leads to the **removal of uncertainty** (Shannon and Weaver, 1964). For illustration, consider a simple children's game: A says "I am thinking of a number between 1 and 16." B has to find the number by asking A yes/no questions. At the outset, B is uncertain about the number. With every question (if appropriately posed), B gains more information until there is no remaining uncertainty, and B knows the number exactly.

How many questions does B need to ask? An inefficient strategy would be: "Is it 1?," "Is it 2?," and so on. On average, this requires about eight questions to achieve success. A more streamlined approach is to ask each question so that it splits the remaining range of possible answers in half, such as starting with "Is it 8 or less?" This strategy requires only four questions to achieve success. In general, if A's number ranges from 1 to 2^n, then n questions are needed to nail it down precisely. In this case, we say that B had an uncertainty about A's number equal to n bits. During the question-and-answer game, that uncertainty was completely removed, so A transferred n bits of information to B.

8.6.1 Entropy

This leads us to a quantitative definition of uncertainty: The uncertainty about a random variable X is equal to the minimal number of yes/no questions required to determine X precisely. The uncertainty is also called **entropy**, denoted as $H(X)$ and is measured in units of bits.

If $X \in \{x_1, \ldots, x_n\}$ is a discrete random variable and all the outcomes are equally likely a priori, as in the guessing game, then

$$H(X) = \log_2 n. \tag{8.123}$$

Generally, X does not follow a uniform distribution. For example, X might be the outcome of a roll of two dice (as discussed in section 6.3.1). If X follows the probability mass function $P(X)$, then

$$H(X) = -\sum_i P(x_i) \log_2 P(x_i). \tag{8.124}$$

Note that equation (8.123) is just a special case of equation (8.124).

This choice for measuring **entropy** is the only mathematical expression that satisfies two common-sense expectations of such a measure: (1) It should be positive. (2) If two variables X and Y are statistically independent, the uncertainty about both should be the sum of the uncertainties about each individually.

Example 8.7 (Bernoulli distribution) Remember that a Bernoulli random variable is one that can take two outcomes, with probability p and $1-p$, respectively, such as the outcome of a coin toss (as in section 6.3.5). If $X \sim \text{Bern}(p)$, then the entropy is (figure 8.28)

$$H(X; p) = -p \log_2 p - (1-p) \log_2 (1-p). \tag{8.125}$$

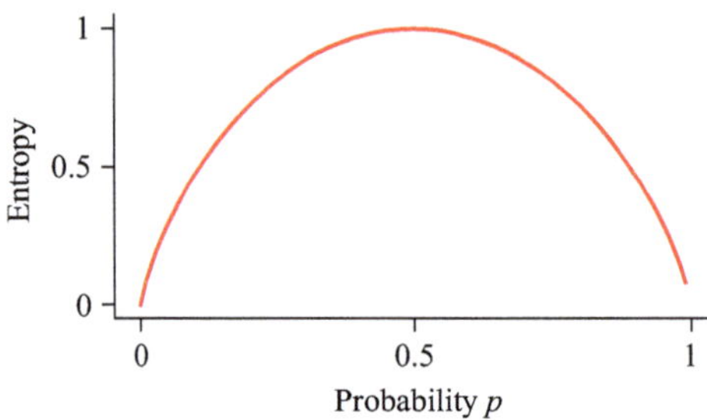

Figure 8.28
The entropy of a Bernoulli variable with bias p.

For what value of p is this entropy a maximum? Of course, one can look for the extremum of $H(X;p)$ with respect to p. For a simpler argument, note that the entropy must be symmetric under the exchange of p with $1-p$. Therefore, the maximum has to be at $p=0.5$, when the two values are equally likely. On the other hand, when $p=0$ or $p=1$, then the value of X is certain and the entropy $H(X)=0$. For these calculations, it is useful to remember that

$$x \log x \xrightarrow[x\to 0]{} 0. \tag{8.126}$$

$\square$

Example 8.8 (What is the entropy of English?) It is well worth reading Shannon's paper on this topic (Shannon, 1951). He asks: When reading an English text, what is the uncertainty about the next character on the page?

A naive estimate goes as follows: Ignoring spaces and puctuation, there are 26 letters, so using equation (8.123), the entropy $H(C)$ of the next character C is

$$H_0(C) = \log_2 26 = 4.70 \text{ bits.} \tag{8.127}$$

However, the characters don't appear at equal frequency, so one should measure those frequencies and use the more general expression in equation (8.124) to get

$$H_1(C) = -\sum_{i=1}^{26} p_i \log_2 p_i = 4.08 \text{ bits.} \tag{8.128}$$

As it turns out, consecutive characters are not independent of each other: Certain letter pairs (like "QU") happen much more often than expected from the product of their individual frequencies. By tabulating the frequencies of letter pairs, one gets to $H_2(C) = 3.56$. Shannon pursues this further, estimating the frequencies of tri-grams and words, and with each step obtains a lower entropy estimate. Eventually, he engages human subjects in a guessing game: after reading 100 characters in a book, they must guess the next one. From the number of guesses required, Shannon estimated the true entropy of English as somewhere between 0.6 and 1.3 bits/character:

$$0.6 < H_\infty(C) < 1.3. \tag{8.129}$$

This exercise foreshadows two interesting insights: First, the entropy of a symbol string depends not only on the frequency of each symbol, but on the correlations across symbols. Second, it should be possible to store English text in a very efficient way: the calculations suggest that one only needs about 1 bit per character. Instead, the popular ASCII code for English uses 8 bits per character—a great waste of bits. $\square$

Example 8.9 (The entropy of DNA) Genetic material is stored in chromosomes, each one consisting of a long macromolecule of double-stranded DNA. Each strand of DNA is a long sequence of nucleotides that have one of four possible values: adenine (A), thymine (T), cytosine (C), or guanine (G). What is the entropy per base-pair of a long DNA sequence?

As in the case of English, we can start with the naive estimate, assuming that all 4 nucleotides appear at equal frequency, $p_i = \frac{1}{4}$, where $i \in \{A, T, G, C\}$:

$$H_0 = -\sum p_i \log_2(p_i) = 2 \text{ bits per base-pair.} \tag{8.130}$$

In actuality, the four bases do not appear at equal frequency. For example, in human chromosome 11, one finds that $p_i = [0.289, 0.289, 0.211, 0.211]$. With that knowledge, the entropy is

$$H_1 = -\sum p_i \log_2(p_i) = 1.9822 \text{ bits per base-pair.} \tag{8.131}$$

Further, it turns out that two successive nucleotides (di-grams) are not statistically independent. Again, one can estimate the frequencies of di-grams from the human chromosome 11 data to find a lower estimate:

$$H_2 = 1.9350 \text{ bits per base-pair.} \tag{8.132}$$

As in the case of English, knowledge of the statistical structure of the signal serves to reduce the uncertainty.

In coding regions of the chromosome, DNA carries the instructions for assembling amino acids into proteins. You can further explore the information-theoretic aspects of this genetic code in exercise 10.21. $\square$

8.6.2 Communication Channel

Shannon (1948) formalized the process of communication between a source and a destination as follows (figure 8.29):

On the source side, the message gets encoded into a signal to be conveyed on a channel. During transmission on the channel, that signal may be corrupted by noise. On the destination side, the received signal gets translated back into an interpretable message. The context of Shannon's work was telecommunications, so the channel in question was typically a telephone transmission line or a wireless connection. Each of these channels suffers from a different kind of noise corruption. Ideally, the transmitters and receivers must be adapted to the kind of noise encountered, so as to allow error-free transmission regardless.

As it turns out, this framework lends itself to illuminate a vast number of phenomena, including many cases of signaling and communication in biology.

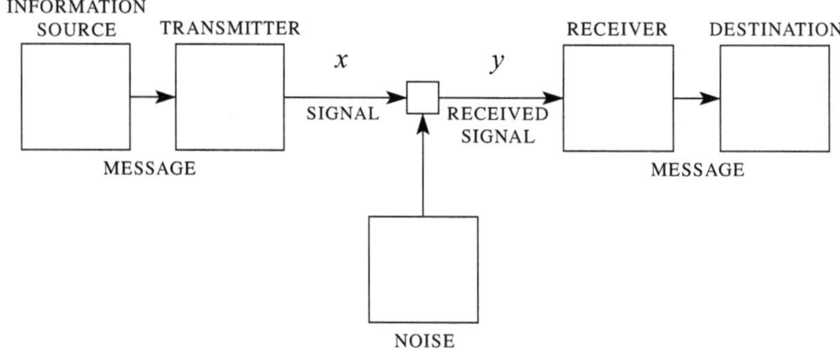

Figure 8.29
Shannon's framework for communication.

8.6.3 Mutual Information

Within this framework, suppose that the transmitter puts signal X on the channel and the receiver observes signal Y. Because of transmission noise, it is possible that $X \neq Y$. Suppose that X and Y follow a joint distribution $P_{XY}(X, Y)$. How much information about X does the receiver get from observing Y?

As discussed previously, information corresponds to a reduction of uncertainty. Prior to receiving signal Y, the receiver's uncertainty about X is the entropy $H(X)$, which depends only on the marginal probability distribution $P_X(X)$:

$$H(X) = -\sum_{x \in X} P_X(x) \log_2 P_X(x). \tag{8.133}$$

After the receiver sees the particular symbol $Y = y$, the probability distribution of X shifts from $P_X(X)$ to $P_{X|Y}(X|Y = y)$—namely, the probability conditional on observation of y. Now the remaining entropy is

$$H(X|y) = -\sum_{x \in X} P_{X|Y}(x|y) \log_2 P_{X|Y}(x|y) \tag{8.134}$$

and the information gained in the process is

$$I(X|y) = H(X) - H(X|y). \tag{8.135}$$

To assess the average gain over many transmissions, one averages this expression over all possible outcomes of Y:

$$I(X, Y) = \sum_{y \in Y} P_Y(y) I(X|y)$$
$$= -\sum_{x \in X, y \in Y} P_{XY}(x, y) \log_2 \left(\frac{P_{XY}(x, y)}{P_X(x) P_Y(y)} \right). \tag{8.136}$$

The quantity $I(X, Y)$ is called the **mutual information** between the random variables X and Y. It tells us how much uncertainty about X is removed by measuring Y. Note that the expression for $I(X, Y)$ is symmetric in X and Y. So the information gained by the receiver about the transmitted message is equal to the information that the transmitter has about what appears at the receiver.

Note the special case in which X and Y are statistically independent. Then the joint probability factors into the product of the marginal probabilities as in equation (6.5.4), so the log term vanishes and the mutual information is zero. This is the extreme of a lousy communication channel, in which noise completely dominates the signal.

8.6.3.1 Mutual information for continuous variables Suppose that the symbol X placed on the channel and the receiver symbol Y are both continuous random variables, such as an electric voltage. Now the joint distribution of X and Y is a probability density function $P(x, y)$. The mutual information extends in a straightforward way by simply converting the sum to an integral:

$$I(X, Y) = - \int_{x,y} P_{XY}(x, y) \log_2 \left(\frac{P_{XY}(x, y)}{P_X(x)P_Y(y)} \right) dxdy. \tag{8.137}$$

8.6.4 Channel Capacity

The **capacity** of a communications channel is the maximum rate of information that can be transmitted down the channel, measured either in bits per symbol or bits per unit time. This maximum value of the mutual information between output and input is taken over all possible distributions of the signals, given some constraint.

8.6.4.1 Example: Binary channel with noise For example, consider the transmission of binary symbols $x \in \{0, 1\}$ across a noisy channel that occasionally changes a 0 into a 1 or vice versa. Suppose that the probability of error (of either kind) is q. So the probability of the output Y conditional on the input X is

$$P_{Y|X}(y|x) = \begin{cases} 1 - q, & \text{if } y = x \\ q, & \text{if } y \neq x. \end{cases} \tag{8.138}$$

To optimize the use of this channel, we have only one degree of freedom—namely, the fraction of time we use 0 and 1 for X. Because the channel properties are symmetric with respect to swapping 0 and 1, the only plausible optimum is when we use both symbols at equal frequency[10]. In that case, the channel capacity becomes

$$C = I(X, Y)_{\text{opt}} = 1 + q \log_2 q + (1 - q) \log(1 - q). \tag{8.139}$$

In this case, the constraint is that the signal can be either 0 or 1, and the free parameter is the fraction of 0s and 1s that are present. You can explore what happens if the channel's errors are asymmetric in exercise 10.20.

8.6.4.2 Example: Gaussian channel with noise Now consider a continuous channel with Gaussian noise. The signals X and Y are continuous variables, and the channel adds a random noise with Gaussian distribution, such that $y = x + n$ with $n \sim \mathcal{N}(0, N)$. Also, we suppose that the transmitter has limited signal power, so the variance of X is fixed: $\text{Var}[X] = S^2$. Then one can show that the capacity is

$$C = \frac{1}{2} \log_2 \left(1 + \frac{S^2}{N^2} \right). \tag{8.140}$$

---

10. You can save a lot of effort with symmetry arguments like this.

To transmit at this limit, the optimal symbol distribution of X is Gaussian with variance S: $X \sim \mathcal{N}(0, S)$.

Note the result in equation (8.140) is logarithmic in the signal-to-noise (SNR) ratio on the channel. It has a simple interpretation: For large $\frac{S^2}{N^2}$, $C \approx \log_2 \frac{S}{N}$. But $\frac{S}{N}$ is the number of signal levels that are separated by the noise amplitude. So C is simply the $\log_2$ of the number of distinguishable signals (i.e. the number of bits needed to specify the signal to within the noise amplitude).

8.6.4.3 Redundancy The redundancy R of a communication link is the degree to which it fails to use the full channel capacity. Redundancy is generally a consequence of nonoptimal symbol use, namely, when the transmitter fails to encode the message appropriately for the channel. Redundancy is expressed as a fraction of the capacity wasted:

$$R = 1 - \frac{I(X, Y)}{C}. \tag{8.141}$$

For example, the ASCII code that uses eight binary digits to encode a character is a rather inefficient representation of English. Using Shannon's estimate that the entropy of English is about 1 bit/character, one concludes that the ASCII code has a redundancy of $R \approx 7/8$.

8.6.5 The Channel Coding Theorem

So far, we have mostly engaged in definitions of information-theoretic quantities. But what is the payback for using this way of measuring information? One powerful result is the **channel coding theorem**: Given a noisy channel with capacity C, one can use it to transmit error-free messages at an information rate up to C.

It seems counterintuitive that one can use a noisy channel for error-free communication at all. Obviously, this requires an ingenious encoding and decoding scheme that makes the message robust to the kinds of disruption that occurs on the channel. Notably, the theorem does not spell out how to achieve this, and much engineering effort goes into devising clever encoders and decoders to match messages to a channel. However, the theorem tells you when to stop trying. In many cases, it is easy to compute the capacity of the channel (see sections 8.6.4.1 and 8.6.4.2), and you can stop improving your encoders once the information rate that they support gets close to C.

In a number of biological applications, it is possible to compute the capacity C for a given signaling pathway, starting from an understanding of signal and noise in the system. Then one can ask how much information actually passes through that channel. In a few cases, the information rate seems to approach the capacity of the channel, suggesting a certain optimization of the encoding and decoding mechanisms (Tkacik and Bialek, 2016).

8.6.6 The Data-Processing Inequality

Consider a signaling chain from X to Z through an intermediate signal Y:

$$X \to Y \to Z. \tag{8.142}$$

Along such a chain, the mutual information can only decrease. In particular,

$$I(X, Z) < I(X, Y) \quad \text{and} \quad I(X, Z) < I(Y, Z). \tag{8.143}$$

In other words, along a signaling chain information can only be lost, not created. This has consequences for the capacity: if a signaling chain should have capacity C, then each individual link must have capacity $\geq C$.

In biology, we find that a signal frequently changes its physical identity along the way. For example, communication in the nervous system alternates between electrical voltage across the membrane, calcium concentration at a synapse, neurotransmitter concentration in the synaptic cleft, ionic current into the dendrite, and back to membrane voltage. Information theory is agnostic to the physical embodiment of a signal, and this is one of its chief attractions. At every stage along the way, one can measure rates and capacities in the universal unit of bits, and interpretation of those results is supported by the theorems of information theory.

8.6.7 Further Reading

The founding document of information theory is still one of the most readable introductions. Shannon and Weaver (1964) present the original papers with additional didactic material. A good technical reference is Cover and Thomas (2012), and a survey of biological applications can be found in Tkacik and Bialek (2016). Nelson (2022) explains many of the physical mechanisms by which living organisms gain information about their surroundings.

9.1 Luria-Delbrück Revisited

To illustrate some of the concepts around statistics and inference, we will revisit Luria and Delbrück (1943). Recall the basic structure of their experiment presented in section 6.1.1:

1. Grow a culture of bacteria to a total number of $\sim 10^8$.
2. Expose the culture to a virus.
3. Plate the culture and count the colonies. Each colony on the plate corresponds to a single mutant cell in the culture that has become resistant to the virus.

The authors considered two basic hypotheses to account for the observed mutant counts:

- H_0: The resistance mutations happen at the time of exposure to the virus. Each bacterium has a small chance of mutating and thus surviving and forming a colony.
- H_1: The mutations happened earlier throughout the growth of the culture, independent of the virus. Mutant cells pass the mutation on to their offspring.

9.1.1 The Null Hypothesis

The null hypothesis H_0 predicts that the number of mutant cells in the culture should follow a Poisson distribution. If N is the number of cells in the culture and β is the probability of mutation per cell, then the number of mutants m follows

$$m \sim \mathrm{Poiss}(v), \tag{9.1}$$

where

$$v = \beta N \tag{9.2}$$

is the mean number of mutants per culture. If one were to repeat the same experiment many times, the observed number of mutants should vary from trial to trial, and the variance of that result should equal the mean:

$$\mathrm{Var}[m] = \mathrm{E}[m]. \tag{9.3}$$

TABLE 3

Distribution of the numbers of resistant bacteria in series of similar cultures.

EXPERIMENT NO.	22		23	
Number of cultures	100		87	
Volume of cultures, cc	.2*		.2*	
Volume of samples, cc	.05		.2	
	Resistant bacteria	*Number of cultures*	*Resistant bacteria*	*Number of cultures*
	0	57	0	29
	1	20	1	17
	2	5	2	4
	3	2	3	3
	4	3	4	3
	5	1	5	2
	6– 10	7	6– 10	5
	11– 20	2	11– 20	6
	21– 50	2	21– 50	7
	51– 100	0	51– 100	5
	101– 200	0	101– 200	2
	201– 500	0	201– 500	4
	501–1000	1	501–1000	0
Average per sample	10.12		28.6	
Variance (corrected for sampling)	6270		6431	
Average per culture	40.48		28.6	
Bacteria per culture	2.8×10^8		2.4×10^8	
Mutation rate	2.3×10^{-8}		2.37×10^{-8}	
Standard deviation { exp.	7.8		2.8	
Average { calc.	1.5		1.5	

* Cultures in synthetic medium.

Figure 9.1
A data table from Luria and Delbrück (1943).

Instead, Luria and Delbrück found that the variance was much larger. For example, in their experiment 23 (figure 9.1), across a sample of 87 cultures, the mean number of mutants was 28.6 and the variance was 6,431, about 225 times greater than the mean. This is obviously inconsistent with the Poisson distribution, so one can reject the null hypothesis with confidence.

How high is that confidence? We can perform a chi-square test to see how well the observed distribution matches the best-fit Poisson distribution. Recall that the maximum-likelihood estimate for the parameter v is the mean count (section 7.1.4):

$$\hat{v} = \langle m \rangle_X = 28.6. \tag{9.4}$$

Under the null hypothesis using the maximum likelihood estimator (MLE) parameter, the expected number of cultures with m resistant bacteria is

$$E_m = n \frac{\hat{v}^m}{m!} e^{-\hat{v}}. \tag{9.5}$$

Then Pearson's chi-square becomes

$$X^2 = \sum_m \frac{(X_m - E_m)^2}{E_m} \approx 10^{241}. \tag{9.6}$$

This is an astronomically large number[1] and close to the largest number possible in 64-bit floating-point arithmetic. Computing the associated p-value would require special numerical techniques, but that is a pointless exercise. Certainly, $p \ll 10^{-100}$, which for all intents and purposes is zero.

9.1.1.1 An aside on the meaning of tiny p-values

An interesting question is whether scientific authors should ever quote a ridiculously small p-value, like $p < 10^{-10}$. Recall that the p-value is the probability of obtaining a result that lies this far from the null prediction by chance, assuming that the null hypothesis is true. Legitimately, one needs to consider all the paths by which the results may have ended up in the research report. This includes the possibility of human error. For example, Luria and Delbrück might have mislabeled some vials or made mistakes in copying mutant counts from one table to another. Every scientist has seen errors like these occur in the laboratory, and sometimes they are caught before publication. Some scientists seem to be unusually error-prone. For example, when an accusation of data fabrication arises, the explanation is invariably that the figures were mixed up due to honest error. These researchers should be forbidden from ever quoting a p-value again. The rest of us could adopt an honest estimate of the error rate in the laboratory (maybe 10^{-3}), and use that as a lower bound on p-values.

9.1.2 The Alternative Hypothesis

Here, we want to see what Luria and Delbrück's alternative hypothesis H_1 would predict. Suppose that the culture starts from a single bacterium and goes through i successive rounds of mutation and cell division. In round $j = 0, \ldots, i-1$, there are $n_j = 2^j$ bacteria, and of these, m_j are already mutant. The remaining $n_j - m_j$ bacteria each experience a mutation with probability α. We will call the number of new mutants q_j. Then all the bacteria divide to complete round j. Formally,

$$m_{j+1} = 2\,(m_j + q_j), \tag{9.7}$$

where

$$q_j \sim \text{Binom}(2^j - m_j, \alpha) \tag{9.8}$$

is binomially distributed. Starting with $m_0 = 0$ we want to understand the distribution of m_i, namely the final number of mutants.

This model is somewhat complex because the distribution of q_j depends on the number of mutants in all preceding generations, and therefore on all the preceding $q_{k<j}$. Surely, it is easy to simulate the process, but one would like to get some analytical insight. A useful approximation is that the number of mutants in the culture is almost always very small. For example, in experiment 23, the average was $\bar{m}_i = 28.6$ mutants

---

1. The number of particles in the observable universe is only $\sim 10^{80}$.

out of a total of $n_i = 2.4 \times 10^8$ cells. So let us approximate $2^j - m_j \approx 2^j$, which means that all the bacteria present can undergo mutations (not just the nonmutants). Then

$$q_j \sim \text{Binom}(2^j, \alpha), \tag{9.9}$$

which makes all the q_j independent random variables, thus simplifying the problem.[2]
   Now we can write

$$m_i = \sum_{j=0}^{i-1} q_j 2^{i-j} \tag{9.10}$$

because the mutants arising in round j then go through another $i - j$ rounds of division. This is a sum of independent random variables, so the expectation value and the variance simply sum. Note that

$$\begin{aligned}
\text{E}[q_j] &= \alpha\, 2^j \\
\text{Var}[q_j] &= \alpha(1 - \alpha)\, 2^j.
\end{aligned} \tag{9.11}$$

So the final number of mutants has expectation

$$\text{E}[m_i] = \sum_{j=0}^{i-1} \text{E}[q_j]\, 2^{i-j} = i\,\alpha\, 2^i \tag{9.12}$$

and the variance

$$\begin{aligned}
\text{Var}[m_i] &= \sum_{j=0}^{i-1} \text{Var}[q_j] 2^{2(i-j)} = \sum_{j=0}^{i-1} 2^j \alpha(1-\alpha) 2^{2(i-j)} \\
&= \alpha(1-\alpha) 2^{2i} \sum_{j=0}^{i-1} 2^{-j} = \alpha(1-\alpha)\, 2^{2i+1}(1 - 2^{-i}).
\end{aligned} \tag{9.13}$$

We can further approximate $1 - \alpha \approx 1$ because the mutation rate is small, and $1 - 2^{-i} \approx 1$ because there are many cell divisions, so

$$\text{Var}[m_i] \approx \alpha\, 2^{2i+1}. \tag{9.14}$$

Then the variance-to-mean ratio becomes

$$\frac{\text{Var}[m_i]}{\text{E}[m_i]} \approx \frac{2^{i+1}}{i}. \tag{9.15}$$

Using the numbers from experiment 23 (figure 9.1), $n_i = 2.4 \times 10^8$, so there were $i = \log_2 n_i \approx 28$ cell divisions. One therefore expects a variance-to-mean ratio of

$$\frac{\text{Var}[m_i]}{\text{E}[m_i]} \approx \frac{2^{i+1}}{i} \approx 2 \times 10^7. \tag{9.16}$$

---

2. This approximation leads to a significant error only if two mutations both hit in an early round while the population is small—a very rare event.

```
10000 Experiments with 87 cultures each
28 Cell divisions
Mutation rate 1.0e-08
Expected mean: 75.2
First 10 experiments:

      Mean        Var  Var/Mean
      37.1    16389.6     441.7
      22.4     3516.3     156.9
      22.2     3446.3     155.4
      36.6     9533.4     260.3
      22.3      840.6      37.7
      29.4     4695.5     159.7
      27.9     1161.7      41.6
      30.1     4653.5     154.8
      20.4     1291.7      63.2
      75.3   192515.6    2555.5
```

Figure 9.2

Simulations of experiment 23 as described in table 3 in Luria and Delbrück (1943), using the model of cell divisions and mutations described here. The experiment was simulated 10,000 times, and the outcomes of the first 10 are listed in the table. For each trial, this shows the mean number of mutants across the 87 cultures, the variance, and the ratio of variance to mean.

Instead, the authors report a sampled variance-to-mean ratio of only ~225. So are the data really consistent with the alternative hypothesis of spontaneous mutation? It seems that the observed variance-to-mean ratio violates the expectation from H_1 by an even bigger factor than the expectation from H_0.

To gain some insight, it helps to experiment a bit, numerically of course. In simulating the process, we can perform the authors' experiment 23 not once, but many times. Figure 9.2 shows the results.

Note that both the mean and the variance/mean ratio scatter a lot. Also, most of the time the sample mean is below the expectation. And the typical variance/mean ratio is reasonably close to the value of ~225 reported by the authors. Why is this so different from the analytical expectation of $\sim 2 \times 10^7$?

Figure 9.3 shows the empirical distributions of the sample mean and the sample variance-to-mean ratio from these simulations of 10,000 Luria-Delbrück studies. They are plotted as cumulative distributions with a logarithmic x-axis. Note that both these statistics have a highly skewed distribution, with a thin long tail extending to very large values. Those values hardly ever happen. For example, if the very first bacterium gets mutated, that leads to a huge number of mutants in the final culture, but this will happen only once in 10^8 cultures. Obviously, one cannot sample all parts of this distribution experimentally.

Returning to the original question: Are the data consistent with hypothesis H_1? Note that the value of 225 that Luria and Delbrück reported for variance/mean lies right in the bulk of the distribution, at a cumulative of ~0.6 (figure 9.3B). So one concludes that the observed results are in fact entirely compatible with the alternative hypothesis H_1.

Beyond the issue of hypothesis testing, this example also illustrates the care required when dealing with extremely skewed distributions, where the mean and variance do not reflect where most of the values are located. Recall that for this distribution,

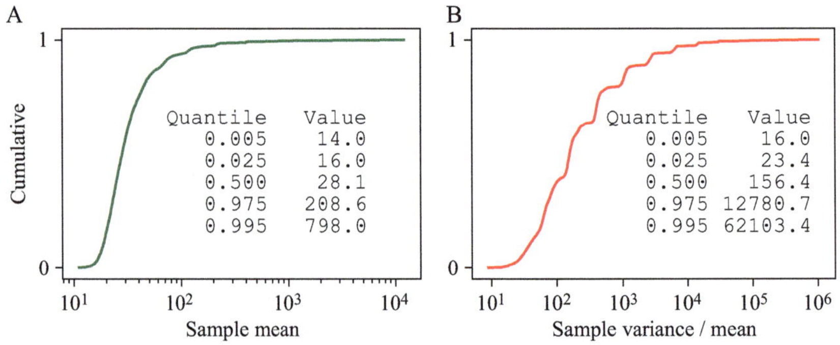

Figure 9.3
Analysis of the simulations of experiment 23. Shown here are the cumulative distribution of the mean number of mutants per culture, along with the relevant quantiles (A), and the same for the ratio of variance to mean (B).

$$\frac{\mathrm{Var}[m_i]}{\mathrm{E}[m_i]} \approx \frac{2^{i+1}}{i} \approx 2 \times 10^7 \tag{9.17}$$

and yet 99 percent of the time, the experiment will produce a variance/mean ratio in the interval [16,62000].

9.2 Signal Processing

9.2.1 Quantization of Continuous Variables

A ubiquitous element of measurement devices is the quantizer, which maps a continuous input value to an output value from a predefined set, typically a set of integers. Quantization introduces noise because two distinct values that got mapped to the same integer can no longer be distinguished. Engineers say that each additional bit of resolution for the quantizer improves the signal-to-noise (SNR) ratio by 6 dB. We will explore what this means and where it comes from in the following short example.

The input-output relationship of a quantizer may look like figure 9.4.

As you can see, an analog value within some range (e.g. 0–1 V) is mapped to a fixed value (0.5 V) by the quantizer. The resolution of a quantizer is the number of bits it uses to represent the signal in its input range. An N-bit quantizer chops its operating range to 2^N levels. Figure 9.4, for example, depicts a 3-bit quantizer. The choice of quantizer resolution can be important—if it is too small, then you will not be able to resolve the signal you are interested in, whereas quantizers with high resolution are sometimes prohibitively expensive or waste power and storage space. Figure 9.5 depicts an example of passing a signal through the quantizer of figure 9.4.

We start with the simple assumption that the signal (denoted by the random variable X) will be *uniformly distributed* across the entire input range $[-x_{\max}, x_{\max}]$[3]. The noise due to quantization is the difference between X and its quantized version, which we will call $Z = X - \mathrm{Quant}(X)$. Our quantizer has N bits, or $M = 2^N$ levels. What is the

---

3. If we know nothing about our signal, this is a reasonable assumption; one might say that it is the distribution with the *maximum entropy*.

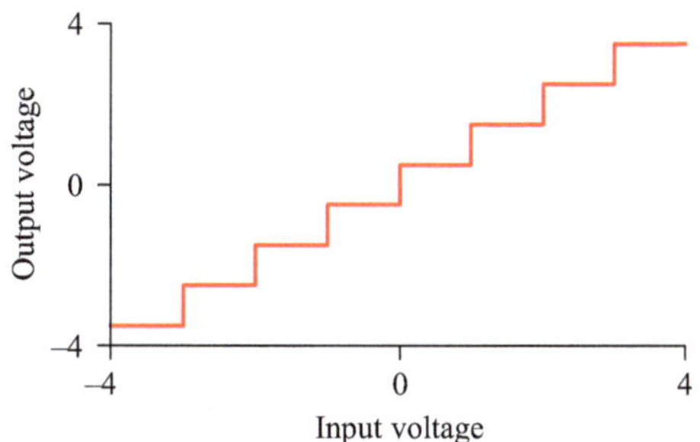

Figure 9.4
Input-output relationship of a quantizer.

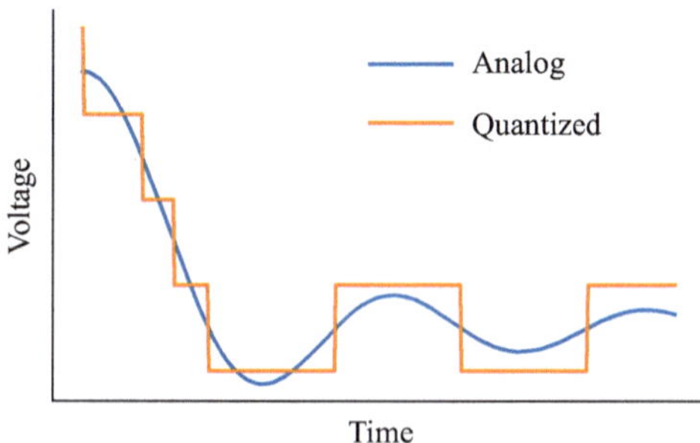

Figure 9.5
A signal is passed through a quantizer.

distribution of Z? In this example, if the input were 0.5 V, the noise would be zero, as it would get mapped to 0.5 V. If the input were close to 0 V or 1 V, on the other hand, the noise would be large because it would also get mapped to 0.5 V. This applies to not just the range from 0 to 1 V, but to every range—in every case, the quantizer noise is uniformly distributed in $[-x_{\max}/M, x_{\max}/M]$. Finally, recall from section 6.4.5 that the variance of the uniform distribution over $[a, b]$ is $\frac{1}{12}(b-a)^2$.

Now, the SNR ratio, measured in decibels, is defined as

$$
\begin{aligned}
\text{SNR} &= 10\log_{10}\frac{\text{Var[Signal]}}{\text{Var[Noise]}} \\
&= 10\log_{10}\frac{\text{Var}[X]}{\text{Var}[Z]} \\
&= 10\log_{10}\frac{(2x_{\max})^2/12}{(2x_{\max}/M)^2/12} \\
&= 20\log_{10} M \\
&= 20N\log_{10} 2 \\
&\approx 6.02N.
\end{aligned}
\tag{9.18}
$$

Indeed, under these assumptions, each additional bit gives you a roughly 6-dB increase in the SNR.

9.2.2 Fluctuation Correlation Analysis

Often a random process arises from the superposition of many independent stochastic processes. Examples include:

- The fluorescence of many particles diffusing in and out of a microscope beam
- The electric current flowing through many ion channels in a neuronal membrane
- A chemical reaction implemented by many independent enzyme molecules

In these cases, one can observe directly only the macroscopic signal that arises from the superposition of many events. Can one derive some clues about the structure of the microscopic events? For example, how many fluorescent molecules are there, and what is their diffusion coefficient? Or how many membrane channels are there, and what are their switching kinetics? Often, the microscopic information can be gleaned from analyzing the fluctuations in the macroscopic quantity. Here, we illustrate this approach with the current fluctuations through a neuronal membrane.

With glass microelectrodes, one can perform very sensitive measurements of the electric current across a neuronal membrane. In fortunate cases, it is possible to observe the current through single channels (Sakmann and Neher, 1995), but more commonly, the current flows through many channels and their individual contributions cannot be resolved. A simple model of this situation is as follows:

- There are n channels in the neuronal membrane.
- Each channel can be open or closed, with transitions between the two states occurring according to a Markov model (figure 9.6), with ON-rate α_{01} and OFF-rate α_{10}.
- When open, the channel passes a current i.
- Different channels switch independently.
- The observed membrane current $I(t)$ is the sum of all the single-channel currents.

The model has four unknowns: $n, i, \alpha_{01}, \alpha_{10}$. Is it possible to derive these microscopic parameters entirely from a macroscopic measurement of the whole-cell membrane current $I(t)$?

Figure 9.7 illustrates the problem by simulation. Figure 9.7A shows a simulation of the current through single channels with parameters $i = 1, \alpha_{01} = 10\,\mathrm{s}^{-1}, \alpha_{10} = 5\,\mathrm{s}^{-1}$. When the current through $n = 1{,}000$ of these channels is summed, one can no longer distinguish the individual events (figure 9.7B). However, an autocorrelation of this summed signal $I(t)$ reveals an interesting exponential decay (figure 9.7C). How should we interpret this result?

The model used here for the individual channel is exactly the random telegraph signal discussed in example 8.5. So the autocorrelation function of the single-channel current is given by equation (8.68):

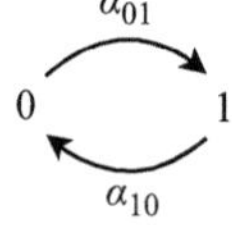

Figure 9.6
Two-state Markov model for a membrane channel.

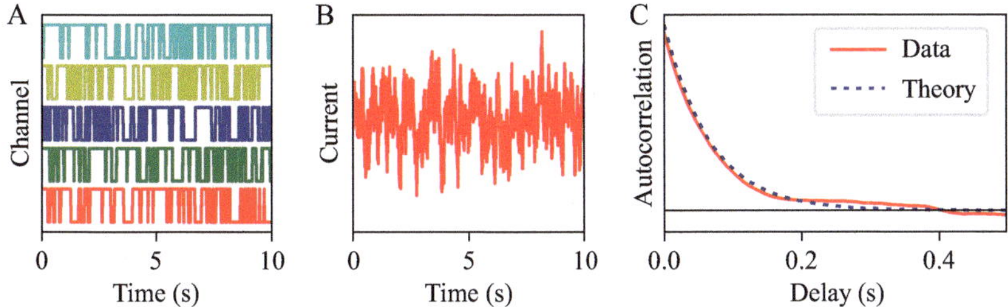

Figure 9.7
Fluctuation analysis of neuronal membrane current. A: Simulation of current through several channels that switch independently between open and closed states. B: The sum of many such single-channel currents. C: The autocorrelation function of the membrane current.

$$c(\tau) = i^2 \frac{\alpha_{01}}{\alpha_{01} + \alpha_{10}} \cdot \left(\frac{\alpha_{01}}{\alpha_{01} + \alpha_{10}} + \frac{\alpha_{10}}{\alpha_{01} + \alpha_{10}} e^{-(\alpha_{01} + \alpha_{10})\tau} \right). \tag{9.19}$$

Because the channels all operate independent of each other, the correlation function of their sum is simply the sum of the correlation functions (see section 6.5.6). So the autocorrelation of the full membrane current is

$$C(\tau) = nc(\tau) = ni^2 \frac{\alpha_{01}}{\alpha_{01} + \alpha_{10}} \cdot \left(\frac{\alpha_{01}}{\alpha_{01} + \alpha_{10}} + \frac{\alpha_{10}}{\alpha_{01} + \alpha_{10}} e^{-(\alpha_{01} + \alpha_{10})\tau} \right). \tag{9.20}$$

By measuring the autocorrelation, one can therefore derive two combinations of the unknown parameters: The value at $\tau = 0$ yields $ni^2 \frac{\alpha_{01}}{\alpha_{01} + \alpha_{10}}$. And the decay constant yields $\alpha_{01} + \alpha_{10}$.

In addition, we can measure the mean current, which is

$$\bar{I} = nip, \tag{9.21}$$

where $p = \frac{\alpha_{01}}{\alpha_{01} + \alpha_{10}}$ is the average open probability of the channel.

That still leaves one missing degree of freedom. Under suitable conditions, one can manipulate the open probability p. For example, for some types of ion channels, p is controlled by the membrane voltage or by the concentration of an external ligand. Suppose that we can manipulate p over some range, but without knowing what it is. For each condition, we measure the mean current $\bar{I}$ and the variance of the current $\mathrm{Var}[I]$. The microscopic model predicts that

$$\mathrm{Var}[I] = C(0) = ni^2 p(1-p) = \bar{I} \left(i - \frac{\bar{I}}{n} \right). \tag{9.22}$$

So from a plot of the variance $\mathrm{Var}[I]$ versus the mean $\bar{I}$, one can derive both i and n. Together with this autocorrelation analysis, this finally yields all four of the microscopic parameters.

Other versions of fluctuation analysis follow a similar theme. In each case, one makes a model of the microscopic processes and predicts how those processes determine the mean and fluctuations in the macroscopic variables.

For further information, Neher and Stevens (1977) reviews these methods as applied to neuronal membranes, and Digman and Gratton (2011) is a more recent review of fluctuation analysis in optical signals.

9.3　Population and Quantitative Genetics

Population genetics is the study of genetic variations in populations and the forces that drive them. Quantitative genetics specifically focuses on the study of *quantitative* traits, or traits that vary continuously within a population (e.g. height). Historically, these two fields have had close interactions with mathematics and statistics. Here, we explore some well-known results from these fields to study these connections. Our aim is not to cover every detail comprehensively, but rather to offer concise examples demonstrating how the mathematics developed in earlier chapters is relevant and applicable in this context.

9.3.1　What Are the Frequencies of Different Genotypes in a Population?

Consider an ideal[4] population with the following properties:

- Diploid (cells carry two copies of each chromosome, unless they are gametes)
- Monoecious (hermaphroditic, or both the male and female sex cells are produced by a single individual, with the possibility of self-fertilization)
- Infinite size and random mating
- No mutation, selection, or migration
- Discrete, nonoverlapping generations

We are interested in genetic variation at a single *locus* (a region of a chromosome). In the simplest case, the gene at the locus has two alleles, which we will call B and R. Since the organism is diploid, this implies that an individual organism has one of three possible genotypes: BB, BR, and RR. Let us denote the frequencies of the B and R alleles p and $q = 1 - p$, respectively. What are the frequencies of all possible genotypes in the next generation? Diploid organisms produce gametes (sex cells) that are haploid (contains one copy of the chromosome)—each having B or R. Because mating is random and the organism is monoecious, an individual in the next generation essentially comes from a process in which two gametes are drawn from a large pool that contains gametes from every individual in the parental generation. Put another way, we can think of the population as an urn full of balls, p of which are blue (B) and $q = 1 - p$ red (R). We then ask: What are the probabilities of drawing with replacement (1) two blue balls (BB); (2) one blue ball and one red ball (BR); and (3) two red balls (RR), given two draws? The answer is given by the binomial distribution (section 6.3.6):

$$P(BB) = \binom{2}{2} p^2 q^0 = p^2$$

$$P(BR) = \binom{2}{1} p^1 q^1 = 2pq \tag{9.23}$$

$$P(RR) = \binom{2}{0} p^0 q^2 = q^2.$$

---

4. We say ideal, but we often find that these conditions apply well enough to real populations.

Indeed, these probabilities add up to 1: $p^2 + 2pq + q^2 = (p+q)^2 = 1$.

What happens to the genotypic frequencies in the next generation? Again, the assumptions we have listed here about the population lead us to the same calculation, so the genotypic frequencies remain unchanged. This situation is known as the *Hardy-Weinberg equilibrium.* Deviation from this equilibrium would imply that at least one of the assumptions is violated. For example, the presence of selection, mutation, migration, or nonrandom mating could drive the genotypic frequencies to change over time.

What if the gene had three alleles (A, B, C)—what are the frequencies of the possible genotypes under these assumptions? In this case, we use the multinomial distribution, a generalization of the binomial. Again, the intuition is that we are reaching into an urn, this time filled with balls of three different colors, and drawing out two balls with replacements.

$$P(AA) = \frac{2!}{2!0!0!}P(A)^2 = P(A)^2$$

$$P(AB) = \frac{2!}{1!1!0!}P(A)P(B) = 2P(A)P(B)$$

$$P(AC) = \frac{2!}{1!0!1!}P(A)P(C) = 2P(A)P(C) \tag{9.24}$$

$$P(BB) = \frac{2!}{0!2!0!}P(B)^2 = P(B)^2$$

$$P(BC) = \frac{2!}{0!1!1!}P(B)P(C) = 2P(B)P(C)$$

$$P(CC) = \frac{2!}{0!0!2!}P(C)^2 = P(C)^2.$$

Armed with this theoretical knowledge, you now go and empirically measure the frequencies of genotypes at a diallelic locus in a population (e.g. by sequencing a sample of the population). From the frequencies of genotypes, you can estimate the frequency of the alleles—for example, $P(B) = P(BB) + \frac{1}{2}P(BR)$. You can then calculate what the genotypic frequencies *should* be under the assumptions of the Hardy-Weinberg equilibrium, a common first step in many analyses like genome-wide association studies (GWAS). How would you test if the measured genotypic frequencies deviate significantly from the expected values? A popular method is to apply the chi-square test (section 7.6). Specifically, sum the normalized squared difference between measured and expected frequencies and it should follow the chi-square distribution with 2 degrees of freedom (since we have three genotypes whose frequencies must sum to 1). Then compute the p-value to estimate the probability that you would get a result at least this extreme; if this is larger than a predefined threshold, you would not be able to reject the null hypothesis (in this case, that the population is in Hardy-Weinberg equilibrium).

9.3.2 What Are the Effects of Genetic Drift?

We now remove the assumption that the population is infinitely large; instead, there are N individuals ($2N$ copies of the gene in total). This leads to a phenomenon known as *genetic drift,* or the change in the allele frequencies due to random sampling. We will explore a simple model called the *Wright-Fisher model* to understand this phenomenon.

The frequency of the B allele in the population at generation t is $p = i/2N$. What is the probability that this will change to $j/2N$ in the next generation, $t+1$? Again, this is like sampling with replacement from an urn, this time with $2N$ balls, i of which are blue (B) and $2N - i$ red (R). The probability that j out of the $2N$ draws will be blue is

$$P_{t+1}(j) = \binom{2N}{j}(i/2N)^j(1 - i/2N)^{2N-j}. \tag{9.25}$$

Given finite sampling, it is clear that the allele frequency can now change just from random fluctuations. What happens as time goes by? Intuitively, the allele frequencies would fluctuate around the initial frequency p, resembling a random walk (section 8.1). The longer we wait, the more opportunity it will have to drift far from p. Eventually, it may reach 0 or 1, at which point it is said to be *fixed*, as there is no more variation in the population—all individuals are now homozygous for one allele. The following argument uses the concept of the *inbreeding coefficient* to make this intuition more precise.

We define the inbreeding coefficient (f_t) at generation t as the probability of the two alleles at the locus to have descended from a common allele. For example, a population with two individuals has four copies of the gene, so the inbreeding coefficient of the next generation is 1/4. The inbreeding coefficient follows this recursive relationship:

$$f_t = \frac{1}{2N} + \left(1 - \frac{1}{2N}\right)f_{t-1}. \tag{9.26}$$

Here, the first term is the probability of having two gametes containing the same allele from the parental generation (i.e. one just before current). The second term is the complement of the first, multiplied by the inbreeding coefficient from the previous generation. Put differently, the first term corresponds to the inbreeding that occurs in this generation, and the second term corresponds to the inbreeding accumulated from previous generations. Subtracting 1 from both sides of equation (9.26) and multiplying by -1, we get

$$
\begin{aligned}
1 - f_t &= 1 - \frac{1}{2N} - \left(1 - \frac{1}{2N}\right)f_{t-1} \\
&= \left(1 - \frac{1}{2N}\right)(1 - f_{t-1}) \\
&= \left(1 - \frac{1}{2N}\right)^t,
\end{aligned}
\tag{9.27}
$$

with the last line following from assumption that the initial inbreeding coefficient (f_0) is 0. Note that if there is inbreeding (i.e. two gametes containing the same copy of the gene unite in the offspring), then the offspring is by definition homozygous. Thus the expected frequency of heterozygotes (assuming that the frequency of one allele is p) at generation t is $2p(1-p)(1-f_t)$ (can you see why?). Now equation (9.27) implies that $1 - f_t \to 0$ as $t \to \infty$, so $2p(1-p)(1-f) \to 0$ as well. In other words, given enough time, genetic drift will cause heterozygotes to disappear from the population and an allele to be fixed. The smaller N, the faster this happens. Figure 9.8 demonstrates this with simulations.

There are other interesting questions to be asked about such a model. For example, what is the probability that an allele will be fixed, given that its initial frequency is

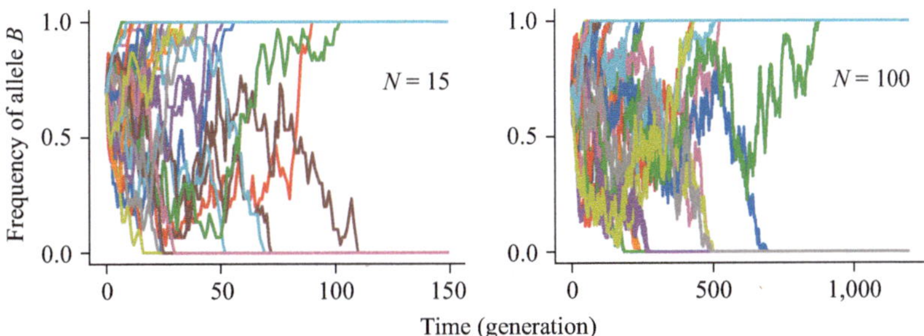

Figure 9.8
Genetic drift leads to the fixation of an allele. A total of 30 trajectories, all starting from the same initial frequency of allele B (0.7), are simulated in a small (left, $N = 15$) and a larger (right, $N = 100$) population. Over time, the frequency of allele B goes to either 0 or 1, and this happens faster for the smaller population (note the different scaling in the time axis).

$i/2N$? On average, how long does it take for an allele to be fixed? These questions are explored in exercise 10.24.

9.3.3 What Are the Effects of Natural Selection?

We will now eliminate another one of our assumptions and consider the effects of selection, which drives differential reproductive success based on the phenotype. Specifically, we will study how applying selection to a population changes its mean phenotype over time.

The rough picture is illustrated in figure 9.9. In the main panel, we plot the phenotype z of parents[5] against that of their offspring. Here, z is a quantitative phenotype that varies continuously (e.g., height). We assume that the two populations have the same mean phenotype μ. Furthermore, we assume that the phenotype is *heritable*—the variation in the phenotype is influenced by genetics, so parents with above-average z tend to produce offspring with above-average z, for example. Although this relationship can take many functional forms, here we will look at the case in which it is roughly linear.

9.3.3.1 Selection in the parental population   We will first examine the effect of selection in the parental population. Let us define some terminology. $P(z)$ is the frequency of phenotype z. *Directional selection differential S* is the change in the mean phenotype of the population after selection within the same generation. If the mean phenotype before selection is μ and the mean phenotype after selection is μ_s, then $S = \mu_s - \mu$ (figure 9.9). $W(z)$ is *fitness*, which we will simply define as the probability that an individual with phenotype z will survive the selection pressure. This can take many forms, and figure 9.9 illustrates the case in which it is a simple step function. The mean individual fitness in the population is $\bar{W} = \mathrm{E}[W(z)] = \int P(z)W(z)\mathrm{d}z$. When this is used to normalize fitness, we get *relative fitness $w(z) = W(z)/\bar{W}$*.

What is $P_s(z)$, or the frequency of phenotype z after selection (i.e. the orange distribution in the bottom panel of figure 9.9)? Clearly, $P_s(z) \propto P(z)W(z)$. With proper normalization,

$$P_s(z) = P(z)W(z) / \int P(z)W(z)\mathrm{d}z$$

---

5. Specifically, the midparent phenotype (the average phenotype of the parents).

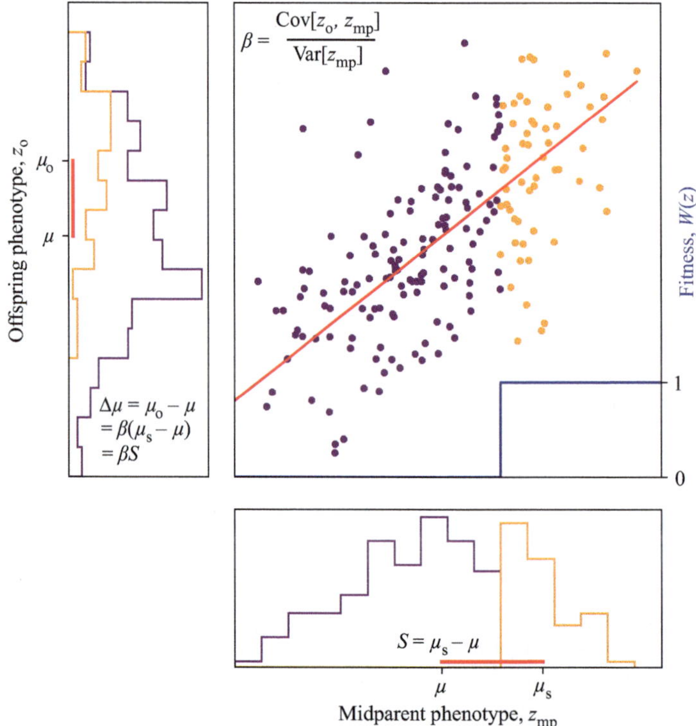

Midparent phenotype, z_{mp}

Figure 9.9
Selection drives intergenerational change in the mean phenotype.

$$= P(z)W(z)/\bar{W}$$

$$= P(z)w(z). \tag{9.28}$$

From this, we can see that μ_s, the mean phenotype after selection, is

$$\mu_s = \int zP_s(z)\mathrm{d}z = \int zw(z)P(z)\mathrm{d}z = \mathrm{E}[zw(z)]. \tag{9.29}$$

Thus

$$S = \mu_s - \mu$$

$$= \mathrm{E}[zw(z)] - \mathrm{E}[z]\,\mathrm{E}[w(z)] \tag{9.30}$$

$$= \mathrm{Cov}[z, w(z)].$$

We use the fact that $\mu = \mathrm{E}[z]$ (by definition) and $\mathrm{E}[w(z)] = \int w(z)P(z)\mathrm{d}z = \int W(z)P(z)/\bar{W}$ $\mathrm{d}z = \bar{W}/\bar{W} = 1$ to go from the first to the second line. In other words, *the change in the mean phenotype due to selection is governed by how correlated the phenotype and the relative fitness are*. This is known as the *Price equation*.[6] One might think it is obvious, but the emphasis here is on the word *relative*—the kind of fitness that matters is that

---

6. The full version includes a non selection term as well.

which gives an individual an advantage over others in the population. Furthermore, this relationship holds regardless of the functional form of relative fitness $w(z)$—it can be any arbitrary nonlinear function of z.

9.3.3.2 Propagation of selection effects to the offspring population

After applying selection, we let the parental generation reproduce and examine what happens to the phenotype in the next generation. As described previously, plotting the phenotype of the parents on the horizontal axis and the phenotype of the offspring on the vertical axis yields a cloud of points that roughly resembles a line with nonzero slope β (figure 9.9). Our goal is to gain a theoretical understanding of β, which translates the phenotypic change S from the parental generation to $\Delta\mu$ in the offsping generation in what is known as the *breeder's equation*: $\Delta\mu = \beta S$ (left panel, figure 9.9).

The theory of linear regression (section 7.9) tells us that the MLE for slope β is S_{XY}/S_{XX}, or put in terms of statistical operators in our case, $\text{Cov}[z_{mp}, z_o]/\text{Var}[z_{mp}]$. How do these quantities relate to genes? To establish this connection, we specify a simple model that relates the genotype to the phenotype. Specifically,

$$z = G + E; \tag{9.31}$$

that is, phenotype z is a sum of the genotypic value G (a function of the genotype) and everything else (e.g. environment) represented by E. Here, we will assume that G and E are statistically independent. Although this is not true in general, it will let us get away with not specifying a third term for the interaction between G and E.

Based on this additive model, we write the following expressions for the mother, father, and offspring phenotypes, respectively:

$$\begin{aligned}
z_m &= G_m + E_m \\
z_f &= G_f + E_f \\
z_o &= G_m + G_f + E_o.
\end{aligned} \tag{9.32}$$

Our goal is to express $\text{Cov}[z_{mp}, z_o]$ and $\text{Var}[z_{mp}]$ in genetic terms from these expressions. To do so, we will make additional assumptions (1) no relationship between genetic and environmental components (i.e. $\text{Cov}[G_i, E_j] = 0$ for any $i, j \in \{m, f, o\}$); (2) no relationship between environmental components ($\text{Cov}[E_i, E_j] = 0$ for any $i, j \in \{m, f, o\}$); and (3) no relationship between parental genetic components (i.e. $\text{Cov}[G_m, G_f] = 0$).[7] Then we have

$$\begin{aligned}
\text{Cov}[z_{mp}, z_o] &= \text{Cov}\left[\frac{1}{2}(G_m + G_f + E_m + E_f), G_m + G_f + E_o\right] \\
&= \frac{1}{2}\text{Cov}[G_m + G_f + E_m + E_f, G_m + G_f + E_o] \\
&= \frac{1}{2}(\text{Var}[G_m] + \text{Var}[G_f])
\end{aligned} \tag{9.33}$$

---

7. Technically, this is not true if the mother and the father is the same organism, in which case it reduces to a variance of parental G, which may not be zero; but here, we will assume that such monoecious reproduction is negligible.

and

$$\mathrm{Var}[z_{\mathrm{mp}}] = \mathrm{Var}\left[\frac{z_{\mathrm{m}} + z_{\mathrm{f}}}{2}\right]$$

$$= \frac{1}{4}\,\mathrm{Var}[z_{\mathrm{m}} + z_{\mathrm{f}}] \qquad\qquad (9.34)$$

$$= \frac{1}{4}\left(\mathrm{Var}[z_{\mathrm{m}}] + \mathrm{Var}[z_{\mathrm{f}}]\right)$$

$$= \frac{1}{2}\,\mathrm{Var}[z].$$

Therefore,

$$\beta = \frac{\mathrm{Cov}[z_{\mathrm{mp}}, z_{\mathrm{o}}]}{\mathrm{Var}[z_{\mathrm{mp}}]}$$

$$= \frac{\mathrm{Var}[G_{\mathrm{m}}] + \mathrm{Var}[G_{\mathrm{f}}]}{\mathrm{Var}[z]}. \qquad\qquad (9.35)$$

Equation (9.35) is also known as *narrow-sense heritability* and is sometimes denoted as h^2. It is the ratio of the sum of the genetic variance in parents to the phenotypic variance.

Going back to the breeder's equation, we have

$$\Delta\mu = \beta S = \frac{\mathrm{Var}[G_{\mathrm{m}}] + \mathrm{Var}[G_{\mathrm{f}}]}{\mathrm{Var}[z]}\,\mathrm{Cov}[z, w(z)]. \qquad\qquad (9.36)$$

Therefore, the response to selection depends on the covariance between the phenotype and the relative fitness (second term), and its efficiency of propagation to the next generation depends on the ratio of genetic to phenoytpic variance (first term).

9.3.4   Further Reading

Lynch, M. and Walsh, B. (1998). *Genetics and Analysis of Quantitative Traits.* Sinauer Associates.

Walsh, B. and Lynch, M. (2018). *Evolution and Selection of Quantitative Traits.* Oxford University Press.

Griffiths, A. J. F., Wessler, S. R., Lewontin, R. C., and Carroll, S. B. (2020). *Introduction to Genetic Analysis.* 12th ed. W. H. Freeman.

9.4   Vision at the Quantum Limit

9.4.1   Humans Can See Single Photons

Following the recognition in the early 1900s that light comes in tiny, discrete packages called "photons," the question arose whether humans can in fact see single photons. Hecht and colleagues (1942) found an answer through a clever method of analysis. The experiments were rather simple: They showed a human subject light flashes of varying intensity and asked each time whether the subject saw the flash or not. Suppose that at least n photons need to be absorbed by the retina to create the sensation of a flash. Then by dialing in a flash intensity close to that threshold, one should observe random fluctuations in the percept: The number of photons m delivered by each flash

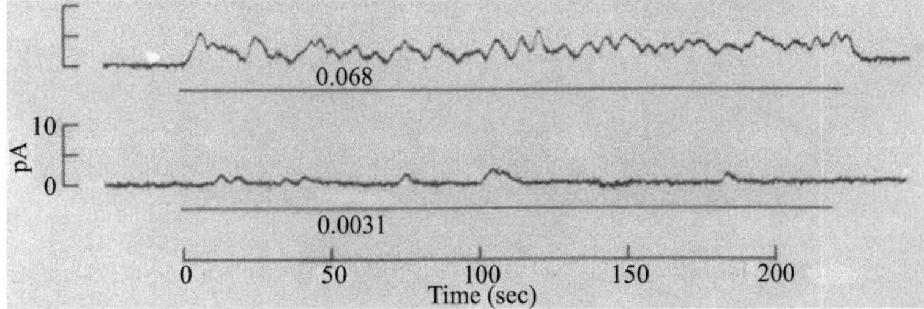

Figure 9.10
Current responses of a toad rod to prolonged very dim light. Lower trace: The small blips correspond to single photons. Upper trace: At 20x greater intensity, the photon responses merge into continuous shot noise. From figure 1 of Baylor et al. (1979).

should be distributed around the mean μ according to the Poisson distribution (section 6.3.7), $m \sim \mathrm{Poiss}(\mu)$. On some trials, one will get $m > n$, and on others, $m < n$. Hecht and colleagues found just such fluctuations, and by varying μ, showed that the results were consistent with a threshold of $n \approx 5$. Some years later, Sakitt (1972) improved on the methods to show that even single photons can be detected.

Note that the photon absorption in the retina causes just about the smallest event possible in a biological system—namely, rotation of one carbon bond in a small molecule called "retinal." This event can trigger a cascade of reactions that within 1 s takes over the whole human body, to the point where the subject jumps up and shouts, "I saw it!" The degree and speed of amplification implemented by the nervous system are truly remarkable!

From a purely methodological perspective, it is intriguing that Luria and Delbrück (1943) were working at the same time, and using the Poisson distribution to find new understanding in an entirely different field of biology.

9.4.2   The Light Response of Photoreceptor Cells
In 1979, Baylor and colleagues recorded electrical signals from rod photoreceptor cells in response to single photons (Baylor et al., 1979; Pugh, 2018). The responses were remarkably large: a transient change in the membrane current by about 5 percent of the maximal value, which lasted a few seconds. Flashes including two photons produced twice that response. To good approximation, the rod photoreceptor acts as a linear system, whose input/output relationship is completely characterized by the impulse response $h(t)$—namely, the response to a single photon (see section 3.1.4).

Under illumination with steady light, the photons absorbed by the photoreceptor act like a Poisson process: The interval between successive photons follows an exponential distribution (section 6.3.7), and the number of photons in any given interval follows a Poisson distribution (section 6.4.6). The superposition of their effects on the membrane current is a noisy-looking waveform in which one can no longer recognize the individual events (figure 9.10, top). Assuming that the events all superpose linearly, this becomes an example of **shot noise**, as explained in section 8.4.7. By measuring the power spectrum of the noise, one can derive the power spectrum of the individual single-photon event. Indeed, this prediction gave a beautiful match to the single-photon responses observed in isolation, confirming the notion of linear superposition (see figure 10 of Baylor et al. (1979)).

Clearly, this linear summation cannot persist over the whole range of light intensities. If 20 photons arrived at the same time, they would completely saturate the rod cell, and no further encoding of light intensity would be possible. By contrast, we know that animals can see very well over about 10 log units (a factor of 10^{10}) of light intensity, in conditions ranging from a moonless night to high noon (Meister and Tessier-Lavigne, 2021). Nature's answer to this challenge is an automatic gain control: a collection of cellular mechanisms collectively called "light adaptation." If the photoreceptor is exposed to steady light, its impulse response will adjust in both amplitude and duration (Schneeweis and Schnapf, 2000), such that variations around the new light level can again be encoded in a linear fashion (see also exercise 5.28).

Light adaptation is powerful and fast; in fact, some signature of the gain control can already be observed during the single photon response itself. Mathematical models with very different styles have been developed to capture this interaction between locally linear processing and dynamically nonlinear adaptation (Clark et al., 2013). There is also a remarkably detailed understanding of the biochemical processes that implement these gain-control mechanisms (Fain et al., 2001).

9.5 Neural Coding

9.5.1 The Linear-Nonlinear Model of Neural Encoding

How do neural signals convey information? As elaborated in section 13.6.3, most nerve cells produce stereotyped electrical pulses called **action potentials**. Every action potential has the same shape (to good approximation), so the message conveyed from one neuron to another must reside entirely in the timing of the pulses. What is the code by which that timing reflects this message?

This question of neural coding has been studied most effectively in sensory systems, where the experimenter can specify what the message is (a physical stimulus) and then measure how the neuron's spiking depends on the message. In some cases, the relationship seems straightforward. For example, so-called **touch receptors** respond

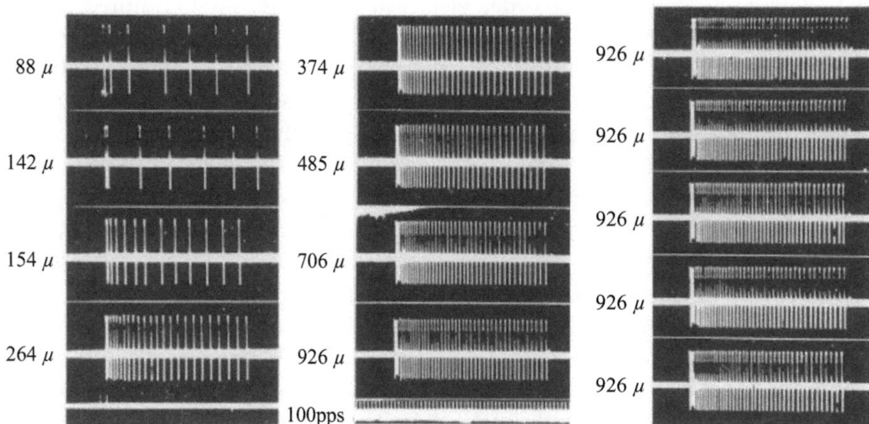

Figure 9.11
Action potentials from a touch receptor in the skin of a cat. Each trace spans 0.55 s, with the indentation applied for 0.4 s. The depth of indentation is listed to the left of each trace. From Werner and Mountcastle (1965).

to pressure on the skin (figure 9.11). With a small skin indentation, the neuron produces an almost perfectly periodic train of spikes. With more indentation, the spike rate increases. One can draw up a lookup table of spike rate versus skin indentation, and that would be a first sketch of the neural code for that receptor cell (Werner and Mountcastle, 1965).

Things get more complicated when the stimulus varies in time, such as when the skin rubs over a textured surface. The receptor cell has some memory of indentation, in part because the skin deforms plastically. So now the spike time depends not only on the instantaneous skin indentation, but also on the recent history. Another complication arises when the stimulus extends in space, such as when the skin gets indented at multiple places: Now there are multiple stimulus variables. Finally, one often finds that the neural response is somewhat stochastic: Even when the experimenter repeats the exact same physical stimulus twice, the resulting spike train may differ. Whether this reflects an irreducibly stochastic influence—like the stochastic arrival of single photons discussed in section 9.4.2—or a deterministic contribution from uncontrolled influences on the neuron's activity is often unclear.

With these extensions, a general formulation of a sensory neuron's response is as a stochastic point process. The instantaneous firing probability per unit time R is a function of the history of the stimulus and the history of the prior spike train; namely,

$$\text{Probability that spike } k \text{ happens in } [t, t+dt] = R\left(s(t' < t), \{t_{i<k}\}\right) \, dt. \qquad (9.37)$$

For practical purposes, one needs to parametrize R somehow and then fit the parameters to experimental results. A popular approach to the response function R is the so-called **linear-nonlinear model**, often abbreviated as "LN model." This model (figure 9.12) makes the following simplifying assumptions:

- The instantaneous rate $R(t)$ is independent of the prior history of the spike train.
- R depends on the preceding stimulus history only through a linear functional of the stimulus:

$$R(t) = N\left(\int_{t'=-\infty}^{t} s(t')L(t-t')dt'\right). \qquad (9.38)$$

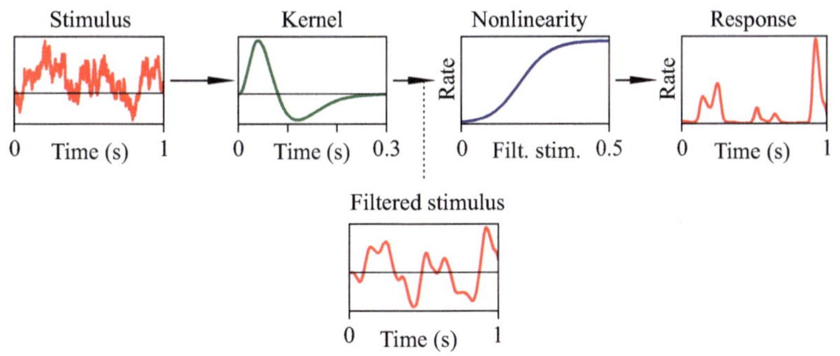

Figure 9.12
Example of an LN model defined by the kernel and the nonlinearity. Red traces show examples of a stimulus $s(t)$, the filtered stimulus $f(t)$, and the resulting firing rate $R(t)$.

The convolution of $s(t)$ with $L(t)$ can be seen as a linear filtering of the stimulus (see sections 3.1.5 and 4.5):

$$f(t) = \int_{t'=-\infty}^{t} s(t')L(t-t')\mathrm{d}t' \tag{9.39}$$

where $L(t)$ is the impulse response of the filter, sometimes called the **kernel** of the model (section 3.1.4). That filtered stimulus gets passed through a function $N(f)$, commonly called the **nonlinearity.** Typically, $N(f)$ is a sigmoid function that tends to zero for small f (firing rates cannot be negative) and saturates at some maximum firing rate at large f.

Suppose now that we have measured the neuron's firing rate $R(t)$, such as by repeating the same stimulus $s(t)$ in many trials, counting the spikes in short time bins, and averaging over all the trials. How can we estimate the kernel $L(t)$ and nonlinearity $N(f)$ of an LN model to fit these data?

9.5.1.1 Reverse correlation analysis A popular approach to doing this is called **reverse correlation** or **spike-triggered averaging.** This takes advantage of the experimenter's ability to freely design the stimulus $s(t)$. To do this, divide time into discrete bins of length Δt and define s_t to be the stimulus presented in the interval $[t\Delta t, (t+1)\Delta t]$. Then proceed with the following steps:

- Suppose that the stimulus time series s_t is Gaussian (i.e. all the s_t are independent draws from a normal distribution):

$$s_t \sim \mathcal{N}(0, \sigma). \tag{9.40}$$

- Suppose that the response time series R_t actually follows the LN-model—namely,[8]

$$R_t = N\left(\sum_{k=1}^{m} s_{t-k}L_k\right). \tag{9.41}$$

It helps to express this in terms of a scalar product:

$$R_t = N\left(\mathbf{s}_t \cdot \mathbf{L}\right), \tag{9.42}$$

where $\mathbf{s}_t = [s_{t-1}, \ldots, s_{t-m}]^{\top}$ is an m-dimensional vector with the stimulus history preceding time t, and $\mathbf{L} = [L_1, \ldots, L_m]^{\top}$ is an m-dimensional vector representing the kernel weights applied to those stimulus coefficients.

- Then the kernel **L** is proportional to the **reverse correlation C**:

$$\mathbf{L} \propto \mathbf{C}, \tag{9.43}$$

---

8. Note that this is a discrete convolution, the analog of equation (9.39) for discrete time, where m is the length of the filter and $m\Delta t$ is the length of the stimulus history that contributes to the response.

where

$$C = \frac{1}{n} \sum_{t=1}^{n} s_t R_t, \tag{9.44}$$

and n is the number of bins that cover the whole experiment, which lasts a total of $n\Delta t$ seconds.

If R_t is a binned spike train, with values of 0 (no spike) or 1 (spike), then the reverse correlation can also be seen as the **spike-triggered average stimulus**—namely, the average stimulus preceding a spike:

$$C = \langle s_t \rangle_{R_t=1}. \tag{9.45}$$

There is a simple geometric proof for the result in equation (9.43) that relies only on symmetry. By inserting equation (9.42) into equation (9.44), we get

$$C = \frac{1}{n} \sum_{t=1}^{n} s_t N (s_t \cdot L). \tag{9.46}$$

The quantity on the left is a vector. On the right, the only vectors are s_t and L. But s_t has a spherically symmetric probability distribution:

$$P(s_t) \propto \exp\left(-\frac{|s_t|^2}{2\sigma^2} \right). \tag{9.47}$$

This distribution has no preferred direction in the m-dimensional space. So the only remaining vector that points in a specific direction is L. Therefore, C must point in the same direction as L.

With this result, one can simply use equation (9.43) to compute the reverse correlation C from the data, set $L = C$, compute the filtered stimulus $f_t = s_t \cdot C$, and then plot R_t against f_t to obtain the nonlinearity function $N(f)$. You can explore this approach in exercise 10.26.

Note that the argument relies entirely on spherical symmetry of the stimulus distribution $P(s_t)$. This is one major reason why experimenters often draw sensory stimuli from a Gaussian ensemble. Of course, there are more intricate methods of fitting the LN model that also apply to non spherical stimulus distributions.

For more on this and related topics, see Chichilnisky (2001), Paninski (2004), and Pitkow and Meister (2014).

9.5.2 The Information Capacity of a Neuronal Spike Train

How much information can be conveyed by a spiking neuron? As discussed in section 13.6, a train of spikes from a neuron can be idealized as a point process. All the spikes are identical events, and the information is carried entirely by the time of arrival of the events $\{t_i\}$. How much information can be transmitted this way?

Clearly, the information capacity of this channel (see section 8.6.4) depends on the time resolution of the system (namely, the encoder mechanisms that produce the spike train and the decoder mechanisms that receive it). Suppose that the encoder and the decoder have a temporal resolution of Δt, meaning that the encoder can reliably place spikes to within Δt, and the decoder can reliably distinguish spike times to within Δt. If Δt is very small, then a single spike time can take on many distinguishable values, thus conveying a great deal of information.

A second constraint on the capacity is the average rate of spiking, r. Obviously, if there are more spikes per unit time, they will convey more information per unit time.

For a simple approximation, let us quantize time into bins of width Δt. The system cannot interpret the position of a spike within a bin, so the only relevant coding variable is the number of spikes per bin:

$$n_i = \text{number of spikes in the interval } [i\Delta t, (i+1)\Delta t]. \tag{9.48}$$

Now consider the case where the firing rate is low compared to the time resolution, such that

$$r\Delta t \ll 1. \tag{9.49}$$

Then each time bin contains either 0 or 1 spikes (to good approximation), so the system acts like a Bernoulli variable that uses the symbol 1 with probablity $p = r\Delta t$. That binary channel is used optimally if subsequent symbols are statistically independent, in which case the information rate per unit time becomes (see example 8.7):

$$C = \frac{-1}{\Delta t} \left(p \log_2 p + (1-p) \log_2 (1-p) \right), \quad p = r\Delta t. \tag{9.50}$$

For small p, this is

$$C \approx \frac{-1}{\Delta t} p \log_2 p = -r \log_2 r\Delta t. \tag{9.51}$$

As expected, the capacity increases with the firing rate r and with the timing precision $1/\Delta t$.

What is the statistical nature of such a spike train that achieves maximal information rate? We had to assume that successive symbols n_i are statistically independent, meaning that spikes are produced independently with probability p during each small time interval Δt. This defines a homogeneous Poisson process with intensity r (see section 8.4.3). So a neuronal spike train used to full capacity should look like the most random possible point process.

This is an instance of a more general principle that states, "Pure information looks like noise." When a channel is used to full capacity, its symbol stream should offer no recognizable patterns anymore. If such a pattern existed, one could exploit it to recode the symbol stream and communicate the same information with fewer symbols.

Now we can add some biologically realistic numbers. A timing precision of $\Delta t = 1$ ms is not uncommon. For example, neurons in the visual system will produce spike times to that resolution when presented repeatedly with the same visual stimulus (Berry et al., 1997). However, there are also instances where a spike precision down to $\Delta t = 1\,\mu$s is important for the decoder of the neural message (Kawasaki et al., 1988). The mean firing rate of neurons varies from ~ 1 to ~ 100 spikes/s depending on the brain region. Note that all these cases satisfy the assumption $r\Delta t \ll 1$. So, taking the extremes of these two parameters, the capacity of neuronal spike trains ranges roughly from

$$C = 10\,\text{bits/s}, \quad \text{if} \quad r = 1\,/\text{s}, \ \Delta t = 1\,\text{ms} \tag{9.52}$$

to

$$C = 1300 \, \text{bits/s}, \quad \text{if} \quad r = 100 \, /\text{s}, \, \Delta t = 1 \, \mu\text{s}. \tag{9.53}$$

Alternatively, one can express the capacity in terms of information per spike:

$$C/r = -\log_2 r\Delta t = \begin{cases} 10 \, \text{bits/spike}, & \text{if } r = 1 \, /\text{s}, \, \Delta t = 1 \, \text{ms} \\ 13 \, \text{bits/spike}, & \text{if } r = 100 \, /\text{s}, \, \Delta t = 1 \, \mu\text{s}. \end{cases} \tag{9.54}$$

This time, the outcome in the two extreme cases is much more similar, and one can simply remember the **ballpark number of 10 bits/spike** for the capacity of neuronal spike trains.

9.5.3 The Information Rate of Neuronal Spike Trains

Now that we have a sense of how much information a spike train *could* convey, it becomes interesting to figure out how much it actually *does* convey. The best answers come from studies in sensory systems (namely, spike trains of neurons that are as close to sensory receptors as possible). The reason is that here, the investigator can specify the message that the neuron should convey, by presenting a sensory stimulus, like a movie to the eye, or a melody to the ear. Then one can record the spike train of a sensory neuron and ask how much information that carries about the sensory stimulus.

One popular method for performing this estimate uses the tool of neural decoding (Borst and Theunissen, 1999). Here, one constructs a decoder that converts the spike train back into an estimate of the sensory stimulus. The decoder is an algorithm— often no more complex than a linear transform—and the parameters of the decoder are optimized so the stimulus estimate comes as close as possible to the actual stimulus over many instantiations of the experiment. Then one measures signal and noise in the

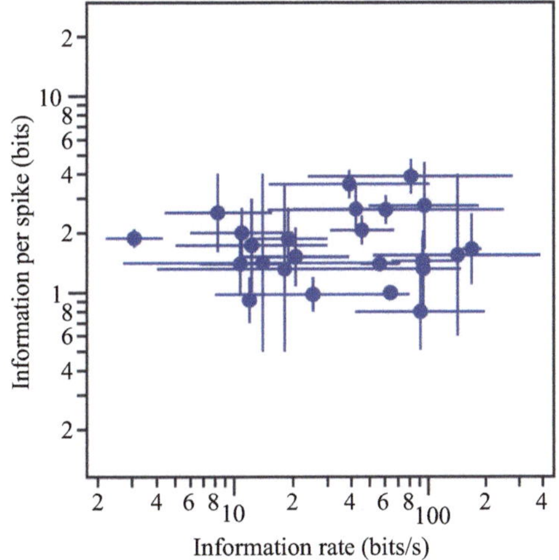

Figure 9.13

Information rate of neuronal spike trains from many studies in different species and brain areas: information per spike plotted against information per unit time. Bars indicate the range in each study.

stimulus estimate and converts that to an information rate (see section 8.6.4.2). Because there is no guarantee that the researcher found the best possible decoder, this typically yields a conservative estimate, meaning a lower bound on the actual information rate.

Figure 9.13 summarizes results from two dozen different studies, covering sensory neurons in insects, crustaceans, amphibians, birds, and mammals; some of these examples are taken from Borst and Theunissen (1999). It plots the measured information rate per spike against the information rate per unit time. The horizontal and vertical bars indicate the range reported in each study. Note that the information rates cluster rather tightly, in the range of 1 to 4 bits/spike, even though the firing rates vary by a factor of 100. If the capacity of a spike train is 10 bits/spike (see section 9.5.2), then these neurons transmit at about 10–40 percent of capacity, or equivalently with a redundancy of 60–90 percent.

Obtaining information rates that are this high requires some effort in designing the stimulus. Obviously, it needs to be matched to the capabilities of the sensory organs and receptors. For example, a melody in the ultrasound range will not be well encoded by human auditory neurons. Often one finds that stimuli that carry strong behavioral meaning, such as the mating call of a frog, will elicit the highest information rates in sensory neurons (Rieke et al., 1995).

For more on this topic, read Rieke et al. (1997).

Exercise 10.1 (Linda problem) Linda is 31 years old, single, outspoken, and very bright. She majored in philosophy. As a student, she was deeply concerned with issues of discrimination and social justice, and she also participated in anti-nuclear demonstrations. Which is more probable? Why?

(a) Linda is a bank teller.
(b) Linda is a bank teller and is active in the feminist movement.

Exercise 10.2 (The case of Sally Clark) Sally Clark, a mother in England, was convicted of murdering her two elder sons in 1999. Her first son died suddenly within a few weeks of his birth, and two years later, her second died in a similar manner. A month later, she was arrested and tried for both deaths. The prosecution's case relied significantly on the statistical evidence presented by a pediatrician, Professor Sir Roy Meadow. From Wikipedia:

Professor Sir Meadow stated in evidence as an expert witness that "one sudden infant death in a family is a tragedy, two is suspicious and three is murder unless proven otherwise" (Meadow's law). He claimed that, for an affluent non-smoking family like the Clarks, the probability of a single cot death was 1 in 8,543, so the probability of two cot deaths in the same family was around "1 in 73 million" (8543 × 8543). Given that there are around 700,000 live births in Britain each year, Meadow argued that a double cot death would be expected to occur once every hundred years.

Do you agree with Professor Sir Meadow? Why?

Exercise 10.3 (Boys and girls) For (a) and (b) below, assume that boys and girls are equally frequent.

(a) Family Jones has two children. One is a boy. What is the probability that the other child is a girl?
(b) Family Miller has two children. The older one is a girl. What is the probability that the other child is a boy?
(c) A more careful study is done on families with two children, and the following (imaginary!) probabilities are found for their gender distribution, with the older child listed first:

$$P(BB) = b = 0.30 \quad P(BG) = m = 0.21.$$

$$P(GB) = s = 0.23 \quad P(GG) = g = 0.26.$$

Given these statistics, what would be the revised answers to (a) and (b)?

Exercise 10.4 (Hair and eye color) There are 43 people in your class. Of these, 15 have blue eyes. Out of these 15, 10 are blonde. The probability that a student has blue eyes given that the student is blonde is 2/3. What is the probability that a non-blue-eyed person is not blonde?

Exercise 10.5 (The binomial distribution is normalized) Convince yourself that this is true; namely, that

$$\sum_{m=0}^{n} P(m) = 1,$$

where

$$P(m) = \binom{n}{m} p^m (1-p)^{n-m}.$$

Exercise 10.6 (The variance is the mean of the square minus the square of the mean) Convince yourself that the statement in this exercise is true, as claimed in equation (6.14).

Exercise 10.7 (The mean and variance of the binomial distribution) Convince yourself that the mean and variance of the binomial distribution are as claimed in equation (6.23).

Exercise 10.8 (Run lengths) Consider successive flips of a coin that delivers "1" with probability p. Define n as the number of consecutive flips of "0" before the first appearance of "1." This is sometimes called the "length of a run" of zeroes. We are interested in the probability distribution of this run length. It plays a big role in sports statistics, as well as many more other human endeavors, such as assessing the time between failures of an airplane or the distance between random mutations on a chromosome.

(a) Show that

$$P(n) = p(1-p)^n.$$

This is called the "geometric distribution." Sketch the function. What is the most probable run length?

(b) Confirm that $P(n)$ is normalized; that is, $\sum_{n=0}^{\infty} P(n) = 1$.

(c) Show that the average run length is

$$\langle n \rangle = \sum_{n=0}^{\infty} nP(n) = \frac{1-p}{p}. \tag{10.1}$$

Hint: Explain why

$$\sum_{n=0}^{\infty} n(1-p)^n = -(1-p)\frac{\mathrm{d}}{\mathrm{d}p} \sum_{n=0}^{\infty} (1-p)^n$$

and then use what you know about $\sum_{n=0}^{\infty}(1-p)^n$. This is a popular trick for figuring out certain infinite sums starting from a series with a known sum.

(d) Suppose that a chromosome has been exposed to point mutations with probability p per base. According to equation (10.1), the mean distance between mutations is $(1-p)/p$. Imagine the chromosome strung out in a line left to right, and pick a random location on the chromosome. What is the mean distance to the nearest mutation on the right? What is the mean distance to the nearest mutation on the left? What is the mean distance between the nearest mutations on the left and on the right? Is there a paradox, and if so, how do you resolve it?

Exercise 10.9 (Uncorrelated is not independent) If two random variables are statistically independent, then they have zero covariance (i.e. they are uncorrelated). However, the reverse is not true. Construct an example of two random variables that are uncorrelated but not independent. Specifically, constrain the two variables X and Y to be discrete and ternary, meaning that they can only take one of three possible values.

(a) Find a probability distribution $P(X, Y)$ under which X and Y are uncorrelated but statistically dependent. Demonstrate both properties.

(b) Bonus question: Is this possible if X and Y are binary variables? Why or why not?

Exercise 10.10 (Covariance) Suppose that you have two statistically independent random variables: X_1 and X_2. They have means μ_1 and μ_2 respectively, and standard deviations σ_1 and σ_2.

(a) What is the covariance of X_1 and X_2?

(b) Now define two new random variables Y_1 and Y_2 as the sum and the difference of X_1 and X_2, respectively. What are the means and standard deviations of Y_1 and Y_2?

(c) What is their covariance?

(d) What is their correlation?

Exercise 10.11 (Games to win) Two basketball teams, A and B, play a series of games that ends when one team wins 2 games (draws are not allowed). Team A wins each game with probability p. What is the expected number of games? For what value of p is this maximized?

Exercise 10.12 (Polya's urn) A classic problem, known as "Polya's urn," is relevant to genetic drift in a population with a selectively neutral polymorphic trait.

(a) **A growing population with two alleles of a gene:** Consider an urn with initially one black and one white ball in it. A ball is drawn at random and replaced by two balls of the same color as the one drawn. This process is repeated many times.
  (i) Simulate this process using a programming package.
  (ii) Plot the fraction of black balls as a function of time (the number of draws from the urn). Try this several times.
  (iii) Discuss your observations. For example: What happens at early times? What happens at late times? Does each simulation lead to the same result?

(b) **A population of fixed size:** The urn now starts with N black and N white balls. Two balls are drawn randomly. If they are the same color, they are put back in. If they are different colors, one of the colors is selected randomly, and two balls of that color are put back in. This process is repeated many times, with the number of balls in the urn remaining at $2N$.

  (i) Simulate this process using a programming package.

  (ii) Plot the fraction of black balls as a function of time (the number of draws from the urn). Try this several times, starting with $N = 10$ or with $N = 20$.

  (iii) Discuss your observations and contrast with part (a). Speculate on the biological significance of this process, particularly the probability of one of the colors going "extinct" after some time.

Exercise 10.13 (MLE for the binomial distribution) Convince yourself of the claim in equation (7.4); namely, that the maximum likelihood estimator (MLE) for the bias of a binomial variable is simply the fraction of ones in the data sample.

Exercise 10.14 (How large can a tumor grow before it must develop a network of blood vessels?) Assume a spherical tumor of radius a embedded in surrounding tissue. It needs to consume oxygen at a rate of

$$R = 0.3\mu L/s/cm^3 \text{ of tissue.}$$

The concentration of oxygen at the surface of the tumor is

$$C_S = 7\mu L/s/cm^3$$

stabilized at that value by the surrounding vascularized tissue. Oxygen diffuses through the tissue of the tumor with a diffusion coefficient of

$$D = 2 \times 10^{-5} cm^2/s.$$

We want to find the largest radius a compatible with these constraints.

(a) Consider a sphere of tissue inside the tumor with radius $r < a$. What is the required total oxygen consumption per unit time inside the sphere? Translate that to a required flux of oxygen per unit time and area across the surface of the sphere, $J(r)$. Give your results in terms of the variables r and R.

(b) For a spherically symmetric diffusion problem in three dimensions, the flux $J(r)$ and concentration $C(r)$ depend only on radius r, and are related by

$$J(r) = -D\frac{\partial}{\partial r}C(r).$$

Using the results from (a), derive the form of the steady-state oxygen concentration $C(r)$ inside the tumor. Give your results in terms of r, R, and D.

(c) Find the radius a such that $C(a) = C_S$ at the surface of the sphere. From a simple constraint on $C(0)$, derive the maximal possible radius of the organism. First, give your answer in terms of R, D, and C_S. Then substitute the values of those parameters to answer the question in this exercise.

Exercise 10.15 (Inhomogeneous Poisson process) As discussed in section 8.4.4, this is a point process defined by the instantaneous event rate $\lambda(t)$:

$$\text{Probability of an event in the interval } [t, t+\mathrm{d}t] = \lambda(t)\mathrm{d}t.$$

(a) The text claims that this process is equivalent to a homogeneous process with intensity 1.0 in warp time τ, which is related to t by

$$\frac{\mathrm{d}\tau}{\mathrm{d}t} = \lambda(t).$$

Prove to yourself that this statement is correct.

(b) Using the warp time trick, simulate events from a Poisson process with a sinusoidally varying rate:

$$\lambda(t) = a(1 + m\sin\omega t)$$

(c) Generate many instantiations of this, and test the claim in the text, that the number of events in any given time interval $[t_a, t_b]$ is Poisson distributed (equation 8.92).

Exercise 10.16 (Mountain biking and big data) An aging mountainbiker, let's call him $\mathfrak{M}$, tracks his physical decline by riding the same trail every few weeks and recording the time that he takes to do it. He also compares his fitness to the general population by recording how many riders he passes and how many pass him. On a recent ride, $\mathfrak{M}$ passed 16 riders, and he was passed by zero riders.

(a) Propose an analysis to test how well $\mathfrak{M}$ does relative to the general population. Is he faster or slower? What confidence can we have in the result?

(b) While celebrating his ride over lunch, $\mathfrak{M}$ browses results from the population on Strava.com (Google it). His time on that specific ride ranked 4105 among 8729 total entries (remember that short times are better). Is there a conflict between the two results? List some potential confounding variables that hamper the comparison.

Exercise 10.17 (Linear regression with variable noise) Consider the problem of multiple regression in equation (7.98) for a data set where each measurement Y_i has a different associated uncertainty σ_i. Now the model is

$$Y_i = \sum_j b_j X_{ij} + E_i,$$

where

$$E_i \sim \mathcal{N}(0, \sigma_i^2),$$

and the values of σ_i^2 are known. Find the maximum likelihood estimators for the parameters b_j.

Exercise 10.18 (Principal component analysis derived) Prove the claim in section 8.5.1 that the first principal component of a data set is the eigenvector of the covariance matrix with the largest eigenvalue.

Here are some hints:

For a single principal component, equation (8.106) with $D=1$ simplifies to

$$\mathbf{x}_j = \mathbf{m} + c_j \mathbf{w} + \mathbf{e}_j. \tag{10.2}$$

The residual error of equation (8.107) is

$$E = \frac{1}{T} \sum_{j=1}^{T} \mathbf{e}_j^{\top} \mathbf{e}_j, \tag{10.3}$$

where

$$\mathbf{e}_j = \mathbf{x}_j - \mathbf{m} - c_j \mathbf{w}. \tag{10.4}$$

Now minimize equation (10.3) with respect to the parameters $\mathbf{m}$, $\mathbf{w}$, and c_j by differentiation:

$$\frac{\partial}{\partial \mathbf{m}} E = \frac{\partial}{\partial \mathbf{w}} E = \frac{\partial}{\partial c_j} E = 0.$$

Use this to prove that $\mathbf{w}$ must be an eigenvector of the covariance matrix $\mathbf{C}$, as seen in equation (8.109) with eigenvalue $\lambda > 0$.

Then show that the error is

$$E = V - \lambda, \tag{10.5}$$

where V is the total variance of the data set; see equation (8.113). Use this to argue why $\mathbf{w}$ should be the eigenvector with the largest eigenvalue.

Exercise 10.19 (COVID and vitamin D) During the COVID-19 pandemic, many theories were suggested to explain why the virus made some people sicker than others. One idea was that vitamin D deficiency weakens the immune system. A study was conducted (Entrenas Castillo et al., 2020) comparing the incidence of admission to the ICU (indicating serious illness) between a group that was given calcifediol (vitamin D3) and an untreated control group. The following results were obtained:

Outcome	Untreated	Treated with Calcifediol
Admitted to ICU	13	1
Not admitted to ICU	13	49

(a) Consider the null hypothesis that calcifediol has no influence on whether a patient will end up in the ICU. What is the probability that you would get this result or worse?

(b) Given your answer to part (a), do you think we should administer calcifediol to COVID patients?

Exercise 10.20 (Capacity of an asymmetric binary channel) Expanding on the symmetric binary channel in section 8.6.4.1, consider a binary channel with asymmetric errors. Specifically, $X=1$ gets turned into $Y=0$ with probability q, but $X=0$ is always faithfully transmitted into $Y=0$. What is the capacity of this channel?

(a) Assuming that $X=1$ gets used with probability p, what is the mutual information $I(X,Y)$?
(b) Maximize that mutual information with respect to p to find the capacity.

Exercise 10.21 (Capacity of the genetic code)

(a) What is the capacity of the human genome, namely, the maximal amount of information it could encode, assuming optimal use of all symbols? How does that compare to the capacity of a USB stick that you carry in your pocket?
(b) Within exons, we know that the genetic code maps triplets of bases into twenty amino acids plus a stop signal. Given these constraints, what is the capacity of the genetic code, again assuming optimal use of all symbols (bases and amino acids).

Exercise 10.22 (Capacity of the brain) It is commonly believed that memories are stored in the brain by the strength of synaptic connections between neurons.

(a) What is the memory capacity of the brain? Assume that there are 10^{14} synapses, and each synapse can take on 26 distinct strength values (Bartol et al., 2015).
(b) How does that compare to the capacity of the genome (Exercise 10.21)?
(c) Use the data-processing inequality to set a limit on the information stored in the brain as a result of genetically specified developmental processes. Where would the additional information come from?

Exercise 10.23 (Mutual information of two Gaussian variables) Show that if x and y are jointly Gaussian variables (see section 6.5.5) with a correlation coefficient (see (6.70)) of c, then their mutual information is

$$I(x,y) = -\frac{1}{2}\log_2\left(1-c^2\right).$$

Exercise 10.24 (Moran process for genetic drift) Consider a population of n bacteria, of which m are mutant. We will implement a single birth-death process that keeps the population size the same: Randomly choose one bacterium to divide, and another bacterium to die.

$$m(t+1) = \begin{cases} m(t)+1, & \text{with probability } \frac{m}{n}\frac{n-m}{n-1} \\[2mm] m(t), & \text{with probability } \frac{m}{n}\frac{m-1}{n-1} + \frac{n-m}{n}\frac{n-m-1}{n-1} \\[2mm] m(t)-1, & \text{with probability } \frac{n-m}{n}\frac{m}{n-1} \end{cases} \qquad (10.6)$$

(a) Explain equation (10.6). Hint: $\frac{m}{n}$ is the probability of picking a mutant cell for division, and $\frac{n-m}{n-1}$ is the probability of picking a nonmutant for death.
(b) As we learned in section 9.3, the probability that every bacterium will become a mutant at time t (i.e. $m(t)=n$) is nonzero. Find $R(m)$, the probability that starting with m mutants, the entire population will become mutant.

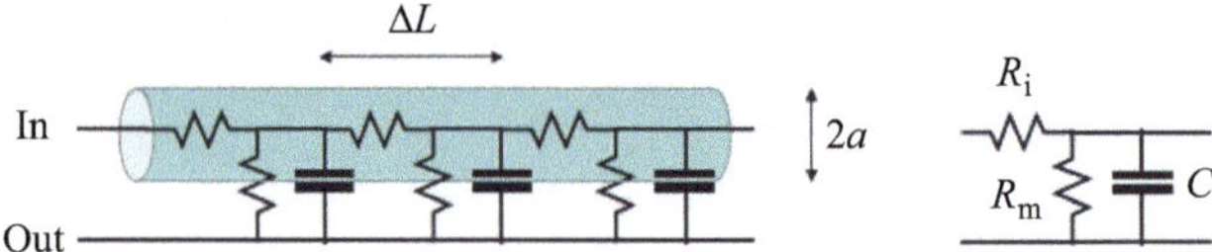

Short segment of length ΔL:

$$R_i = r_i \frac{\Delta L}{\pi a^2} \qquad R_m = \frac{r_m}{2\pi a\,\Delta L} \qquad C = c_m\,2\pi a\,\Delta L$$

Figure 10.1
Electric circuit of a passive dendritic cable.

Hint: Given $P(m) = \frac{m}{n}\frac{n-m}{n-1}$, note that $R(m) = P(m)R(m+1) + (1 - 2P(m))R(m) + P(m)R(m-1)$. Start with this to derive $R(m)$.

(c) We want to know how long it takes for the entire population to become mutant. Find $T(m)$, the mean number of steps to fixation at either $m = 0$ or $m = n$.

Hint: Start with $T(m) = 1 + P(m)T(m+1) + (1 - 2P(m))T(m) + P(m)T(m-1)$ and turn it into a differential equation by nothing that as $n \to \infty$, $x = m/n$ and $T(m+1) - 2T(m) + T(m-1) \to \frac{1}{n^2}\frac{d^2}{dx^2}T(x)$.

Exercise 10.25 (Dynamics of dendritic conduction) A passive dendritic cable can be represented by the electrical circuit of figure 10.1.

R_i represents internal axoplasmic resistance along a short stretch of length L. R_m is the membrane resistance over that same short stretch and C is the membrane capacitance. By using Kirchhoff's Laws for currents and voltages, and by letting ΔL go to zero, one can derive the **cable equation** for the voltage along the cable:

$$\lambda^2 \frac{\partial^2 V}{\partial x^2} = \tau \frac{\partial V}{\partial t} + V, \qquad \lambda = \sqrt{\frac{a\,r_m}{2\,r_i}}, \qquad \tau = r_m\,c_m.$$

(a) Derive this equation.

(b) What is its steady-state solution after all the voltages have settled so that $\frac{\partial V}{\partial t} = 0$? Interpret that voltage profile in words.

(c) Now we would like to know the full time-dependent solution. In particular, suppose the voltage at $t = 0$ is a brief pulse that injects some charge into the cable. How does that voltage pulse propagate down the cable?

(d) Analytical solution. Show that

$$V(x, t) = \frac{1}{\sqrt{4\pi t/\tau}} \exp\left(-\frac{\tau x^2}{4\lambda^2 t}\right) \exp\left(-\frac{t}{\tau}\right)$$

satisfies the cable equation. This is called the Green's function for the problem.

(e) Compute and plot the spatial profile $V(x, t)$ at some times t, and do the same for the time course $V(x, t)$ at some locations x. Describe in words what is going on.

(f) How fast does the pulse propagate down the dendrite? For a given location x, at what time t_{max} does the pulse reach a maximum? How does that t_{max} depend on location x?

For more on the topic, read Dayan and Abbott (2005).

Exercise 10.26 (The LN model applied to responses of a visual neuron) Here, we analyze the sensory responses of a visual neuron in the fly that is called "H1." This neuron is part of the flight control system of the fly. To stabilize its flight, the fly must measure its own rotation relative to the world. It does that largely by analyzing full-field motion on its retina, also called "optic flow." Neuron H1 seems to encode the optic flow in the horizontal direction. More information about the H1 neuron can be found here: https://en.wikipedia.org/wiki/H1_neuron.

We will work with a neural recording from neuron H1. In this experiment, the fly was fixed in the lab frame and presented with a visual pattern that moved back and forth in the horizontal direction. The velocity of that pattern was modulated with a Gaussian distribution. The data file contains a record of the stimulus velocity and a record of all the spikes fired by neuron H1.

Your task is to find out how neuron H1 encodes the velocity stimulus. In particular:

- What is the delay between the visual input and the neuron's response? This is a critical chracteristic of a real-time control system.
- What is the time resolution of the neuron? Are there fluctuations in the stimulus that are too fast for the neuron to resolve?

Apply the LN model to the neuron's responses, as discussed in section 9.5.1. In particular, compute the filter $L(t)$ using reverse correlation between the stimulus and response. The shape of $L(t)$ will offer answers to these questions.

Details:

(a) Load the data from the directory associated with this book (https://mitpress.mit .edu/mathinbio). You will get two arrays: One contains a time series of the stimulus (in arbitrary units), with one sample per ms. The other contains the corresponding spike train from H1 in bins that are 1 ms wide. Each bin contains at most 1 spike.

(b) Compute a histogram of the stimulus. Compare that to a Gaussian. Recall that a Gaussian stimulus is useful for interpretation of the reverse correlation.

(c) Compute the reverse correlation function. Because the response is sparse (i.e. only a small fraction of the time bins contain a spike), you may benefit from converting the time series to a point process, and using the "spike-triggered average" formulation of the reverse correlation. For a start, use a stimulus window of $[-0.5 \text{ s}, +0.5 \text{ s}]$ relative to the spike.

(d) Interpret the shape of the reverse correlation to answer these questions.

(e) Bonus question: Estimate the nonlinearity $N(f)$ in the LN model of this neuron's response.

For further information on this topic, see de Ruyter van Steveninck et al. (1988).

Exercise 10.27 (Poisson distribution of photon counts) Recall the study by Hecht et al. (1942), claiming that humans can see flashes of light containing just a few photons (see section 9.4.1). The entire argument relies on the assumption that flashes of light produced in the identical way will deliver a number of photons to the eye that follows the Poisson distribution. Explain why they could be so confident in that assumption.

(a) Imagine creating a flash of light by placing a shutter in front of a light bulb and opening the shutter for a very brief period. Explain why the number of photons that get through the shutter is Poisson-distributed.

(b) On its way to the subject's eye, lots of other things happen to this flash of light. It has to pass through various optical filters, bounce off mirrors, pass through apertures, get through the cornea of the eye to the retina, and ultimately be absorbed by molecules of retinal (a low-probability event). Explain why–regardless of the precise chain of events–the number of photons absorbed by retinal molecules is still Poisson-distributed.

III NONLINEAR DYNAMICS

11 Basics of Dynamical Systems

11.1 Motivation

In biology, change is the only constant. Organisms adapt, cells differentiate, and genes orchestrate a complex range of activities, all as part of an ever-evolving tapestry of life. Understading these dynamic processes often requires mathematical models. As we have seen in part I, linear models are frequently the first tool in the mathematical biologist's toolbox. They are exceptionally useful for describing systems whose responses are proportional to the input. In these cases, the mathematical treatment is often straightforward, allowing easy analytical solutions and interpretations. But biological systems are rarely so accommodating. Many respond nonlinearly, involve feedback loops, or are inherently cyclical. These reveal the limitations of linear models, which often cannot capture the complex dynamics that are fundamental to biological systems.

This is where nonlinear dynamics comes into play, providing a more accurate framework for describing systems where the output is not proportional to the input and the sum is more than its parts. The realm of applications for nonlinear dynamics in biology is vast, encompassing everything from population dynamics and the spread of disease to the firing patterns of neurons and the intricate interplay of transcription networks.

11.1.1 Oscillations in Biology

One example of commonly observed nonlinear dynamics in biology is oscillations. Biological oscillators cover a wide range of timescales, from a day in the case of circadian rhythms to a millisecond in the electric organ discharge of the electric fish (figure 11.1). What are the underlying mechanisms that cause these oscillations? How do they maintain their frequencies? Properly understanding these oscillations requires nonlinear models. In this part of the book, we will introduce fundamental concepts that underlie the mathematical analysis of nonlinear systems and discuss how these have been used to gain a quantitative understanding of entrainment in biological oscillators (section 13.5) and even to create synthetic oscillators in living cells (section 13.1), among other applications.

11.2 What Is a Dynamical System?

We will start by considering problems that are defined by a set of first-order differential equations of the type

$$\frac{\mathrm{d}}{\mathrm{d}t}x_i = f_i(x_1, \ldots, x_n),$$

$$(11.1)$$

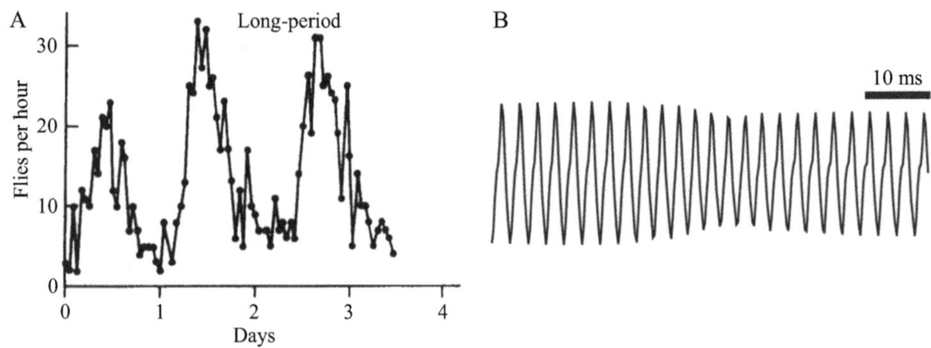

Figure 11.1
The timescale of oscillations in biology span many log units. (A) Eclosion rhythm in *Drosophila melanogaster*, adapted from Konopka and Benzer (1971). (B) Waveform of electric organ discharge in a male *Apteronotus leptorhynchus* fish, adapted from Bastian et al. (2001).

where $x_i, i = 1, \ldots n$ represents the set of variables that fully characterizes the state of the system. One can write this more concisely using vector notation as

$$\frac{\mathrm{d}}{\mathrm{d}t}\mathbf{x} = \mathbf{f}(\mathbf{x}). \tag{11.2}$$

11.2.1   Linear and Nonlinear Dynamics

We have already encountered linear dynamical equations, meaning that the term on the right side of equation (11.2) is a linear function of the system variables:

$$\frac{\mathrm{d}}{\mathrm{d}t}\mathbf{x} = \mathbf{A}\mathbf{x}. \tag{11.3}$$

As discussed previously (in section 1.4 and example 2.19), these systems have particularly simple solutions, of the type

$$\mathbf{x} = \mathbf{x}_0 e^{\mathbf{A}t}. \tag{11.4}$$

That encompasses a rather limited set of behaviors, restricted to exponential growth or decay and sinusoidal oscillation.

Here, we will attack the more general case of nonlinear dynamics, where the right side of equation (11.2) is some arbitrary function. This is an exceedingly common occurrence in biological problems. Take, for example, the most elementary case of a bimolecular chemical reaction:

$$A + B \rightarrow AB, \tag{11.5}$$

which leads to this dynamic equation for the concentrations:

$$\frac{\mathrm{d}}{\mathrm{d}t}[AB] = k[A][B]. \tag{11.6}$$

Cell biology abounds with biochemical networks in which the reactions are nonlinear functions of the reactants. The same applies to the multicellular networks encountered in neuroscience and the multiorganism networks found in population biology and evolution. Occasionally, we will encounter a system whose overall behavior appears linear, but this arises only through the peculiar interaction of its nonlinear components. Clearly, then, we need some methods to deal with the generic nonlinear case of the dynamic equations. What are the options?

11.2.1.1 Analytical solutions Occasionally, a nonlinear differential equation will have an analytical solution. The previously described bimolecular reaction (equation 11.6) is an example (see exercise 14.1). However, stringing together even a small system of such reactions quickly leads to intractable differential equations. When one finds an analytical solution, even to a simplified version of the original problem, that should be seen as a fortuitious accident. Certainly, there is no correspondence between interesting questions in biology and analytically tractable differential equations.

11.2.1.2 Numerical solutions Alternatively, one can simulate the dynamical system numerically by stepping through the differential equations with sufficiently small time steps. This is often a useful approach, and its domain of applications expands as our computers become faster and their memory larger. On the other hand, one quickly discovers challenges: Often the exact numerical parameters in the equations are unknown, and extending the numerical simulations over a combinatorial space of many parameters becomes computationally expensive. Worse still, the form of the equations themselves is sometimes uncertain. Finally, even when the numerical scheme converges, it may provide little insight into what happened and why.

11.2.1.3 Qualitative solutions Frequently, we ask questions that accept approximate or qualitative answers. We don't always need the time course of the system variables to 64-bit precision. Instead, we might simply want to know how a system responds to some perturbation: Will it blow up? Oscillate? Or quickly return to baseline? What is the gain of its response?

In this book, we will explore a way of thinking about dynamical systems that delivers these qualitative answers. Frequently, this answer is an end in itself. On other occasions, it is an essential first step in deciding how to approach the system by other means, like numerical simulation. Working out the problems in this way involves a lot of sketching because important conclusions can often be reached by just looking at shapes on paper. So gather some paper and colored pencils, or whatever the current substitute is for those tools.[1]

---

1. When the historian Charles Weiner interviewed Richard Feynman, they got to talking about Feynman's notebooks. Weiner: "So this [notebook] represents the record of the day-to-day work." Feynman: "I actually did the work on the paper." [...] Weiner: "Well, the work was done in your head but the record of it is still here." Feynman: "No, it's not a record, not really, it's working. You have to work on paper and this is the paper. OK?" (American Institute of Physics, 1973).

11.2.2 Definitions

We already encountered the **standard form** of dynamical system for n variables $x_1, \ldots, x_n$ here:

$$\frac{\mathrm{d}}{\mathrm{d}t} x_i = f_i(x_1, \ldots, x_n), \tag{11.7}$$

and the associated vector notation:

$$\frac{\mathrm{d}}{\mathrm{d}t} \mathbf{x} = \mathbf{f}(\mathbf{x}). \tag{11.8}$$

As it turns out, most ordinary differential equations can be put in standard form. The insistence on a first-order derivative on the left side appears like an undue restriction, but most of the problems that we care about can be adapted to this formulation (see exercises 14.2 and 14.3). In this book, we will restrict ourselves to systems that are deterministic, but methods exist to deal with equations that include stochastic terms.

The **phase space** is simply the space in which the dynamic variables $x_1, \ldots, x_n$ live. The **dimensionality** of that space is the number of dynamic variables n. The state of the dynamic variables, $\mathbf{x} = (x_1, \ldots, x_n)$ is called a **phase point**. As the phase point moves over time, it traces a **trajectory** through the phase space. The time derivative of that phase point, $\frac{\mathrm{d}}{\mathrm{d}t}\mathbf{x}$, is called the **velocity**. So the dynamic equations $\frac{\mathrm{d}}{\mathrm{d}t}\mathbf{x} = \mathbf{f}(\mathbf{x})$ specify the velocity at every point in phase space. That vector field $\mathbf{f}(\mathbf{x})$ is sometimes called the **flow**. The analogy here is to phase points as particles in a flowing liquid. These concepts are depicted in a two-dimensional phase space in figure 11.2.

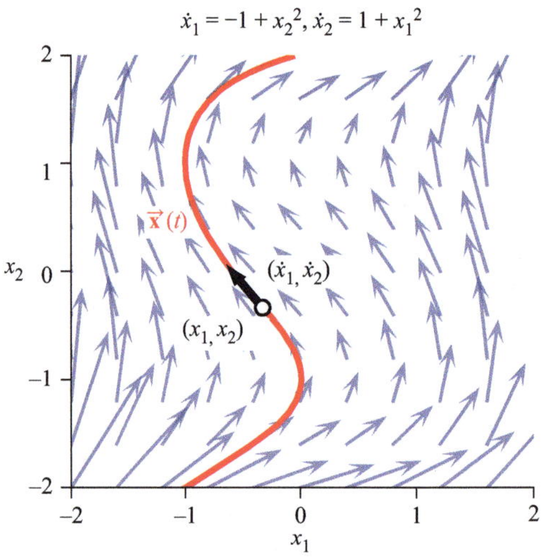

Figure 11.2
Some terminology for dynamical systems. A phase point (x_1, x_2) (circle), the local velocity vector $(\dot{x}_1, \dot{x}_2)$ (black arrow), a trajectory $\vec{\mathbf{x}}(t)$ (red), and the flow field (blue).

11.3 Flows, Fixed Points, and Bifurcations

11.3.1 Illustrative Example: The Fish Farm

Let us begin with a classic example that illustrates the use of these concepts for the graphical analysis of a dynamics problem. Imagine a population of fish growing in a pond. The change in the number of fish is a balance of birth and death. When there are only a few fish, we expect that both birth and death rates are proportional to the number of fish. Therefore, defining $x(t)$ as the number of fish at time t, we have

$$\frac{\mathrm{d}x}{\mathrm{d}t} \equiv \dot{x} = rx, \tag{11.9}$$

where the per-capita growth rate r is the difference between the birth and the death rates. If r is positive, this leads to an exponential increase of the population. Of course, that has to come to an end when the fish exhaust the resources in the pond. So suppose that the growth rate declines with the size of the population, such that

$$\dot{x} = f(x) = rx(1 - x/k). \tag{11.10}$$

In other words, $r(1 - x/k)$ is the new effective growth rate.

In this example, x is the single **dynamic variable**. The **phase space** is one-dimensional, namely the x-axis. The **dynamic equation** (equation 11.10) spells out the velocity $\dot{x}$ as a function of x. To understand what will happen, it helps to graph that function (figure 11.3), which is a downward-pointing parabola.

11.3.2 Flow and Fixed Points

We can immediately sketch out the **flow** of the system by marking arrows on the x-axis. The **phase point** will travel rightward wherever f is positive and leftward where f is negative.

In two special places on the x-axis, the velocity is zero: at $x^* = 0$ and $x^* = k$. These are called the **fixed points** of the system. When the phase point is at a fixed point, it will remain there forever.

The two fixed points have very different characters, based on what happens to trajectories directly next to the fixed point. For the left fixed point, $x^* = 0$, a phase point some small amount to the right (say, $x = \epsilon$), will move away from the fixed point with increasing velocity. In practical terms, zero fish is very different from two fish (of the opposite sex) because the latter situation will quickly escalate to many fish. We call this an **unstable fixed point** because phase points in immediate vicinity of the fixed

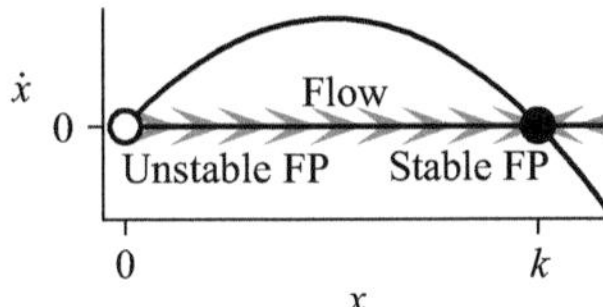

Figure 11.3
Phase diagram of a fish farm, with the unstable fixed point at the open circle and the stable fixed point at the closed circle.

point get driven away from it. By contrast, the right fixed point, $x^* = k$, is a **stable fixed point**. Phase points near the fixed point, whether to the right or to the left, get driven toward this fixed point.

We see now that the stability of a fixed point x^* depends on the slope of the velocity,

$$\frac{\mathrm{d}}{\mathrm{d}x} f \equiv f'(x), \tag{11.11}$$

where it crosses the x-axis:

$$x^* \text{ is } \begin{cases} \text{stable, if } f'(x^*) < 0 \\ \text{unstable, if } f'(x^*) > 0. \end{cases} \tag{11.12}$$

One generally marks a stable fixed point with a closed circle and an unstable one with an open circle (figure 11.3).

11.3.3 Linearization at the Fixed Points

Beyond the qualitative insights about the stability at the fixed points, we can make some quantitative statements about the system's dynamics near the fixed points. Consider a small deviation from the fixed point $\xi = x - x^*$. We want to know how $\xi(t)$ evolves. Using a Taylor series expansion in equation (section 1.5.2.4),

$$\dot{\xi} = f(x^* + \xi) = f(x^*) + \frac{\mathrm{d}}{\mathrm{d}x} f(x^*) \cdot \xi + \frac{\mathrm{d}^2}{\mathrm{d}x^2} f(x^*) \cdot \frac{1}{2}\xi^2 + \cdots . \tag{11.13}$$

For small ξ, we can truncate the Taylor series after the linear term. Also, $f(x^*) = 0$ because x^* is a fixed point. So

$$\dot{\xi} \approx \lambda \xi, \tag{11.14}$$

where

$$\lambda = \frac{\mathrm{d}}{\mathrm{d}x} f(x^*) \equiv f'(x^*) \tag{11.15}$$

is called the **stability parameter** of the fixed point.

Since equation 11.14 is a first order differential equation, the deviation $\xi(t)$ will follow an exponential time course

$$\xi(t) \sim \mathrm{e}^{\lambda t} \tag{11.16}$$

of either growth or decay, depending on the sign of λ. For example, a large negative value of λ means fast decay to the stable fixed point. A small positive value indicates slow deviation away from an unstable fixed point.

Based on this linearization analysis, one can start sketching possible trajectories $x(t)$, as shown in figure 11.4. Near the fixed point at $x^* = 0$, x will grow exponentially. Near the fixed point at $x^* = k$, it will decay exponentially toward k. And in the middle, at $x = k/2$, the velocity is highest and varies slowly, producing an approximately straight section of the curve $x(t)$. This sort of "stitching" of trajectories can be quite

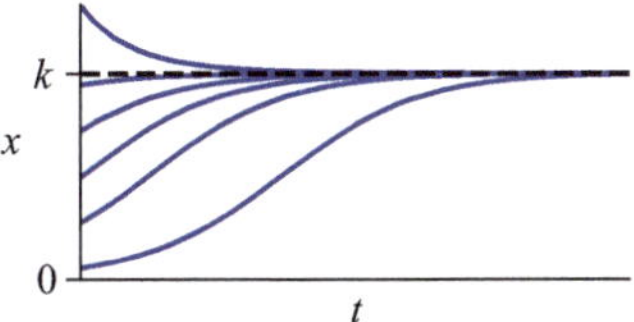

Figure 11.4
Sketch of trajectories for a fish farm.

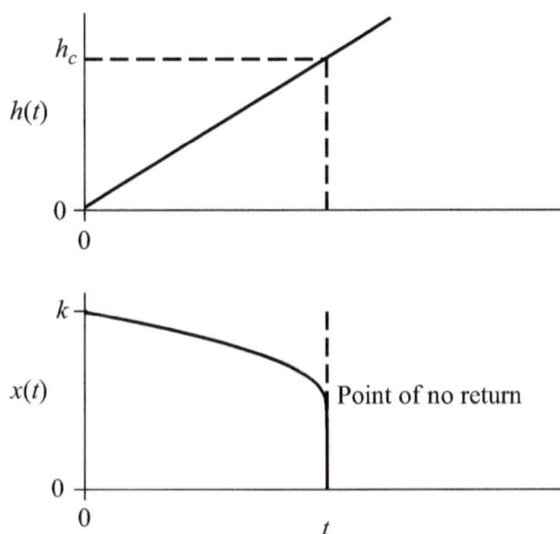

Figure 11.5
The road to ruin for the fish farm. Top: Fishing rate increases gradually over time. Bottom: Fish population suddenly plummets to zero at the point of no return.

useful, combining a analytical solution near the fixed points, with a guess or numerical simulation at places far from the fixed point.

11.3.4 Dependence on a Model Parameter

So far, nothing terribly surprising has happened. Now the farmer starts fishing. They take fish out of the pond every day to sell at the market, say at a constant rate h per unit time. To maximize profits, they increase the fishing rate h a little bit every week (figure 11.5). The business seems to move along swimmingly. Then one morning, the farmer finds the pond empty: no more fish. What happened?

The introduction of fishing changed the dynamic equation slightly, to

$$\dot{x} = rx\left(1 - x/k\right) - h. \tag{11.17}$$

The velocity $\dot{x} = f(x)$ is a parabola, like before, but shifted down by the fishing rate h (figure 11.6). For small h, the parabola still intersects the x-axis in two places, so there still are two fixed points. The left fixed point is unstable and now occurs at $x^* > 0$, which means that a certain minimum number of fish is needed to maintain the population at all. In the long run, you cannot take more fish out of the pond than were born the previous day. The right fixed point is stable and lies at a lower x than before, again due to the effect of fishing.

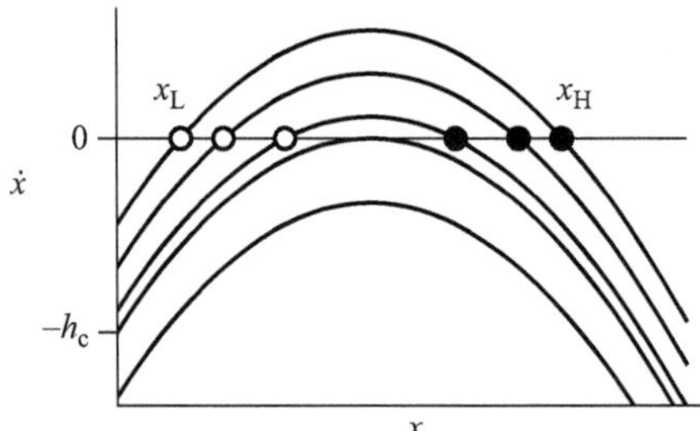

Figure 11.6
Dynamic equation including fishing. The dynamics of the fish population (x) is driven away from the unstable fixed point x_L toward the stable fixed point x_H. The parabola is shifted down by the fishing rate h. When $h < h_\mathrm{c}$, the parabola slips entirely below the x-axis and $\dot{x}$ is negative everywhere, leading to $x = 0$: no more fish!

Something dramatic happens when the fishing rate reaches a certain critical value where the parabola slips entirely below the x-axis. Now there are no fixed points left, and the velocity is negative everywhere. Wherever the phase point starts, it will slide leftward, accelerate, and eventually slam into the limit $x = 0$: no more fish!

This explains the farmer's disconcerting experience: As they gradually increased h, the pond followed along at the higher fixed point x_H. But once h crossed the critical value h_c, the population rapidly declined to zero (figure 11.5).

The key event that precipitated the decline was the disappearance of the fixed points. So we can summarize the behavior of the system by tracking how the fixed points change depending on the parameter h.

For $h < h_\mathrm{c}$ there are two fixed points: one stable, one unstable. At the critical parameter value $h = h_\mathrm{c}$ the two fixed points "merge," leaving nothing behind. So $h = h_\mathrm{c}$ marks the transition between two qualitatively different systems. Such a change in character owing to a change in fixed points is called a **bifurcation**. The particular form encountered here is called a **saddle-node bifurcation** (figure 11.7), a name that might make sense after section 11.4. It is also called a "blue sky bifurcation" because, moving in the direction of decreasing h, two fixed points appear seemingly out of the blue sky.

11.3.5 Bifurcation Types in One Dimension

The saddle-node bifurcation occurs very commonly in 1D systems as a system parameter changes. Whenever an extremum of the velocity $f(x)$ passes through zero, a pair of fixed points is either created or destroyed (figure 11.8).

At that extremum, the first two terms of the Taylor series of $f(x)$ vanish:

$$f(x^* + \xi) = f(x^*) + \frac{\mathrm{d}}{\mathrm{d}x}f(x^*) \cdot \xi + \frac{\mathrm{d}^2}{\mathrm{d}x^2}f(x^*) \cdot \frac{1}{2}\xi^2 + \cdots \tag{11.18}$$

because both the value and the slope are zero, and one can approximate $f(x)$ with a parabola. We can choose a generic parabola to stand for all such instances, like

$$\dot{x} = r - x^2. \tag{11.19}$$

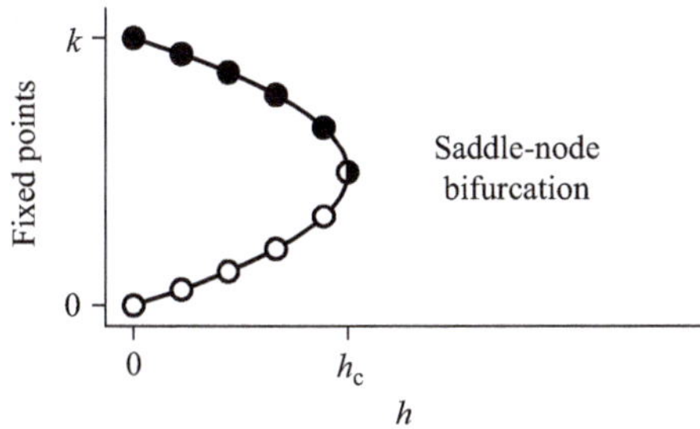

Figure 11.7
Bifurcation plot for the fish farm.

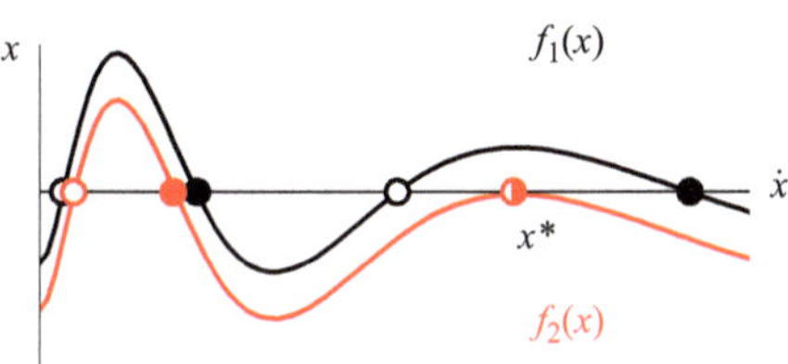

Figure 11.8
How saddle-node bifurcations arise commonly in 1D systems. Here, the velocity function $f(x)$ depends on some system parameter, and when it reaches $f_2(x)$, a pair of fixed points disappear.

This is known as the **normal form** of the saddle-node bifurcation, essentially the simplest dynamic equation that undergoes such a transition. Note the similarity to the fish farm in equation (11.17), where $r = -h$.

Generalizing the lessons from the fish farm, one finds that all 1D saddle-node bifurcations have the following properties (figure 11.9):

- For r above the critical value $r_c = 0$, there are two fixed points, one stable and one unstable.
- As r passes through the critical value $r_c = 0$, the two fixed points approach each other increasingly rapidly, according to

$$\left| x_2^* - x_1^* \right| \propto \sqrt{|r - r_c|}. \tag{11.20}$$

- The stability factors at the fixed points also decrease dramatically as

$$\left| \lambda_{1,2} \right| \propto \sqrt{|r - r_c|}. \tag{11.21}$$

So as the system nears the bifurcation, the dynamics at the fixed points slow sharply. Sometimes this is seen experimentally as unusually slow fluctuations in the system variables, which can be an indicator that a bifurcation point is nearby.

If the function $f(x)$ meets some special conditions, then other kinds of bifurcation graphs can arise. Figure 11.9 shows some examples of these types, each with the normal form of its dynamic equation. Except for the saddle-node bifurcation, these conditions

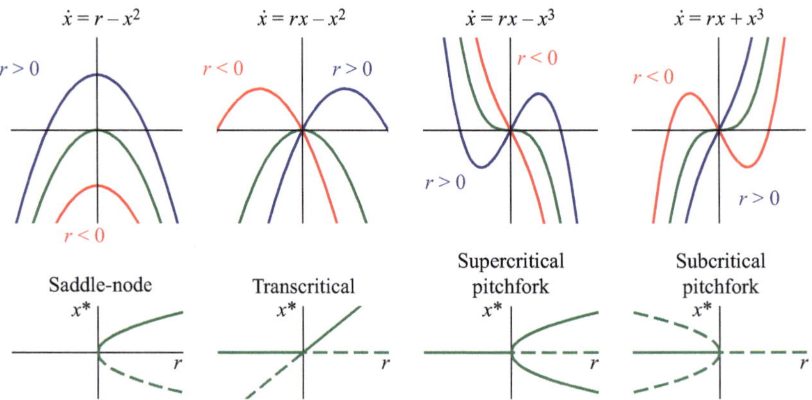

Figure 11.9
Types of bifurcations in 1D systems. Top: Normal form of the dynamic equation with one parameter r. Middle: Sketch of $\dot{x}$ vs. x with r above and below the critical value. Bottom: Bifurcation plot of the fixed point locations as a function of the parameter r; solid = stable fixed points; dashed = unstable fixed points.

usually require some form of symmetry or other constraints on the dynamical system. For more on this, read chapter 3 of Strogatz (2015).

11.3.6 Phenomena in One Dimension

The range of phenomena in 1D dynamics is extremely limited. Fundamentally, the phase space is a line with some alternation of unstable and stable fixed points (figure 11.8). Depending on where the phase point starts, it just travels monotonically to the neighboring stable fixed point or to the end of the range.

In particular, note that *oscillations are impossible in a 1D system*. For an oscillation, the phase point $x(t)$ must be able to reverse and pass through the same x-value twice, once with positive and once with negative velocity. But the dynamic equation $\dot{x} = f(x)$ assigns just one velocity to each phase point, so that can't work.

As we have seen, the qualitative behavior of a system is largely determined by the number and types of its fixed points. Two systems with the same sequence of fixed points act the same way, except for some warping of the time axis. Interesting changes happen when fixed points are lost or gained: these bifurcations are often the focus of attention in nonlinear dynamics.

As we will see, two-dimensional (2D) systems offer a much richer menagerie of behaviors. Nonetheless, we will find occasions where 1D dynamics come in handy in analyzing a more complex system (e.g., because one of the dimensions is decoupled from the other or varies on a very different time scale).

11.4 Dynamics in Two Dimensions

We will now consider dynamical systems with two variables, typically denoted as (x, y). The phase space is now a plane (figure 11.10), and the velocity of the phase point is set by two dynamic equations:

$$\dot{x} = f(x, y) \qquad \dot{y} = g(x, y). \tag{11.22}$$

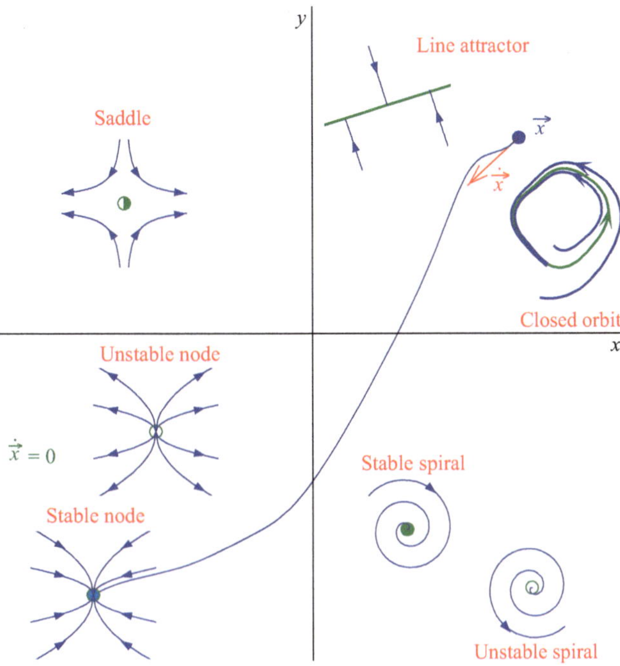

Figure 11.10
A survey of fixed points that occur in 2D systems.

We will sometimes write these in vector form:

$$\dot{\mathbf{x}} = \mathbf{f}(\mathbf{x}), \quad \mathbf{x} = \begin{bmatrix} x \\ y \end{bmatrix}, \quad \mathbf{f}(\mathbf{x}) = \begin{bmatrix} f(x,y) \\ g(x,y) \end{bmatrix}. \tag{11.23}$$

This seemingly minor increase in complexity from one dimension allows for a wonderfully rich set of new phenomena.

11.4.1 Fixed Points and Attractors

As before, it pays to start investigating a 2D system by considering its fixed points, namely, points in phase space where the velocity vanishes:

$$f(x,y) = g(x,y) = 0. \tag{11.24}$$

They come in a wide variety (figure 11.10). An isolated fixed point that attracts all the nearby phase points is called a **stable node**. One that repels nearby phase points is an **unstable node**. These two types resemble the fixed points in 1D systems. In addition, one now finds a type of fixed point that attracts some phase points in its vicinity and repels others: this is called a **saddle**. Trajectories near a saddle will approach it from one direction and then leave in another direction, much like a marble rolling over a horse's saddle. Notice that a saddle is neither stable nor unstable, but some awkward mixture of the two.

Another type of isolated fixed point is called a spiral. A **stable spiral** attracts nearby phase points, but they oscillate around the fixed point as they approach in

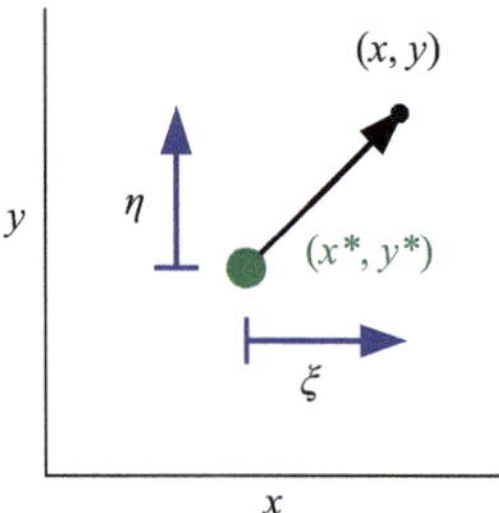

Figure 11.11
Linearization about a fixed point in two dimensions.

an ever-tightening spiral. Similarly, an **unstable spiral** repels nearby phase points as they spiral outward.

Then one may encounter an entire line of fixed points. If the line attracts nearby phase points, it is called a **line attractor**; otherwise, it is a **line repeller**.

Another new feature in two dimensions is a **closed orbit**. This is a trajectory that closes on itself. If a phase point starts on this trajectory, it will remain there forever, even though the velocity doesn't drop to zero. If a closed orbit also attracts nearby phase points into it, it is called a **stable limit cycle**. If it repels nearby phase points, it is an **unstable limit cycle**.

In the following discussion, we will learn techniques to spot and categorize the fixed points of a 2D system.

11.4.2 Linearization

To classify the type of a given fixed point, one needs to inspect how phase points in the immediate vicinity behave. Recall how we determined the stability of a fixed point in one dimension (section 11.3.3): by linearizing the velocity nearby and inspecting the derivative of the velocity at the fixed point, $\frac{\mathrm{d}}{\mathrm{d}x}f(x^*)$. The same approach will serve us again.

Say that we have a fixed point $\mathbf{x}^* = (x^*, y^*)$ and want to consider the immediate neighborhood (figure 11.11). Define small deviations from the fixed point as

$$\xi = x - x^*, \quad \eta = y - y^*. \tag{11.25}$$

Then the Taylor expansion of the velocity around the fixed point is

$$\dot{\xi} = f(x, y) = f(x^*, y^*) + \frac{\partial}{\partial x}f(x^*, y^*) \cdot \xi + \frac{\partial}{\partial y}f(x^*, y^*) \cdot \eta$$

$$+ \frac{\partial^2}{\partial x^2}f(x^*, y^*)\frac{1}{2}\xi^2 + \dots$$

$$\dot{\eta} = g(x, y) = g(x^*, y^*) + \frac{\partial}{\partial x}g(x^*, y^*) \cdot \xi + \frac{\partial}{\partial y}g(x^*, y^*) \cdot \eta$$

$$+ \frac{\partial^2}{\partial x^2}g(x^*, y^*)\frac{1}{2}\xi^2 + \dots. \tag{11.26}$$

Now

$$f(x^*, y^*) = g(x^*, y^*) = 0 \tag{11.27}$$

because (x^*, y^*) is a fixed point. If we cut off the Taylor series after the linear terms, we are left with the approximate dynamic equations:

$$\dot{\xi} \approx f_x \cdot \xi + f_y \cdot \eta$$
$$\dot{\eta} \approx g_x \cdot \xi + g_y \cdot \eta, \tag{11.28}$$

where the partial derivatives of the velocity are defined as

$$f_x = \frac{\partial}{\partial x} f(x^*, y^*), \quad f_y = \frac{\partial}{\partial y} f(x^*, y^*), \quad g_x = \frac{\partial}{\partial x} g(x^*, y^*), \quad g_y = \frac{\partial}{\partial y} g(x^*, y^*). \tag{11.29}$$

So near a fixed point, one can approximate the dynamics with a purely linear system. By inspecting the nature of that linear system, we will identify the type of fixed point.[2]

11.4.3 Linear Dynamics in Two Dimensions

The linearized dynamics in equation (11.28) can be written as

$$\dot{x} = Ax, \tag{11.30}$$

where

$$x = \begin{bmatrix} x \\ y \end{bmatrix}, A = \begin{bmatrix} f_x & f_y \\ g_x & g_y \end{bmatrix}, \tag{11.31}$$

and for purely notational convenience, we have replaced $x - x^*$ with x, so the fixed point is at the origin $x^* = (0, 0)$.

The matrix A is called the **Jacobian matrix** of the flow field $\dot{x}$, and it completely determines the nature of the dynamics around the fixed point. As discussed in section 2.11, your immediate impulse should be to ask about the eigenvectors of A. If we can transform the coordinates (x, y) to a new basis where A is diagonal, then solving the dynamics will be trivial, as it will decouple into two independent 1D linear systems.

We are guaranteed that A has at least one eigenvector (section 2.9), but let us start with the general case, where it has two eigenvectors. Then the two eigenvectors form a basis and we can perform the basis transform. In the new basis, the dynamics look like

$$\dot{x} = \begin{bmatrix} \lambda & 0 \\ 0 & \mu \end{bmatrix} \cdot x, \tag{11.32}$$

where λ and μ are the two eigenvalues. The solution is simply

$$x = x_0 e^{\lambda t}$$
$$y = y_0 e^{\mu t}. \tag{11.33}$$

The nature of these trajectories depends on the values of λ and μ. Let us consider a few distinct cases:

---

2. In some special cases, it happens that all the partial derivatives in equation (11.29) vanish, and then linearization fails and one has to resort to the second-order terms of the Taylor series in equation (section 11.26).

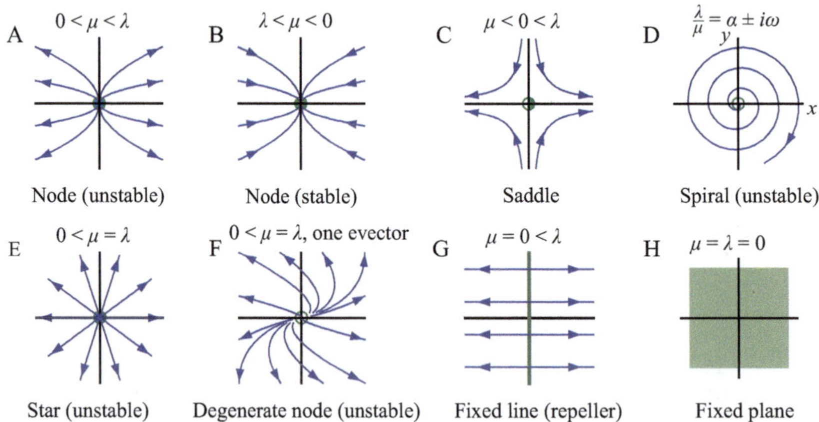

Figure 11.12
Trajectories around various types of fixed points in two dimensions.

- $0 < \mu < \lambda$

Both x and y grow exponentially with time (equation 11.33), moving away from the fixed point. Apparently, we are dealing with an **unstable node**. What shape of trajectories will this produce in the (x, y) plane? We can remove t from equation (11.33) to get

$$y \sim x^{|\mu|/|\lambda|}. \tag{11.34}$$

Because $\lambda > \mu$ and both are positive, these are power functions with a positive power $\mu/\lambda < 1$. They leave the fixed point vertically and then bend over to eventually become horizontal (figure 11.12A). In this case, the y-axis, which has the smaller eigenvalue μ, is called the **slow direction** of the fixed point; and the x-axis with the larger eigenvalue is the **fast direction**. Close to the fixed point, the trajectories move along the slow direction; far from the fixed point, they bend to move along the fast direction.

- $\lambda < \mu < 0$

Now x and y decay exponentially with time as in equation (11.33), approaching the fixed point monotonically. This makes for a **stable node** (figure 11.12B). The shape of the trajectories given in equation (11.34) is a power function as before, with a positive power $\mu/\lambda < 1$. Again, the slow direction is the y-axis because $|\mu| < |\lambda|$.

- $\mu < 0 < \lambda$

This gets more interesting. Now $x(t)$ grows exponentially with time, while $y(t)$ decays with time. The phase point approaches the fixed point along the y-direction but leaves it along the x-direction. The shape of the resulting trajectory is now a negative power function:

$$y \sim x^{-|\mu|/|\lambda|}, \tag{11.35}$$

which looks like a hyperbola. This sort of flow identifies a **saddle** (figure 11.12C). The axis of approach, here the y-axis associated with the negative eigenvalue, is also called the saddle's **stable manifold**. A phase point that finds itself on this line will forever stay there and approach the fixed point. The axis of departure, here the x-axis with the positive eigenvalue, is called the **unstable manifold**. A point on this straight line will remain on the line traveling away from the fixed point.

- $\lambda = \alpha + i\omega,\ \mu = \alpha - i\omega$

What if the eigenvalues are complex? Then one is guaranteed to be the complex conjugate of the other[3]. So the solution to equation (11.33) takes the form

$$
\begin{aligned}
x &= x_0 e^{\alpha t} e^{i\omega t} \\
y &= y_0 e^{\alpha t} e^{-i\omega t}.
\end{aligned}
\tag{11.36}
$$

What will this look like in the basis of the original dynamics problem? Let's call those original phase space coordinates (x', y'). Those coordinates must have real values since they correspond to real trajectories. The most general way to obtain two real coordinates from a basis transform of equation (11.36) will be

$$
\begin{aligned}
x' &= e^{\alpha t}\,(r\cos\omega t + u\sin\omega t) \\
y' &= e^{\alpha t}\,(v\cos\omega t + w\sin\omega t)
\end{aligned}
\tag{11.37}
$$

with some real coefficients r, u, v, w. Now

$$
\begin{aligned}
x' &= r\cos\omega t + u\sin\omega t \\
y' &= v\cos\omega t + w\sin\omega t
\end{aligned}
\tag{11.38}
$$

describes an ellipse revolving around the origin $(0, 0)$. Therefore, equation (11.37) describes an elliptical spiral that revolves around the fixed point and either grows exponentially for $\alpha > 0$ or shrinks exponentially for $\alpha < 0$ (figure 11.12D). It revolves at the angular frequency ω, completing one cycle in a period of $2\pi/\omega$. In the special case $\alpha = 0$, the trajectories are all closed loops. Such a fixed point is called a **center**.

By now, we have covered all the generic situations that will occur regarding the eigenvalues of the matrix **A**. But some interesting special cases remain to be discussed, which may arise from some peculiar constraint on the dynamic equations.

- $\lambda = \mu$

This is a special case of a node. According to equation (11.34), the trajectories follow $y \sim x$, which means that all phase points flow into ($\lambda < 0$) or out from ($\lambda > 0$) the fixed point along straight lines. This is called a **star node**, (figure 11.12E).

- $\lambda \neq \mu = 0$

---

3. In general, this follows from the complex conjugate root theorem because the eigenvalues are the roots of the characteristic polynomial, which has all real coefficients.

When one of the eigenvalues is zero, equation (11.34) says that the corresponding coordinate remains constant regardless of its value, whereas the other one grows or shrinks depending on the sign of λ:

$$x = x_0 e^{\lambda t}, \quad y = y_0. \tag{11.39}$$

Those trajectories are straight lines parallel to the x-axis that terminate on a **line of fixed points** along the y-axis (figure 11.12G): either a **line attractor** ($\lambda < 0$) or a **line repeller** ($\lambda > 0$).

- $\lambda = \mu = 0$

This is a bit of a trivial case where every point is a fixed point and nothing moves ever (figure 11.12H). Needless to say, this rarely arises in living systems.

- Only one eigenvector

Finally, we must consider the possibility that matrix **A** has only one eigenvector. Such a fixed point is called a **degenerate node**. A bit of calculation reveals the shape of the trajectories around the fixed point (see exercise 14.4). The single eigenvector determines one special axis. A trajectory leaves the fixed point along that axis in one direction or the other and then gradually bends 180 degrees, so it ultimately goes to infinity in the opposite direction (figure 11.12F). If $\lambda < 0$, the phase points move inward along these trajectories and the degenerate node is **stable**; if $\lambda > 0$, it is **unstable**.

11.4.4 Classification of Fixed Points

At last, we can return to the question that motivated this whole excursion into linear dynamics: What type is my fixed point? If the Jacobian matrix at the fixed point is

$$\mathbf{A} = \begin{bmatrix} f_x & f_y \\ g_x & g_y \end{bmatrix}, \tag{11.40}$$

then the eigenvalues z are the roots of the characteristic polynomial:

$$\mathrm{Det}\,(\mathbf{A} - z\mathbf{I}) = 0 \tag{11.41}$$

with the solutions $z = \lambda, \mu$ given by

$$\lambda = -\frac{T}{2} + \sqrt{\frac{T^2}{4} - D}$$

$$\mu = -\frac{T}{2} - \sqrt{\frac{T^2}{4} - D}, \tag{11.42}$$

where

$$T = \mathrm{Trace}\,(\mathbf{A}) = f_x + g_y$$

$$D = \mathrm{Det}\,(\mathbf{A}) = f_x g_y - f_y g_x. \tag{11.43}$$

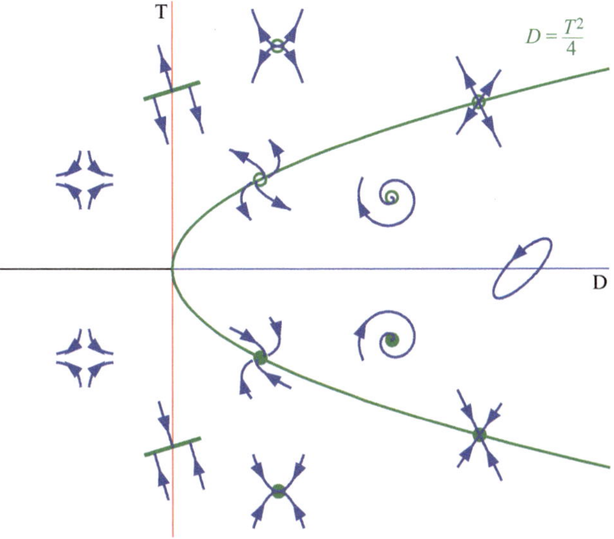

Figure 11.13
Types of fixed points in two dimensions, depending on the determinant D and trace T of the Jacobian matrix.

The type of the fixed point is determined by the values of λ and μ, which in turn are driven by the trace T and determinant D of the Jacobian matrix. Figure 11.13 summarizes those relationships.

11.5 Bifurcation Analysis of 2D Systems

The best way to illustrate the application of principles from the preceding sections is to work through some canonical examples of nonlinear systems. In each case, the goal will be to gain a qualitative understanding of the system's modes of behavior and how they depend on the quantitative parameters.

11.5.1 Fixed Points: Rabbits vs Sheep

This system arises in population biology when two species compete for the same resource. In the case of the fish farm, the fish compete for food in the pond only with other fish. What happens when rabbits and sheep both feed on the same resource (say, a field of grass), but each with a different voracity? Although this problem is commonly phrased in the language of population biology, similar dynamics occur in many other instances of competition.

The so-called **Lotka-Volterra model** describes the growth of each species with a logistic differential equation that includes cross-species terms. If

$$x = \text{number of rabbits}$$
$$y = \text{number of sheep,}$$

$$(11.44)$$

then the dynamic equations are

$$\dot{x} = ux\,(a - x - by)$$
$$\dot{y} = vy\,(c - y - dx).$$
$$(11.45)$$

The first part of equation (11.45), $\dot{x} = ux\,(a - x)$, is already familiar to us from the fish farm example: ua is the maximal per-capita growth rate of rabbits, and a is the capacity of the field of grass (namely, the maximal number of rabbits that it can sustain). That capacity is now reduced by a second term, $-by$, which reflects the fact that sheep are grazing on this field as well. Similarly, the capacity to sustain sheep is limited by the number of rabbits, reflected by the term $-dx$ in equation (11.45).

11.5.1.1 Nondimensionalize the Variables This dynamics problem seems to have a boatload of free parameters–four altogether in equation (11.45). It would be tedious if we had to explain how the system behaves all throughout a four-dimensional parameter space. But on closer inspection, one of these can be eliminated and the equations simplified, as follows:

1. Scale the rabbit number by substituting $\xi = x/a$.
2. Scale the sheep number by substituting $\eta = y/c$.
3. Scale the time variable by substituting $\tau = uat$.

Out of notational laziness, we will call these new variables ξ, η, τ again by the names x, y, t. Then the new dynamic equations become

$$\dot{x} = x\,(1 - x - \beta y)$$
$$\dot{y} = \gamma y\,(1 - y - \delta x),$$
$$(11.46)$$

where

$$\beta = bc/a, \quad \delta = da/c, \quad \gamma = vc/ua, \qquad (11.47)$$

and we must remember that we performed the following substitutions

$$x/a \rightarrow x, \quad y/c \rightarrow y, \quad uat \rightarrow t. \qquad (11.48)$$

The new system (equation 11.46) will behave in all qualitative respects just like the old system (equation 11.45). All the trajectories will be identical except for a scaling along the x-axis, the y-axis, and the time axis. Note also that the new variables x, y, and t are dimensionless: they all have units of 1, whereas the old variables had units of rabbits, sheep, and seconds. The simple act of **nondimensionalizing the variables** almost invariably simplifies the equations. It also brings out the fundamentals of the problem that it shares with those from other domains where the units might be viruses or calcium ions. As a rule, the first step in approaching a dynamical system (or more generally, any applied math problem) should be to get rid of tedious real-world units— but do keep track of what substitutions were needed to accomplish that so the final result can be expressed back in those units.

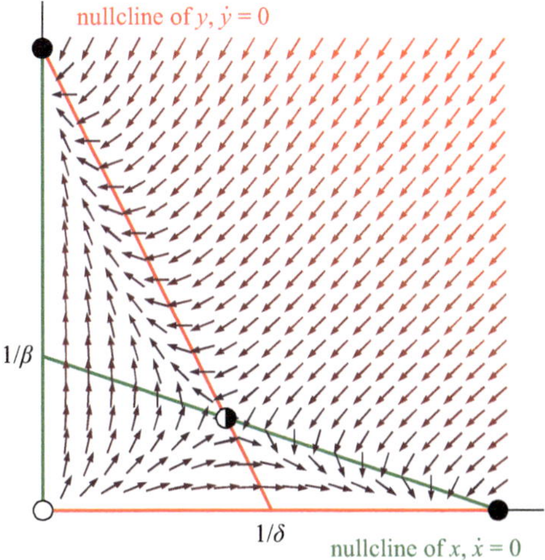

Figure 11.14
A rough phase portrait of the "rabbit vs sheep" system for the case $\beta > 1, \delta > 1$. The vector field has been normalized in length, and the magnitude is color-coded, with brighter shades of red corresponding to faster flow.

11.5.1.2 Create a Rough Phase Portrait The 2D phase space certainly invites graphical exploration by sketching. Here are a few useful steps to follow.

- Find the nullclines and fixed points

Nullclines are those curves in the phase space along which the velocity is zero for one or the other variable. From equation (11.46), one can see directly the following points:

1. $\dot{x} = 0$ if $x = 0$ or $x = 1 - \beta y$. These two straight lines are the nullclines of x.
2. $\dot{y} = 0$ if $y = 0$ or $y = 1 - \delta x$. These are the nullclines of y.

Figure 11.14 plots these nullclines for the case where $\beta > 1$ and $\delta > 1$.
At the intersection of nullclines for the two variables, we find the fixed points, where $\dot{x} = \dot{y} = 0$. In the present example, there are four such fixed points, at

$$\mathbf{x}^* = (0,0)\,,\,(0,1)\,,\,(1,0)\,,\,\left(\frac{1-\beta}{1-\beta\delta}\,,\,\frac{1-\delta}{1-\beta\delta}\right). \tag{11.49}$$

- Sketch some trajectories

A good way to get started is to draw the flow on the nullclines. Here, the velocity arrows are either vertical or horizontal, and it is often straightforward to decide their direction based on the dynamic equations. As the nullclines partition the phase space, we now know where the trajectories flow in or out of these partitions. Based on that, one can start guessing the shape of trajectories. In turn, that allows a preliminary guess as to the character of each fixed point. In this case, for example, it looks as though $(0,0)$ will be an unstable node, $(0, 1)$ and $(1, 0)$ will be stable nodes, and the fourth fixed point in the middle will be a saddle.

Sometimes this rough sketch of the **phase portrait** is already sufficient to answer basic questions. From here, one can proceed more quantitatively, for example by linearizing around the fixed points.

11.5.1.3 Linearize at the Fixed Points Here, the goal is to confirm the character of each of the fixed points and to identify their special axes. To illustrate that process, we will pick a specific set of parameters, such as

$$\gamma = 3, \quad \beta = 2, \quad \delta = 2. \tag{11.50}$$

From equation (11.46), we find the Jacobian matrix as

$$\mathbf{A} = \begin{bmatrix} f_x & f_y \\ g_x & g_y \end{bmatrix} = \begin{bmatrix} 1 - 2x - 2y & -2x \\ -6y & 3 - 6y - 6x \end{bmatrix}. \tag{11.51}$$

We will now find its eigenvectors and eigenvalues at each of the fixed points:

1. $\mathbf{x}^* = (0, 0)$.
   Here,

$$\mathbf{A} = \begin{bmatrix} 1 & 0 \\ 0 & 3 \end{bmatrix}, \tag{11.52}$$

which obviously has the eigenvalues

$$\lambda_1 = 1, \quad \lambda_2 = 3 \tag{11.53}$$

and the corresponding eigenvectors

$$\mathbf{e}_1 = \begin{bmatrix} 1 \\ 0 \end{bmatrix}, \quad \mathbf{e}_2 = \begin{bmatrix} 0 \\ 1 \end{bmatrix}. \tag{11.54}$$

This identifies the fixed point as an unstable node (section 11.4.3), and the nearby trajectories will have the shape $y \sim x^3$ (figure 11.15).

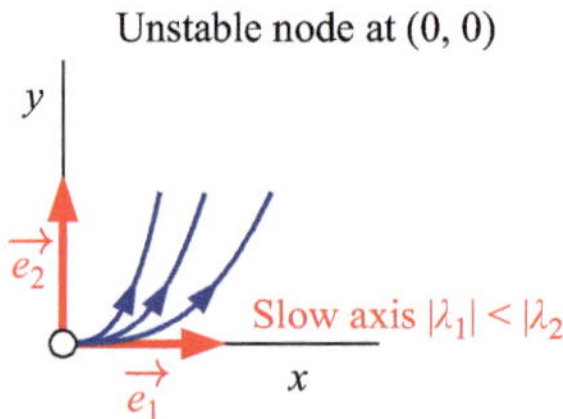

Figure 11.15
Details around the fixed point at $(0, 0)$.

2. $\mathbf{x}^* = (1, 0)$.

Here,

$$A = \begin{bmatrix} -1 & -2 \\ 0 & -3 \end{bmatrix}. \tag{11.55}$$

Recall that the eigenvalues of a triangular matrix are listed along the diagonal, so

$$\lambda_1 = -1, \quad \lambda_2 = -3, \tag{11.56}$$

and a bit of algebra finds the eigenvectors as

$$\mathbf{e}_1 = \begin{bmatrix} 1 \\ 0 \end{bmatrix}, \quad \mathbf{e}_2 = \begin{bmatrix} 1 \\ 1 \end{bmatrix}. \tag{11.57}$$

This is a stable node, with $\mathbf{e}_2$ as the fast axis, so the trajectories will look like figure 11.16.

3. $\mathbf{x}^* = (0, 1)$.

Proceeding in the same way, we find

$$A = \begin{bmatrix} -1 & 0 \\ -6 & -3 \end{bmatrix}, \tag{11.58}$$

with eigenvalues

$$\lambda_1 = -1, \quad \lambda_2 = -3, \tag{11.59}$$

and eigenvectors

$$\mathbf{e}_1 = \begin{bmatrix} 1 \\ -3 \end{bmatrix}, \mathbf{e}_2 = \begin{bmatrix} 0 \\ 1 \end{bmatrix}. \tag{11.60}$$

This is another stable node (figure 11.17).

4. $\mathbf{x}^* = (1/3, 1/3)$.

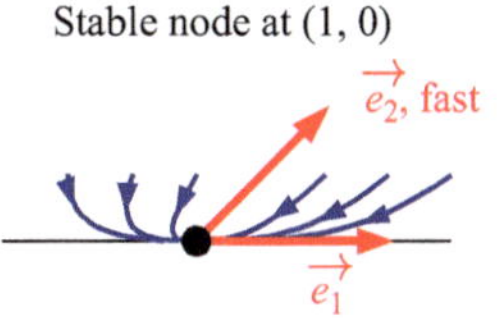

Figure 11.16
Details around the fixed point at $(1, 0)$.

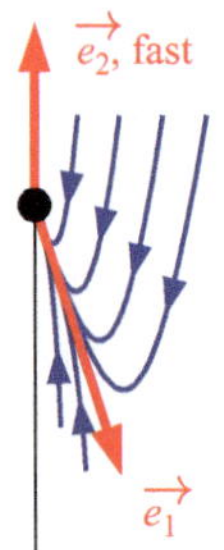

Figure 11.17
Details around the fixed point at $(0, 1)$.

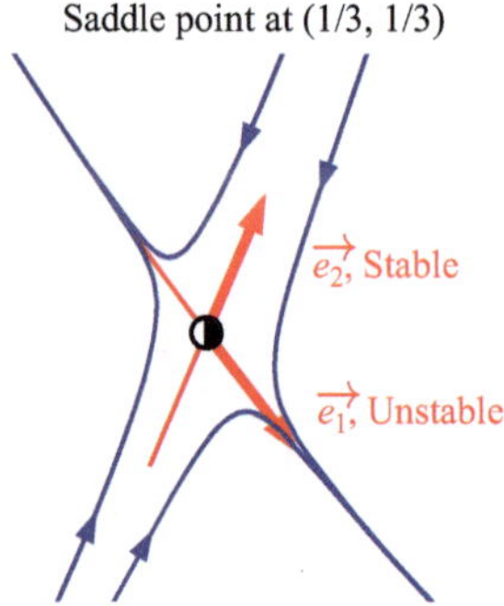

Figure 11.18
Details around the fixed point at $(1/3, 1/3)$.

Here,

$$
\mathbf{A} = \begin{bmatrix} -1/3 & -2/3 \\ -2 & -1 \end{bmatrix}
$$
$$
T = \frac{-4}{3} \tag{11.61}
$$
$$
D = -1,
$$

which leads to

$$
\lambda_1 = \frac{-2 + \sqrt{13}}{3}, \quad \lambda_2 = \frac{-2 - \sqrt{13}}{3} \tag{11.62}
$$

and

$$
\mathbf{e}_1 = \begin{bmatrix} 1 \\ \frac{1 - \sqrt{13}}{2} \end{bmatrix}, \quad \mathbf{e}_2 = \begin{bmatrix} 1 \\ \frac{1 + \sqrt{13}}{2} \end{bmatrix}. \tag{11.63}
$$

Because the two eigenvalues have opposite signs, this is a saddle, with $\mathbf{e}_1$ as the unstable axis and $\mathbf{e}_2$ as the stable axis (figure 11.18).

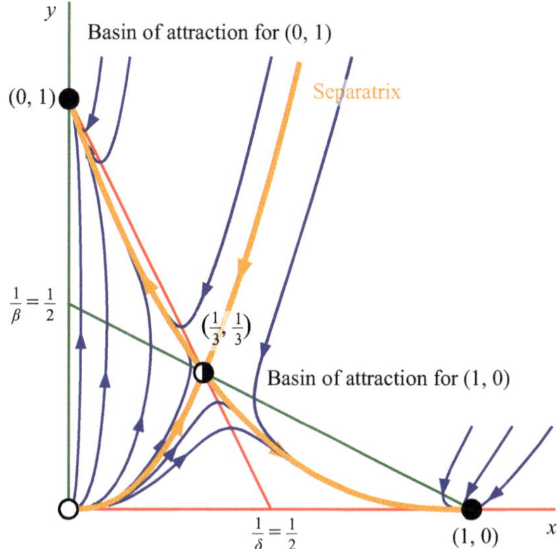

Figure 11.19
A more quantitative phase plot.

11.5.2 Produce a Quantitative Phase Plot

Given the known shapes of trajectories at the fixed points, and the flows on the null-clines, one can now produce a more serious phase portrait (figure 11.19) and stitch together long trajectories. Now we are in a position to interpret the flow and draw conclusions about the behavior of this system.

- There are just two stable fixed points. They correspond to a final state with only rabbits, $\mathbf{x}^* = (1, 0)$, or only sheep, $\mathbf{x}^* = (0, 1)$. So in the end, one species will always completely displace the other from the field of grass.
- Which of the two species wins depends on where the trajectory starts. In fact, the entire phase space is divided into two **basins of attraction**, one for each fixed point. These regions are separated by a curve, called a **separatrix**. Phase points on one side of the separatrix invariably flow toward one of the fixed points, and those on the other side to the other fixed point, much as rain drops falling on either side of the US continental divide will end up in either the Pacific or the Atlantic.

11.5.2.1 Identify the Bifurcations So far, we have only considered the case where $\beta > 1$ and $\delta > 1$, which produces large cross-species terms in equation (11.46). That means that each species is quite voracious, leading to strong competition, and ultimately **each species can eliminate the other**. What happens with different forms of competition?

- $\beta < 1$ and $\delta < 1$: Now the two species compete less strongly, for example because each also has a separate resource available. This leads to a very different phase portrait because the nullclines now intersect differently (figure 11.20). The origin $(0, 0)$ is still an unstable node, but the fixed points on the axes are now saddles. And the fixed point at the interior is a stable node. As a result, all trajectories terminate in the sole

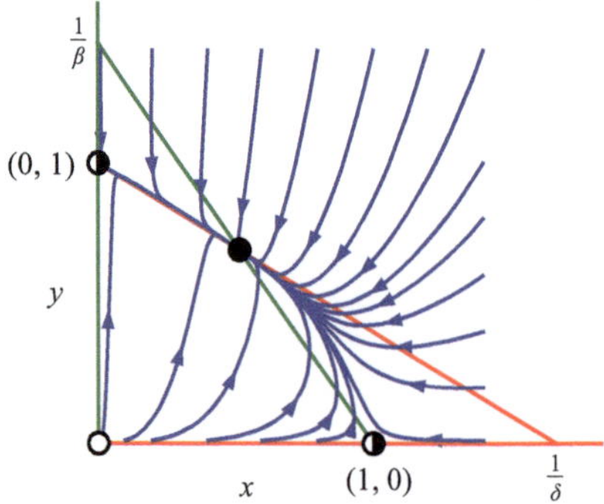

Figure 11.20
Phase portrait for the case $\beta < 1$ and $\delta < 1$.

stable fixed point. This means that the only long-term outcome is **co-existence of the two species** at a specific population ratio.

- $\beta > 1$ and $\delta < 1$: Now the sheep are voracious and affect the rabbits more than the other way round. The phase plot has only two fixed points, and the only stable node lies on the y-axis (figure 11.21). This leads to the guaranteed **elimination of one species**, as sheep invariably displace all the rabbits.[4]

So a change in the model parameters, specifically β and δ, can alter the landscape of fixed points and dramatically affect the dynamics of the system. Figure 11.22 illustrates these bifurcations and how they arise. For example, start from the case $\beta > 1, \delta > 1$ shown in the top right of figure 11.22 and consider what happens when β slowly decreases: the saddle point will gradually slide toward the stable node at $(0, 1)$. At $\beta = 1$, the two fixed points merge; and for $\beta < 1$, only a saddle is left behind. Going around the quadrants of figure 11.22, one sees successive mergers and splits between two fixed points, one a saddle and one a node. This finally explains the term **saddle-node bifurcation**.

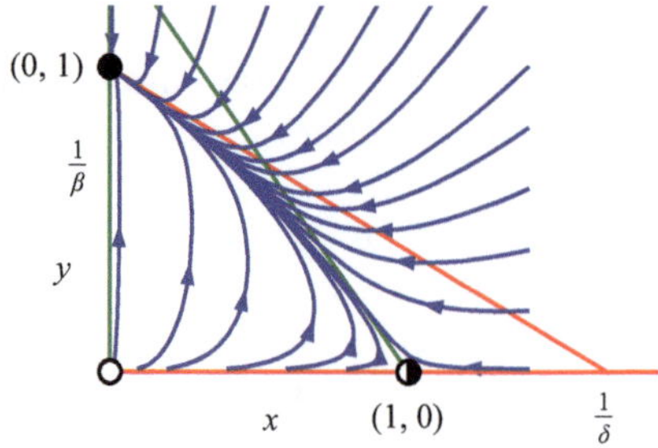

Figure 11.21
Phase portrait for the case $\beta > 1$ and $\delta < 1$.

---

4. Surely you are wondering what happens in the special case of $\beta = 1, \delta = 1$. Find out by doing exercise 14.5.

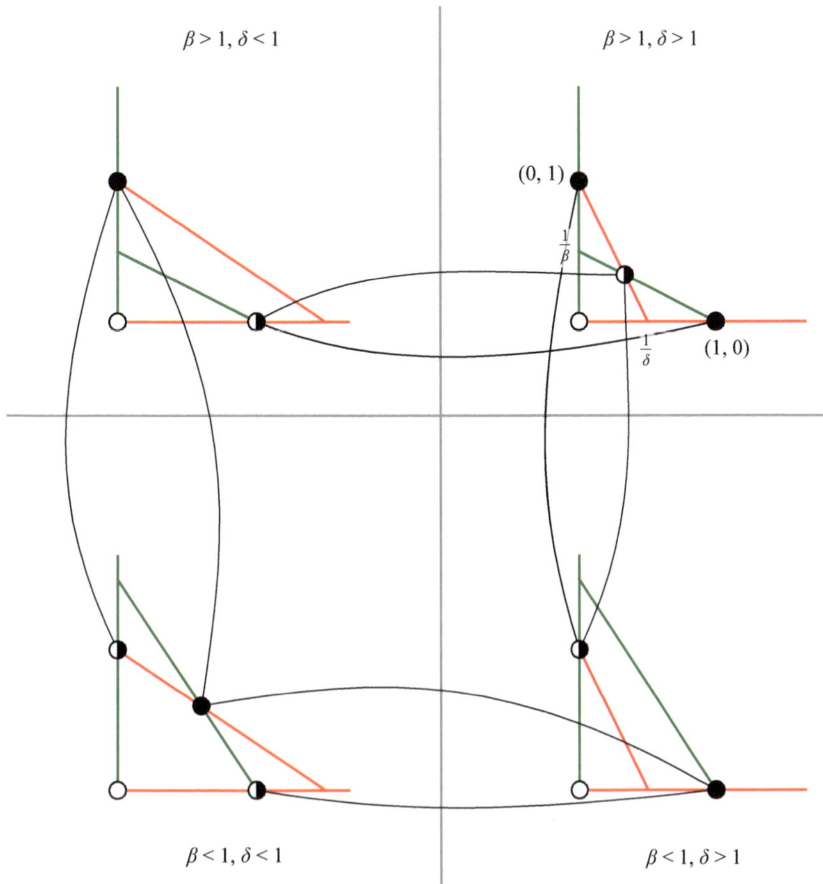

Figure 11.22
Saddle-node bifurcations in the "rabbits versus sheep" system, depending on parameters β and δ. Black arcs indicate the splitting/merging of fixed points at bifurcations.

11.5.3 Closed Orbits: Lynx vs Hare

This next example illustrates the occurrence of trajectories that close on themselves in a loop, which allows the system to oscillate in a periodic fashion (figure 11.23). Such **closed orbits** occur in three possible configurations: A **center** is a fixed point surrounded by a continuous family of closed orbits. By contrast, a **limit cycle** is an isolated closed trajectory. It may be **stable**, such that nearby phase points all approach the limit cycle over time, or **unstable**, such that nearby phase points get repelled.

A dramatic example of interspecies competition involves one species eating the other. For historical reasons to be explained later in this chapter, we will call the prey species "hare" and the predator species "lynx". Perhaps the simplest model of such a predator-prey relationship are the Lotka-Volterra equations:

$$\dot{x} = x\left(a - by\right)$$
$$\dot{y} = y\left(cx - d\right). \tag{11.64}$$

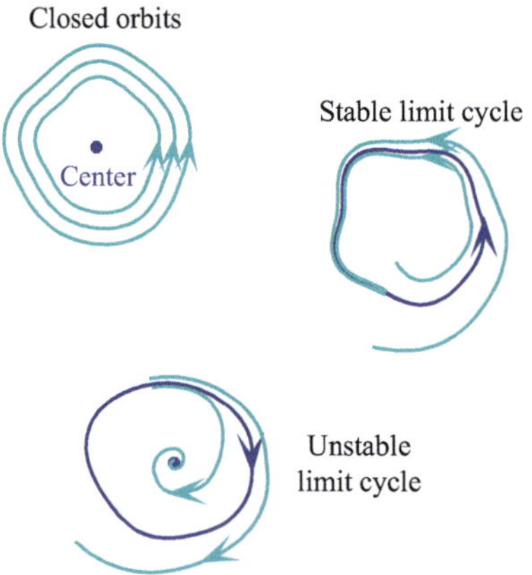

Figure 11.23
Closed orbits in 2D systems.

Here, x denotes the number of hares, and y the number of lynx. The hares are born at the per-capita rate a and die only by predation from lynx at a rate by. The lynx are completely dependent on hares as a food source and thus are born at a rate cx; they die of old age at the rate d.

Again, we start by nondimensionalizing the equations. Using the substitutions

$$x\frac{c}{d} \to x, \; y\frac{b}{a} \to y, \; at \to t, \tag{11.65}$$

we arrive at

$$\dot{x} = x\left(1 - y\right)$$
$$\dot{y} = \gamma y\left(x - 1\right), \tag{11.66}$$

where

$$\gamma = d/a. \tag{11.67}$$

Note that a full three of the four parameters got eliminated!

The nullclines are obvious: $\dot{x} = 0$ at $x = 0$ and $y = 1$; and $\dot{y} = 0$ at $y = 0$ and $x = 1$. The fixed points are at the origin $\mathbf{x}^* = (0, 0)$ and at $\mathbf{x}^* = (1, 1)$. From sketching the flow on the nullclines, it becomes apparent that $(0, 0)$ is a saddle. By contrast, near $(1, 1)$, the flow seems to circle around the fixed point. To check whether those trajectories really form loops instead of a spiral, we have to examine the fixed point more closely.

Linearizing around $\mathbf{x}^* = (1, 1)$, we find the Jacobian matrix is

$$A = \begin{bmatrix} 0 & -1 \\ \gamma & 0 \end{bmatrix}, \tag{11.68}$$

with eigenvalues

$$\lambda_\pm = \pm i\sqrt{\gamma}. \tag{11.69}$$

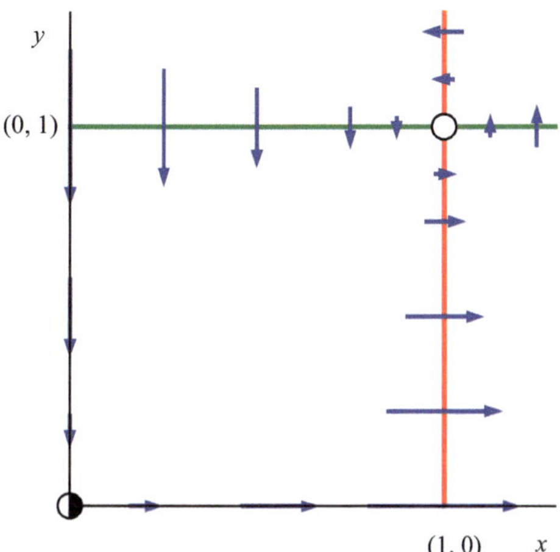

Figure 11.24
Phase portrait of a predator-prey system. Nullclines for x (green) and y (red).

Recall that complex eigenvalues $\lambda_\pm = \alpha \pm i\omega$ indicate a stable spiral for $\alpha < 0$ and an unstable spiral for $\alpha > 0$; see equation (11.37). Here, on the other hand, $\alpha = 0$, so the trajectory neither grows nor shrinks in radius but closes in on itself. Thus we are dealing with a center. Figure 11.24 shows the nullclines, the fixed point, and the flow field.

To be precise, the linearization analysis identifies this point as a **linear center**. It is conceivable that the second-order and higher terms of the Taylor expansion at the fixed point in equation (11.26) change the character of the fixed point to a spiral. Thus one needs to verify by other means whether the fixed point is also a **nonlinear center**. Some theorems are available that guarantee this in special cases (see chapter 6 of Strogatz, 2015 for details).[5]

In the present case, we benefit from the fact that the system has a closed-form analytical solution (see exercise 14.6). This solution confirms that $(1, 1)$ is a nonlinear center (figure 11.25). It is surrounded by closed trajectories; in fact, they cover the entire phase space, and every phase point is on a closed orbit. Near the center, the phase points circle with a period of $2\pi/\sqrt{\gamma}$, just as expected from the imaginary part of the eigenvalue, $\omega = \sqrt{\gamma}$.

11.5.3.1 Oscillations in the fur trade   This treatment shows that a predator-prey system can undergo pronounced periodic oscillations in the population numbers. The basic phenomenon is not a mystery. Suppose that the lynx have grown in to a large population. That year, they will eat most of the rabbits, which leads to fewer rabbits the following season. That in turn will lead to fewer births among the lynx, and so on. One expects the predator population to lag the prey in its oscillations. But does this actually happen in nature?

Figure 11.26 reproduces a historical data set over a century of records on the populations of snowshoe hare and lynx in the Pacific Northwest. The measurements are somewhat indirect: the number of pelts of both species purchased from trappers by the

---

5. More generally, the fixed-point character derived from linearization needs to be verified in all the so-called marginal cases when one of the eigenvalues has zero real part.

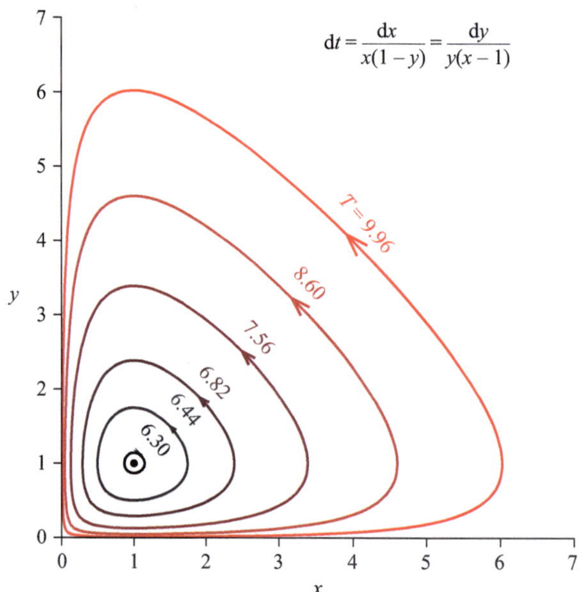

Figure 11.25
Analytical solution of the predator-prey system for $\gamma = 1$. Each curve is labeled with its period T. Note that the period approaches 2π as the orbit approaches the fixed point.

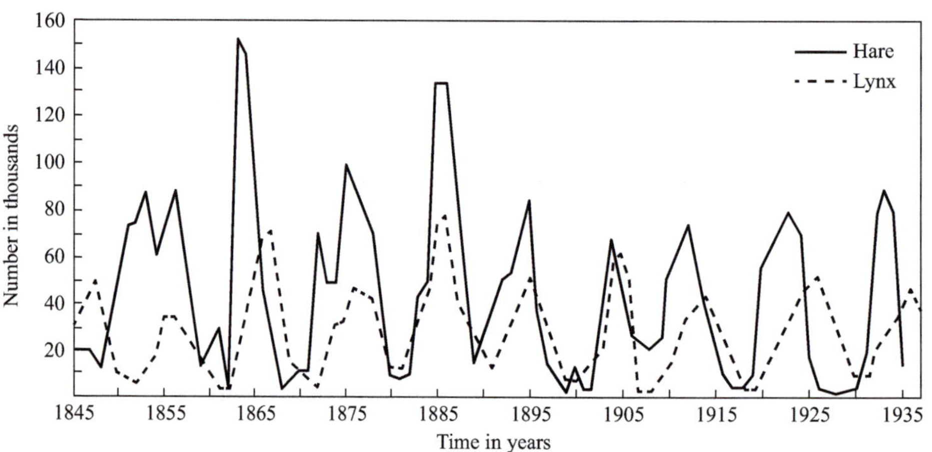

Figure 11.26
Number of pelts from the lynx and the snowshoe hare received by the Hudson's Bay Company. From Murray (2002), redrawn from MacLulich (1937).

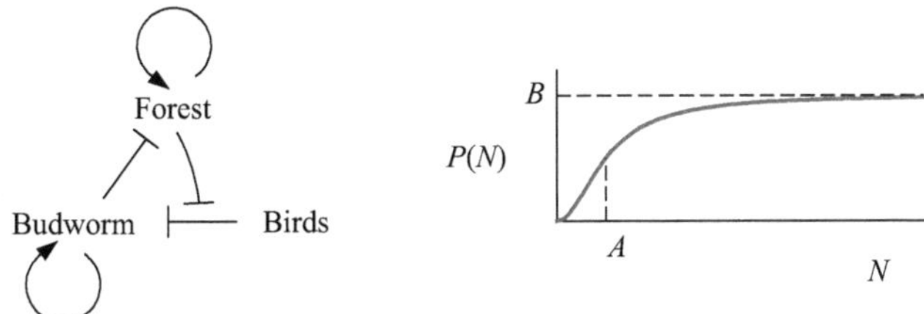

Figure 11.27
Left: Schematic interactions between forest, birds, and budworms. Right: Predation rate vs budworm density.

Hudson's Bay Company.[6] The numbers for both species fluctuate dramatically. The variation is fairly periodic. And the lynx appear to lag the hare by about a quarter cycle, consistent with the prediction of the Lotka-Volterra model (figure 11.25). Still, some other features of the data remain unexplained. For example, there is a pronounced variation of the amplitude from cycle to cycle. A more sophisticated model (Weinstein, 1977) might include the presence of a second predator (the trapper) and—depending on one's political inclinations—a third (the Hudson's Bay Company).

11.5.4 Limit Cycles: Budworm vs Forest

Our last pedagogical example in this section is again drawn from ecology. The spruce budworm is a major forest pest in Canada and the northern United States. Its larvae feed on the buds and needles of the fir trees. Occasionally, the budworm population explodes suddenly and denudes an entire forest, which takes many years to recover. There has been great interest in what causes these periodic outbreaks and how to prevent them. A simple mathematical model (Ludwig et al., 1978) involves three interacting components: needles, budworms, and birds (figure 11.27). Here is the cycle:

- Budworms eat the needles.
- Birds eat the budworms.
- Needles hide the budworms from the birds.

The number of budworms, N, follows the dynamic equation

$$\dot{N} = RN\left(1 - \frac{N}{K}\right) - P\left(N\right). \tag{11.70}$$

We recognize the first few terms as the typical logistic equation for population growth: R is the per-capita reproduction rate, K is the capacity of the forest for budworms. The last term $P\left(N\right)$ represents the rate of predation by birds:

$$P\left(N\right) = B\frac{N^2}{A^2 + N^2}. \tag{11.71}$$

---

6. The Hudson's Bay Company has operated continuously since 1670. It owns, among many other entities, the Saks Fifth Avenue stores in the US.

This is a sigmoidal function of the budworm density N (figure 11.27), and the parameters can be interpreted as follows:

- maximal predation rate at high budworm density. This is where the birds are in "all-you-can-eat" mode.
- half-maximal budworm density. This will vary depending on the density of the forest. Dense trees full of needles will hide the budworms from the birds and thus sustain a high A.

Now, depending on the activity of the budworms, the forest will grow or shrink over the years. As the needle density varies, we assume that both A (the hiding parameter) and K (the capacity for budworms) will vary proportionally, so $k = K/A$ remains constant. We will again postulate a logistic growth equation for the trees:

$$\dot{A} = SA\left(1 - \frac{A}{H} - CN\right), \tag{11.72}$$

where S is the forest's maximal relative growth rate, H reflects the capacity of the land to sustain forest, and C determines the effects of damage from budworms.

We can reduce the parameter set by making the following substitutions:

$$\frac{N}{H} \to x, \quad Rt \to t, \quad \frac{A}{H} \to y, \tag{11.73}$$

which leads to the dynamic equations

$$\dot{x} = x\left(1 - \frac{x}{ky}\right) - b\frac{x^2}{y^2 + x^2} \tag{11.74}$$
$$\dot{y} = ey\left(1 - y - cx\right),$$

where

$$k = \frac{K}{A}, b = \frac{B}{R}, c = CH, e = \frac{S}{R}. \tag{11.75}$$

We can assume that $e = S/R << 1$ because the forest changes much more slowly than the budworm population, on a scale of a decade rather than weeks. Note also that the dynamics of the forest given in equation (11.74) follow the simple form used previously for interspecies competition in equation (11.46), but the dynamics of the budworm in equation (11.74) have a highly nonlinear interaction with the forest, leading to interesting phenomena.

Figure 11.28 illustrates the phase diagram for this system for a particular choice of parameters. The nullclines are given by

$$\dot{x} = 0 : \begin{cases} x = 0, \text{ or} \\ 1 - \frac{x}{ky} - b\frac{x}{y^2 + x^2} = 0 \end{cases}$$
$$\dot{y} = 0 : \begin{cases} y = 0, \text{ or} \\ y = 1 - cx. \end{cases} \tag{11.76}$$

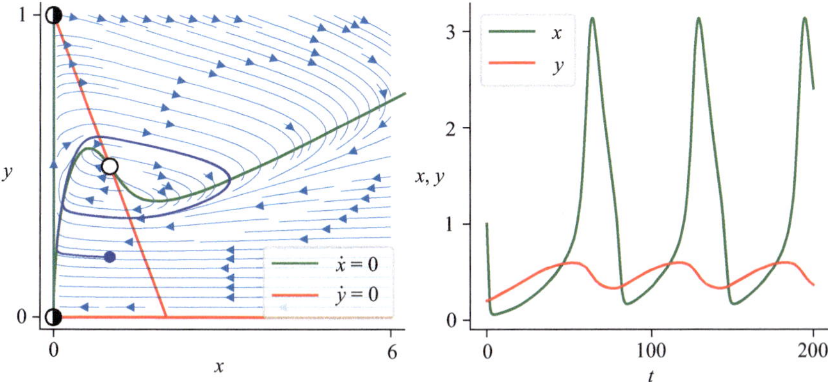

Figure 11.28
Left: Phase diagram for the budworm-vs-forest model using $k = 10, b = 1, c = 0.5$, and $e = 0.05$, including a trajectory that converges on the limit cycle. Right: Time course of x and y during that trajectory.

At the intersection of nullclines on the y-axis, there are two fixed points, and sketching the flow on the nullclines suggests that both are saddles. By linearization, one can show that the third fixed point in the interior space is an unstable spiral. There are no stable fixed points anywhere. And yet the trajectories are naturally confined to the region shown in figure 11.28 because the flow points inward everywhere along the boundary of the region. So where are the trajectories supposed to end up?

11.5.5 Confined Trajectories: The Poincaré-Bendixson Theorem

Here, we get assistance from an important theorem about 2D dynamics:

Theorem 11.1 (Poincaré-Bendixson theorem) If a trajectory is confined to a closed bounded subset of the plane, then it will approach either a stable fixed point or a closed orbit.

For a clean mathematical formulation of this theorem and some of its applications, see chapter 7.3 of Strogatz (2015).

In the present case, there are no stable fixed points, so the system must include a stable limit cycle somewhere. A bit of investigation reveals where it is: as shown in figure 11.28, it follows the N-shaped portion of the nullcline of x.

To understand the dynamics, one must remember that $e << 1$, so the flow is close to horizontal almost everywhere in the plane. Say that a phase point starts somewhere below the N-shaped nullcline (from now on called "the N"). The flow quickly carries it straight left until it crosses the N. Then it gets pushed up and to the right, so it hugs the N from the left, sliding slowly upward. There comes a point when the support from the N disappears and the phase point shoots rightward until it hits the right branch of the N. After crossing the N, it hugs that branch from the right while sliding slowly downward. Eventually, the support from the N disappears again, and the phase point shoots leftward to the left branch of the N. From then on, that cycle repeats almost identically.

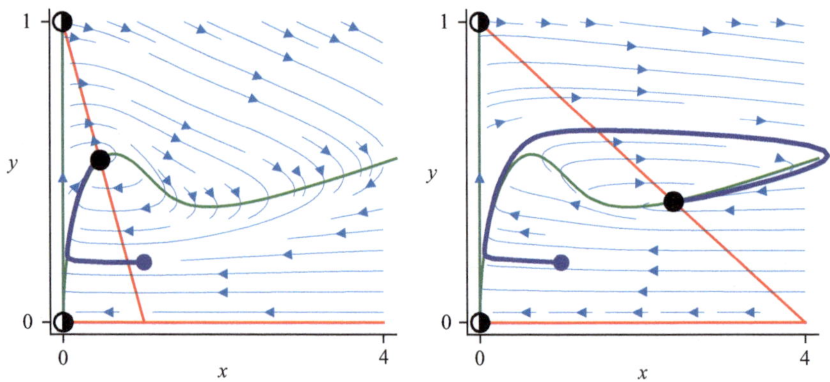

Figure 11.29
Phase diagram for the budworm-vs-forest model using different values of $c = 1$ (left) or $c = 1/4$ (right). In both cases, the interior fixed point is a stable spiral.

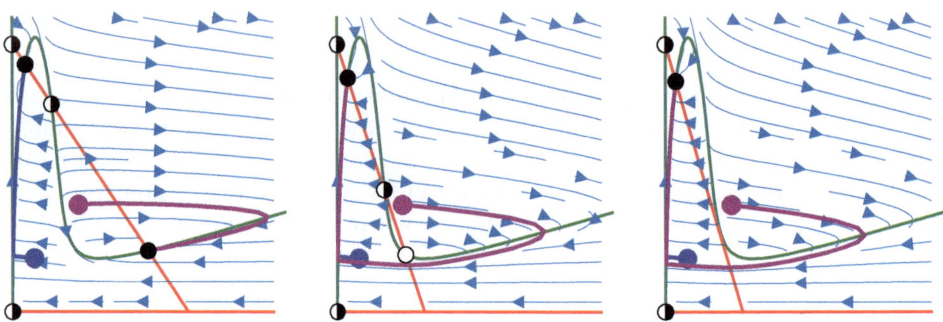

Figure 11.30
Phase diagram for the budworm-vs-forest model using $k = 40, b = 2, e = 0.05$. Display as in figure 11.29. Left: $c = 1/8$, two trajectories converge on the two stable nodes. Middle: $c = 1/4$, the third fixed point has changed to an unstable node. Right: $c = 1/3.5$, two fixed points are lost in a saddle-node bifurcation.

In the ecological context formulated here, the sudden rightward jump of the trajectory corresponds to an explosion of the budworm population, followed by a slow decline of the forest over several years. Eventually, the forest thins to the point where the birds find the budworms efficiently and the budworm population suddenly plummets. Then the forest can recover again slowly until the needles hide the budworms effectively again, and the budworm population explodes.

This kind of behavior—a slow buildup followed by rapid release—is called a **relaxation oscillator**. It generally requires the interaction of two variables with very different time constants.

11.5.6 Bifurcations around a Limit Cycle

The dynamical system described here undergoes a variety of interesting bifurcations, depending on the parameter values. For example, if c is chosen either higher or lower by a factor of 2, the interior fixed point changes to a stable spiral, and there is no limit cycle (figure 11.29). All trajectories end at the single stable fixed point. The pest and the forest remain in equilibrium either at a high or a low insect population. As c varies gradually from 0.5 (figure 11.28) to 1 (figure 11.29), the limit cycle contracts like a noose around the unstable spiral, and it eventually disappears when the spiral converts to a stable one. This is called a **supercritical Hopf bifurcation**.

With a different value of k, one can produce three interior fixed points (figure 11.30). In this case, the middle fixed point is a saddle and both the outer ones are stable nodes. Each of the nodes has its basin of attraction, and two trajectories starting from nearby points may end up at different nodes. The insects and the forest always come to a stable equilibrium, which may be at a high or a low budworm population. As one changes c (namely, the slope of the y-nullcline), the fixed points can change character from a stable node to a stable spiral, to an unstable spiral, and finally to an unstable node. As the slope gets steeper, eventually a saddle and a node coalesce and disappear: we recognize this as another **saddle-node bifurcation**.

12.1   Dynamics in Three or More Dimensions

Clearly, almost any biological system of interest will have more than two interacting components. The main concepts that we encountered in treating one-dimensional (1D) and two-dimensional (2D) systems continue to apply in higher dimensions. For example, the fixed points are always important special locations that organize the phase portrait of the system and constrain what trajectories will do. Linearization near fixed points proceeds in the same way, by finding the eigenvalues of the Jacobian. Of course, this will reveal more complex types: a fixed point may look like a saddle along two dimensions (real eigenvalues of opposite signs) and a stable spiral along another two (complex conjugate eigenvalues with negative real parts). How can one proceed beyond that to predict a system's qualitative behavior?

For 2D systems, we quickly developed some intuition that allows a prediction of a system's behavior, just based on sketching flows on a piece of paper. In higher dimensions, these tools for sketching and visualization suddenly fall away. In three dimensions, one can still produce perspective plots or rotating plots to illustrate a phase diagram. Beyond three dimensions, visualization becomes very difficult. One suspects that this accounts for the dominance of 2D and 3D models in the literature: not because nature favors low dimensions but because humans can't plot anything else with ease.

A good amount of creative energy goes into attacking a high-dimensional system by reducing it to two or three dimensions at a time. This involves sorting the variables into small groups of two or three, and arguing that the groups don't interact with each other. For example, there may be groups of fast and slow variables. For the dynamics of the fast variables, one may treat the slow ones as constant. And for the dynamics of the slow variables, one supposes that the fast ones are in near steady-state equilibrium with the slow ones.

12.2   Chaos

The increased difficulty of visualizion is not the only challenge that one encounters when moving beyond two dimensions. As it happens, 3D systems can engage in a behavior called **chaos** that is impossible in two dimensions. Recall the Poincaré-Bendixson theorem in section 11.1, which determines the asymtotic behavior of a 2D system. At long times, the trajectory ends up in one of two places: a stable fixed point or a closed loop. This just isn't true in three or more dimensions: Instead, the trajectory may converge on a "strange attractor," where it remains in motion within a limited volume but never repeats the same trajectory. This possibility was discovered by Lorenz

(1963) while analyzing a simple 3D model of movements in the atmosphere. It was an astonishing revelation of qualitatively new dynamics, and it is well worth walking through the arguments that led Lorenz to eliminate alternative explanations; see the lucid treatment in chapter 9 of Strogatz (2015).

Perhaps the most consequential characteristic of chaotic dynamics is an **extreme sensitivity to initial conditions**. Two trajectories that start from nearby points in phase space will diverge from each other, with distance increasing as an exponential function of time. The rate-constant of this exponential growth is called a **Lyapunov exponent**, one of the measures that characterize a chaotic system. Given this exponential growth, chaotic trajectories become essentially unpredictable. Suppose that you can measure the initial conditions of a trajectory to some accuracy ϵ. Over time, that just-measurable difference grows to $\epsilon e^{\lambda t}$, where λ is the relevant Lyapunov exponent. Eventually, that uncertainty will encompass everything you care about, such as the size L of the entire phase space. This means that predictions are useless over times longer than $T = \ln(L/\epsilon)$.

The discovery of chaotic dynamics was followed by a flurry of claims in which chaos figured as the explanation for everything, leading up to consciousness and free will. The mystery that came with unpredictability was just too hard to resist.[1] The chaos enthusiasm of the 1980s has now been tempered somewhat. In the life sciences specifically, it is clear that some biological systems can enter a chaotic regime, although they often need to be pushed into extreme conditions (e.g. Crevier and Meister, 1998). What is less obvious is whether chaos plays an essential role in biological systems, in that it allows the system to function in ways that are otherwise impossible. Our own biased opinion here is that evolution has spent more time and effort suppressing chaos in biological systems than exploiting it.

12.3 The Turing Model of Morphogenesis

We end this chapter with an analysis of phenomena in developmental biology that combines the tools of nonlinear dynamics with linear methods like the Fourier transform.

A model proposed by Turing (1952) attacks the question of spatial patterning in development. How do the prominent patterns in living things arise out of a seemingly unpatterned primordium, like an animal egg cell? A photogenic example is the emergence of coat patterns on animal skins. More fundamentally, think of the patterning of the early embryo that ultimately gives rise to all the important specializations like limbs. Turing explained the apparent emergence of something from nothing as a consequence of dynamic instability in a set of chemical reactions that are spatially distributed over the developing organism. By introducing spatial variation, one now has a dynamic variable at each location in space, producing a phase space of potentially infinite dimensionality. The genius of Turing's treatment is that he reduced this problem to two dimensions and solved it by elementary techniques. This is a short tour of Turing's model, largely based on the exposition of the arguments by Murray (2003).

In developmental biology, a chemical that organizes the emergence of spatial structure is called a "morphogen." For example, a spatial gradient in the concentration of a transcription factor may induce dfferent cellular programs at different locations. Of

---

1. ... and it fed the popular logical fallacy of "A is a mystery. B is another mystery. So A and B must be related."

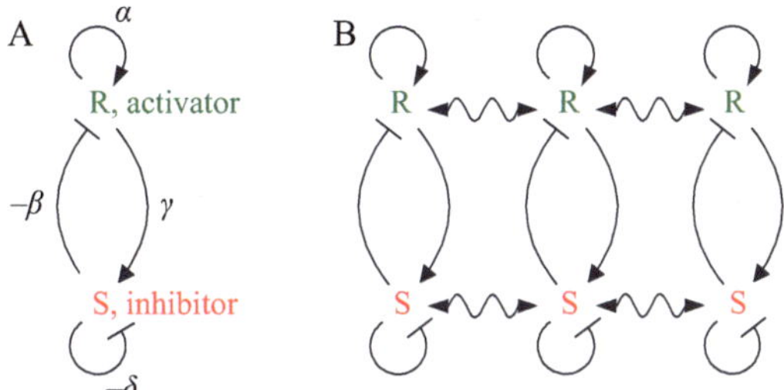

Figure 12.1
A: Schematic of catalysis coupling the activator R and the inhibitor S. B: Spatial system coupling the local reactions between R and S via diffusion between locations.

course, that leaves open the question of how a spatial gradient of any chemical arises within an initially undifferentiated space.

12.3.1   Activator and Inhibitor

We begin with a system of two reacting signaling molecules, R and S, at concentrations $r(t)$ and $s(t)$, which interact through the dynamic equations:

$$\dot{r} = f(r,s), \quad \dot{s} = g(r,s). \tag{12.1}$$

For now, we don't allow any spatial gradients and assume that the reaction vessel is well stirred, so r and s vary only with time.

Suppose that the reaction system has a fixed point (r^*, s^*), whose linearization gives the Jacobian matrix

$$\mathbf{M} = \begin{bmatrix} \partial f/\partial r & \partial f/\partial s \\ \partial g/\partial r & \partial g/\partial s \end{bmatrix} \equiv \begin{bmatrix} \alpha & -\beta \\ \gamma & -\delta \end{bmatrix}, \tag{12.2}$$

where $\alpha, \beta, \gamma, \delta > 0$.

With these signs in the Jacobian matrix, we can identify R as an activator: a small increase in r from the fixed point will catalyze the production of more R and more S. By contrast, S is an inhibitor: a small increase in s will lead to reduction of both R and S (figure 12.1).

The eigenvalues of $\mathbf{M}$ are

$$\lambda_{\pm} = \frac{1}{2}\left(\operatorname{Tr} \mathbf{M} \pm \sqrt{(\operatorname{Tr} \mathbf{M})^2 - 4\operatorname{Det} \mathbf{M}} \right). \tag{12.3}$$

For the fixed point to be stable, we require $\operatorname{Re}(\lambda_{\pm}) < 0$, which in turn means that

$$\operatorname{Tr} \mathbf{M} = \alpha - \delta < 0$$

$$\operatorname{Det} \mathbf{M} = \beta\gamma - \alpha\delta > 0 \tag{12.4}$$

because the determinant is the product of the eigenvalues and this conditions makes sure that neither eigenvalue is zero or purely imaginary. We assume that these conditions obtain and this spatially uniform mixture reaches a stable equilibrium.

12.3.2 Reaction and Diffusion

Suppose now that the chemicals are not stirred but can mix only via lateral diffusion. This allows spatial variation to emerge. For simplicity, we just consider a single spatial dimension, while noting that all the conclusions will be equally valid for 2D and 3D animals. The concentrations $r(x, t)$ and $s(x, t)$ now depend on both time and space. The dynamics are given by the reaction-diffusion equations

$$\frac{\partial r}{\partial t} = f(r, s) + A \frac{\partial^2 r}{\partial x^2}$$

$$\frac{\partial s}{\partial t} = g(r, s) + D \frac{\partial^2 s}{\partial x^2},$$

(12.5)

where A and D are the diffusion coefficients (as discussed in section 8.1.3) of R and S, respectively.

How can one interpret these equations? The local concentration r can change both as a result of the local reactions, through the term $f(r, s)$, and as a result of diffusion of R from or to the neighboring regions, through the term $A \frac{\partial^2 r}{\partial x^2}$. Notice also that we really have an infinite number of dynamic equations in standard form because at every location x, there is a pair of dynamic variables. The second-order differential operator $\frac{\partial^2}{\partial x^2}$ represents the coupling between nearby variable pairs.

Recall that the system has a fixed point (r^*, s^*) that is stable if the two concentrations are spatially uniform. We will now linearize the dynamics around that fixed point and ask whether they remain stable when spatial diffusion is permitted. For small deviations from the fixed point

$$u(x, t) = r(x, t) - r^*, \quad v(x, t) = s(x, t) - s^*,$$

(12.6)

the spatio-temporal dynamic equations are linear—namely,

$$\begin{bmatrix} \frac{\partial u}{\partial t} \\ \frac{\partial v}{\partial t} \end{bmatrix} = \begin{bmatrix} \alpha + A \frac{\partial^2}{\partial x^2} & -\beta \\ \gamma & -\delta + D \frac{\partial^2}{\partial x^2} \end{bmatrix} \begin{bmatrix} u \\ v \end{bmatrix}.$$

(12.7)

as in equation 11.30. We want to know whether this fixed point is stable now that u and v are allowed to vary in space as well as in time.

12.3.3 Fourier Analysis

Recall that the Fourier transform turns linear differential equations into algebraic equations (sections 3.2.4.1 and 4.5.1). For example, if $u(x, t)$ and $\hat{u}(k, t)$ are related by a spatial Fourier transform:

$$u(x, t) \xleftrightarrow{F.T.} \hat{u}(k, t),$$

(12.8)

then

$$\frac{\partial}{\partial x} u(x, t) \leftrightarrow ik \cdot \hat{u}(k, t),$$

(12.9)

and further,

$$\frac{\partial^2}{\partial x^2} u(x,t) \leftrightarrow (ik)^2 \cdot \hat{u}(k,t) = -k^2 \cdot \hat{u}(k,t). \tag{12.10}$$

So after applying a spatial Fourier transform to both dynamic variables, we obtain a much simpler set of dynamic equations:

$$\frac{\partial}{\partial t}\begin{bmatrix} \hat{u}(k,t) \\ \hat{v}(k,t) \end{bmatrix} = \begin{bmatrix} \alpha - Ak^2 & -\beta \\ \gamma & -\delta - Dk^2 \end{bmatrix}\begin{bmatrix} \hat{u}(k,t) \\ \hat{v}(k,t) \end{bmatrix}$$
$$\equiv \mathbf{M}(k)\begin{bmatrix} \hat{u}(k,t) \\ \hat{v}(k,t) \end{bmatrix}. \tag{12.11}$$

Note that we still have infinitely many variables because there is a $\hat{u}$ and $\hat{v}$ for every value of the spatial frequency k. But as is clear from equation (12.11), the variables for different frequencies k_1 and k_2 don't interact at all. So the dynamics can be solved in two dimensions separately for each value of k.

How can we interpret $\hat{u}(k,t)$? Given that k is a spatial frequency, here we are expressing the overall concentration $\hat{u}(k,t)$ as a sum of spatial components $\cos(kx)$. Equation 12.11, then, represents the small deviation in these components from the stable fixed point at which the concentration is uniform everywhere. This deviation can grow or decay with the time course

$$\hat{u}(k,t) \propto e^{\lambda_\pm(k)\cdot t}, \tag{12.12}$$

where $\lambda_\pm(k)$ are the eigenvalues of the Jacobian matrix

$$\mathbf{M}(k) = \begin{bmatrix} \alpha - Ak^2 & -\beta \\ \gamma & -\delta - Dk^2 \end{bmatrix}. \tag{12.13}$$

If one of these eigenvalues has a positive real part, then the fluctuation $\hat{u}(k,t)$ will grow exponentially with time and establish a strong sinusoidal pattern with spatial frequency k.

12.3.4 Dynamic Instability Leads to Pattern Formation

As before, the eigenvalues of the Jacobian matrix are

$$\lambda_\pm(k) = \frac{1}{2}\left(\operatorname{Tr}\mathbf{M}(k) \pm \sqrt{\operatorname{Tr}\mathbf{M}(k)^2 - 4\operatorname{Det}\mathbf{M}(k)}\right). \tag{12.14}$$

We know that

$$\operatorname{Tr}\mathbf{M}(k) = \alpha - \delta - (A+D)k^2 < 0 \tag{12.15}$$

because $\alpha - \delta < 0$ is required for the stability of the spatially uniform system expressed in equation (12.4). Therefore, having a positive eigenvalue requires

$$\operatorname{Det}\mathbf{M}(k) < 0. \tag{12.16}$$

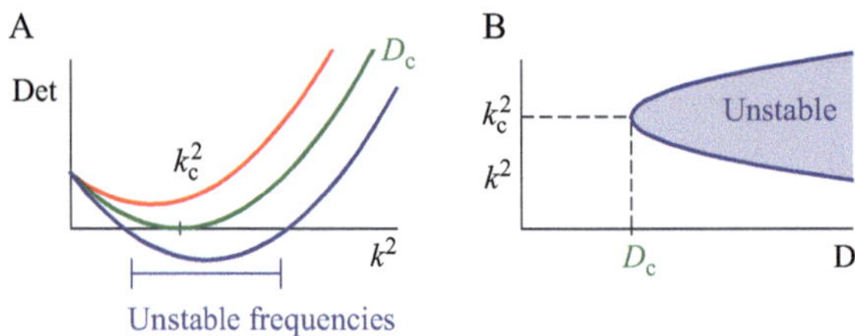

Figure 12.2
A: The function $\mathrm{Det}\,\mathbf{M}(k)$ for various values of D. B: Bifurcation plot indicating the unstable spatial frequencies k.

Let us inspect $\mathrm{Det}\,\mathbf{M}(k)$ as a function of k^2:

$$\mathrm{Det}\,\mathbf{M}(k) = (\beta\gamma - \alpha\delta) + k^2(A\delta - D\alpha) + k^4 AD. \tag{12.17}$$

The first term, $\beta\gamma - \alpha\delta$ is positive, again because of the stability requirement in equation (12.4). What happens if the second term, $(A\delta - D\alpha)$, is negative? Then Det $\mathbf{M}(k)$ becomes an upward parabola in k^2, as plotted in figure 12.2A.

For some value of the parameters, the parabola may be entirely positive, such that all frequencies k are stable and small fluctuations will decay back to the stable fixed point of uniform concentration. However, as one increases D, namely, the diffusion coefficient of the inhibitor, the parabola will dip lower and eventually touch the k^2-axis at some frequency k_c. This is a critical point: If D increases further, the parabola dips into negative territory, frequencies near k_c acquire positive eigenvalues, and those fluctuations will grow exponentially (figure 12.2B). This can give rise to the spontaneous emergence of a spatial pattern.

12.3.5 Conditions for Turing Instability

From the green curve in figure 12.2A one sees that the bifurcation occurs when

$$\mathrm{Det}\,\mathbf{M}(k) = 0 \tag{12.18}$$

and

$$\frac{d}{dk}\,\mathrm{Det}\,\mathbf{M}(k) = 0. \tag{12.19}$$

Using these two conditions, we can derive a condition for the onset of pattern growth:

$$\frac{\alpha}{A} - \frac{\delta}{D} > \frac{2}{\sqrt{AD}}\sqrt{\beta\gamma - \alpha\delta}, \tag{12.20}$$

and the first frequency that goes unstable is

$$k_c = \sqrt{\frac{1}{2}\left(\frac{\alpha}{A} - \frac{\delta}{D}\right)}. \tag{12.21}$$

It is worth restating the additional conditions that create stability in the absence of diffusion:

$$\operatorname{Tr}\mathbf{M} = \alpha - \delta < 0$$
$$\operatorname{Det}\mathbf{M} = \beta\gamma - \alpha\delta > 0. \tag{12.22}$$

Combining equations (12.20) and (12.22), one finds a requirement that

$$D > A. \tag{12.23}$$

Stated in words, the inhibitor S must diffuse more rapidly than the activator R. Often one finds this requirement summarized as "local activation and long-range inhibition."

12.3.6 The Onset of Pattern Growth

Several aspects of this dynamic instability seem surprising at first. For one, the instability is caused and driven by diffusion. If the morphogens were prevented from spreading laterally, they would remain in stable equilibrium. Generally, one thinks of diffusion as a process that "smooths things out" in space. Instead, here it drives the exponential growth of a small perturbation.

Second, the first pattern that appears spontaneously is at a discrete, nonzero frequency, whose value depends on both the reaction rates and the diffusion coefficients as in equation (12.21). This leads to an interesting prediction regarding the sudden onset of pattern formation in a growing organism.

In the treatment shown in figure 12.2, we assumed that some parameter of the reaction-diffusion system changes slowly over time until eventually, one of the eigenvalues acquires a positive real part, leading to sudden pattern development. One can certainly conceive of such a change in molecular parameters, but there is an alternative explanation.

Suppose that the parameters are fixed and positioned such that a range of spatial frequencies is unstable, but this range of frequencies is not accessible for any variations in the morphogens. For example, if the organism has length L, extending over $0 < x < L$, then there can be no spatial flux of the morphogens at $x = 0$ and $x = L$. This means that any sinusoidal variation of the morphogens will be limited to the discrete Fourier components (see section 3.2.6) of the type $\cos jx\pi/L$ with integer j, such as (figure 12.3)

$$u(x) = \sum_{j=0}^{\infty} u_j \cos\frac{j\pi}{L}x. \tag{12.24}$$

So the only frequencies available for pattern formation are

$$k_j = \frac{j\pi}{L}, \quad j = 1, 2, \ldots. \tag{12.25}$$

When the organism is small, the first spatial mode k_1 lies outside the domain of unstable frequencies. As the organism grows, that first mode moves to lower frequencies, and at a critical size L_c it enters the unstable domain. In this way, pattern formation can get triggered purely by growth of the developing organism without a change in the dynamic parameters of the morphogen system (figure 12.3).

Given the expressivity of the Turing mechanism in accounting for several common aspects of biological patterns, this model has long served as a favorite hypothesis

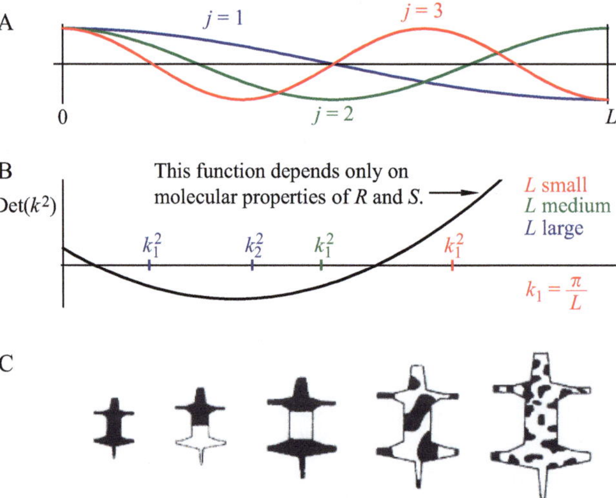

Figure 12.3
A: In an organism of size L, the patterns are restricted to modes with discrete spatial frequencies. B: As the organism grows, the unstable modes enter the allowed domain of frequencies. C: A numerical simulation of coat patterns emerging on the skin of a cow. Only the size of the skin varied between conditions (Adapted from Murray (2003)).

for pattern formation in biological development. More recently, it has become possible to go beyond the phenotype of final patterns and test the model's mechanistic ingredients. For example, the signaling factors Nodal and Lefty serve as an activator/inhibitor pair during patterning of the vertebrate embryo, and Müller et al. (2012) showed that indeed, the diffusion coefficient of the inhibitor far exceeds that of the activator, as predicted by the Turing model in equation (12.23). In other instances, the lateral transport of signaling molecules occurs not via diffusion, but perhaps through cellular movements, and yet the mathematics remain those of the Turing mechanism (Hamada et al., 2014). Still, one does well to remember that there are alternative models that can also account for the formation of periodic patterns (Hiscock and Megason, 2015). For some numerical explorations, see section 13.4.

In this chapter, we will study several applications of nonlinear dynamics in the biological sciences.

13.1 Repressilator

The repressilator (Elowitz and Leibler, 2000) is an artificial genetic circuit made of three repressors: LacI, TetR, and λ cl. The transcription of each repressor is under the control of a promotor that binds another repressor (figure 13.1). As a result, the three repressors cyclically inhibit the transcription of each other, forming a negative feedback loop. Under certain conditions, this gives rise to a sequential oscillation of the repressor concentrations inside a cell (hence the name).

13.1.1 Model

Let us examine the dynamical system used to model the repressilator. It has six coupled differential equations in total: for each of the three repressors, there is one equation for the change in the mRNA concentration, $\dot{m}$, and one for the change in the protein concentration, $\dot{p}$. For instance, here are the equations for the LacI:

$$\dot{m} = -r_{\mathrm{m}} m + \frac{g}{1 + (p_{\mathrm{cl}}/K_{\mathrm{M}})^n} + g_0$$

$$\dot{p} = -r_{\mathrm{p}} p + r_{\mathrm{p}} c m.$$

(13.1)

The first equation in (13.1) governs how the concentration of the $lacI$[1] mRNA transcript changes over time. The first term corresponds to the rate of the breakdown of the mRNA, which is proportional to its concentration with the decay rate r_m. The second and the third terms are about production. Recall that the transcription of $lacI$ is under the control of a promoter that binds the λ cl repressor. Therefore, when the concentration of λ cl is high, the transcription of $lacI$ will be low, though it will be bounded from below by 0; when the concentration of λ cl is low, the transcription of $lacI$ will be high, though it will be bounded from above by some value. Such a relationship is captured by the Hill equation, which expresses the probability that the repressor is bound to the promoter region. For a sanity check, let us consider the case when the concentration of the λ cl repressor (p_{cl}) is high. In that case, the denominator of the second term is large and we are left with g_0, the small rate of transcription of $lacI$ that takes place

---

1. We follow the convention of italicizing the name when referring to the gene / mRNA and capitalizing it when referring to the gene product / protein.

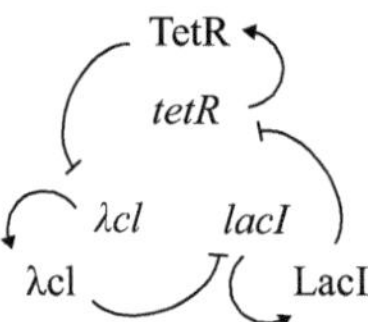

Figure 13.1
The repressilator is a negative feedback network composed of three repressors. The genes are italicized and the gene products (proteins) are capitalized. Lines mean inhibition and arrows mean promotion.

even at maximal repression due to its instrinsic leakiness. On the other hand, when the concentration of λ cl is zero, the denominator of the second term becomes 1, and the growth rate is $g + g_0 \approx g$. This is the maximal transcription rate at zero repression.

The second equation in (13.1) governs how the concentration of the LacI repressor changes over time. Again, the first term is about decay and the second is about growth. We make a simplifying assumption here that both processes have the same rate (r_p). In the second term, the protein production follows the concentration of mRNA (m) multiplied by the translation efficiency (c).

Table 13.1 summarizes the values of these parameters used by the authors.

13.1.2 Dimensional Scaling
We can simplify equation (13.1) by dimensional scaling. There are three units in the model: mRNA concentration, protein concentration, and time. The following change of variables nondimensionalizes the model:

$$M = cm/K_M$$

$$P = p/K_M \tag{13.2}$$

$$\tau = r_m t.$$

What are we doing here? First, the mRNA concentration (m) is converted to protein concentration by multiplying by the transcription efficiency factor, c. That way, the only concentration that we have to deal with is that of the protein. We then rescale protein concentrations by K_M and time by the inverse of mRNA decay rate (r_m). This yields

$$\dot{M}_i = -M_i + \frac{\alpha}{1 + P_j{}^n} + \alpha_0$$

$$\dot{P}_i = -\beta(P_i - M_i), \tag{13.3}$$

where α and α_0 are dimensionless rates of protein production at no repression and maximal repression, respectively, and $\beta = r_p/r_m$ is the ratio of the protein decay rate to the mRNA decay rate. The pairs of subscripts (i, j) are $i =$ (lacI, tetR, cl) and $j =$ (cl, lacI, tetR).

13.1.3 Qualitative Analysis
To analyze the system qualitatively, let us first try to find the fixed points (section 11.3.2) by setting equation (13.3) to zero. It is clear that $\dot{P}_i = 0$ when $P_i = M_i$. Therefore,

Table 13.1
Parameters

Parameter	Value	Unit
r_m	0.5	1/min
r_p	0.1	1/min
g	0.5	Transcripts/s
g_0	5e-4	Transcripts/s
c	20	Protein monomers/transcript
n	2	None
K_M	40	Protein monomers

we obtain

$$M_i = \frac{\alpha}{1 + P_j{}^n} + \alpha_0, \tag{13.4}$$

but since $P_i = M_i$ at the fixed point, we can rewrite this as

$$P_i = \frac{\alpha}{1 + P_j{}^n} + \alpha_0. \tag{13.5}$$

To solve for P_i, we will try to get rid of P_j by substituting equation for P_j in the equations for other repressors. They all take the form of $y = f(x)$, where

$$f(x) = \frac{\alpha}{1 + x^n} + \alpha_0. \tag{13.6}$$

This leads to

$$
\begin{aligned}
P_{\text{lac}} &= f(P_{\text{cl}}) \\
&= f(f(P_{\text{tet}})) \\
&= f(f(f(P_{\text{lac}}))).
\end{aligned}
\tag{13.7}
$$

This is difficult to solve analytically, but we can gain a geometric intuition. We know that $f(x)$ is a monotonically decreasing function, with α controlling the amplitude, α_0 providing vertical shift, and n defining the shape of the curve (i.e. how rapidly it falls). Since $y = P_{\text{lac}}$ (the left hand side of equation 13.7) is a monotonically increasing function, it is clear that it will intersect with $y = f(f(f(P_{\text{lac}})))$ at exactly one point; see an example in figure 13.2 when $\alpha = 10$.

And since the same equation holds for all three repressors, we get the result that the fixed point occurs at $P_{\text{lac}} = P_{\text{tet}} = P_{\text{cl}} = P^*$. This is the unique fixed point for this system.

What about its stability? We learned from section 11.3.3 on linear stability analysis that we can approximate the dynamical system with a Taylor expansion and keep only the first-order term, also known as the **Jacobian matrix**. The Jacobian matrix for this

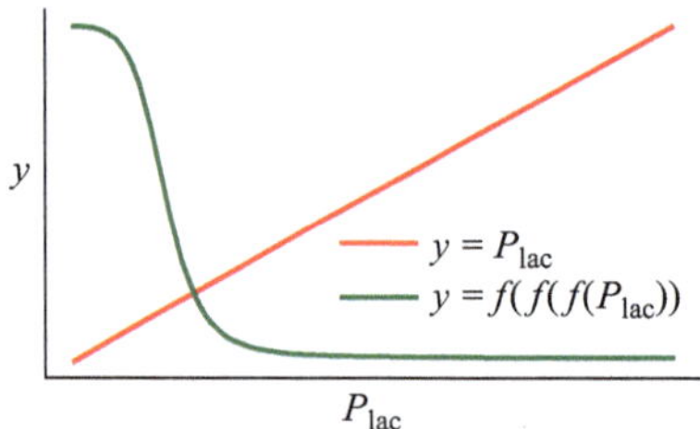

Figure 13.2
Fixed point of the repressilator.

system is

$$J = \begin{pmatrix} -1 & 0 & 0 & 0 & 0 & X \\ 0 & -1 & 0 & X & 0 & 0 \\ 0 & 0 & -1 & 0 & X & 0 \\ \beta & 0 & 0 & -\beta & 0 & 0 \\ 0 & \beta & 0 & 0 & -\beta & 0 \\ 0 & 0 & \beta & 0 & 0 & -\beta \end{pmatrix}, \tag{13.8}$$

where

$$X = -\frac{\alpha n P^{*(n-1)}}{(1 + P^{*n})^2}. \tag{13.9}$$

The next step, as you may recall, is to find the eigenvalues of the Jacobian matrix. When the real parts of all the eigenvalues are negative, the fixed point is stable. In this book, we dealt mainly with one-dimensional (1D) and two-dimensional (2D) systems. The repressilator, however, is six-dimensional, and as such, the eigenvalue analysis is more complicated. It is in fact possible to analytically find the eigenvalues of J (the trick is to think of J as a 2-by-2 block matrix and use the Schur complement to simplify the characteristic polynomial $\det(J - \lambda I)$) and mark stable regions in the α-β plane, as Elowitz and Leibler (2000) do in figure 1b of their paper. This tells us the parameters at which a *bifurcation* occurs. For now, we will skip this and try to infer the dynamics from numerical simulations.

13.1.4 Numerical Simulation

Numerical methods give us a more straightforward way to gain insight into the system's behavior. The code for generating the following figures can be found in the online resources for this book. Once we have solved the system numerically, we can visualize the results in a number of ways. For example, we can examine the trajectory of the system in the phase space. We first plot the trajectory in the 2D spaces spanned by pairs of repressors as in figure 13.3. Note that the trajectory quickly converges to a limit cycle from the inside. This implies that the fixed point we identified in section 13.1.3 is an unstable spiral, with at least one eigenvalue having a positive real part and a nonzero imaginary part. While the system is on the limit cycle, the phase point executes a periodic trajectory. This is verified by plotting the concentrations of the three repressors over time as in figure 13.4, which recapitulates figure 1c from Elowitz and Leibler (2000).

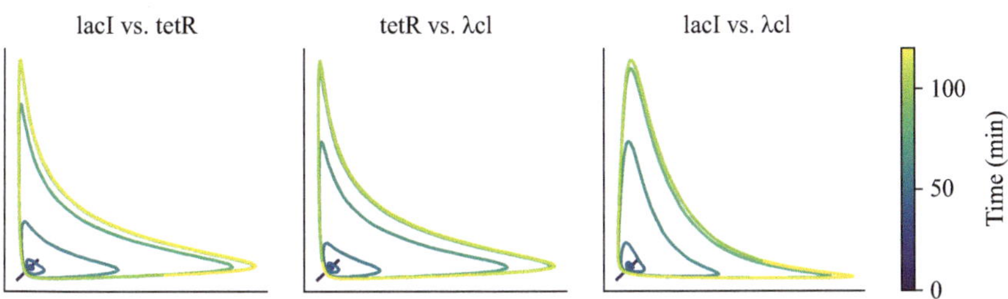

Figure 13.3
Trajectory of the system in the phase space, projected onto 2D subspaces spanned by pairs
of repressors.

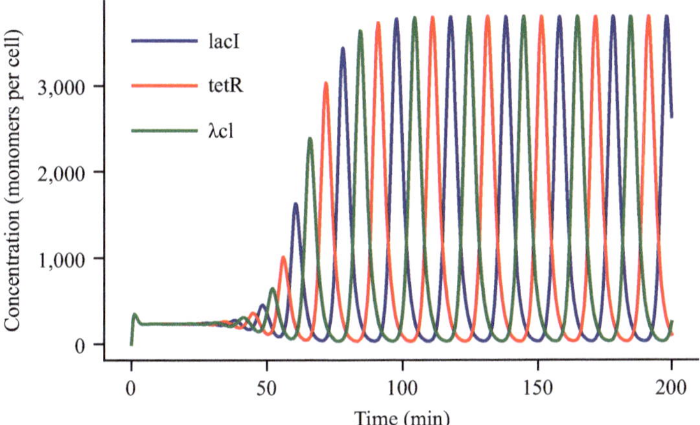

Figure 13.4
Oscillation of the repressors.

This, of course, is just a model; what is remarkable is that this behavior is actually
observed in the living cell! Read the rest of Elowitz and Leibler (2000) for more details
on how they measured the repressilation and how similar this was to the behavior of
naturally occurring biological clocks, such as the proteins involved in the circadian
rhythm. And try to vary the parameters to see whether the system behaves differently.
When does the fixed point become stable?

13.2 Fold Change Detection

Fold change detection is a phenomenon in which a system's repsonse to an input
depends on its relative change against a background, rather than the absolute magnitude of that input (figure 13.4A). Fold change detection is widely observed in biological
phenomena. For instance, *E. coli* move toward an attractive chemical stimulus by sens-
ing the relative change of its concentration against the background (Dahlquist et al.,
1972). Similarly, the smallest change in brightness that the human visual system can
detect increases with the ambient light level, a result known as the Weber-Fechner
law. Here, we will examine how single cells regulate gene expression based on fold
change detection. As shown by Goentoro et al. (2009), this can be achieved with a

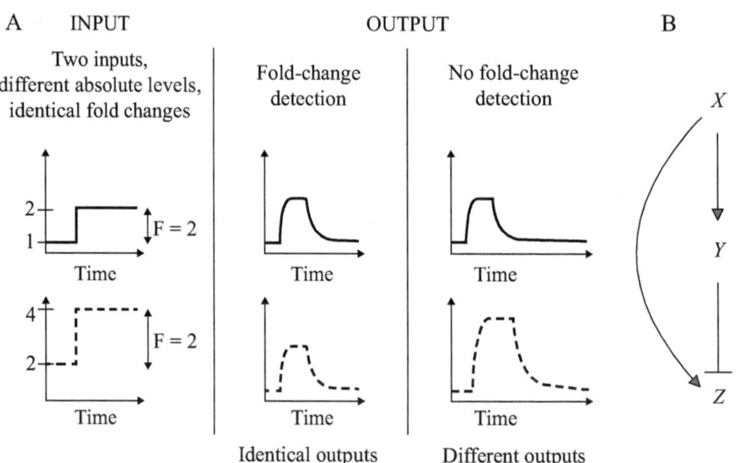

Figure 13.5
(A) Fold change detection (adapted from Goentoro et al., (2009)). (B) Feedforward loop.

transcriptional network motif called a "feedforward loop"—specifically, the incoherent type-1 feedforward loop (Alon, 2019).

13.2.1   Feedforward Loop as a Transcriptional Network Motif

A transcriptional network motif is a frequently appearing pattern in genetic regulatory networks. The feedforward loop is one of the most common such motifs and has the following structure: the transcription factor X activates both the target gene Z and a gene Y that encodes the repressor of Z (figure 13.5B). As we will see, this has the effect of sharpening the time course of the output (Z) by cutting it off early, as well as attenuating the low-frequency components of the time-varying input (X).

Before we continue, we have to note that the activation and repression of Z by X and Y can take many forms. For instance, X and Y may compete for the same sites on the promoter of Z. Or they might have different binding sites and act on Z independently. Finally, they may act cooperatively; that is, the binding of X could facilitate the binding of Y and vice versa. Although Goentoro et al. (2009) considers all three cases, here we will just look at the first case: competition between X and Y for the same site.

13.2.2   Model

We start by identifying the dynamic equations for the system:

$$\dot{Y} = -\alpha_1 Y + \beta_1 \frac{X/K_{XY}}{1 + X/K_{XY}}$$

$$\dot{Z} = -\alpha_2 Z + \beta_2 \frac{X/K_{XZ}}{1 + X/K_{XZ} + Y/K_{YZ}}.$$

(13.10)

The form of equation (13.10) may be familiar from the repressilator example (section 13.1). As before, the first term is about decay and the second term is about growth. The growth term includes a Hill function, with constants K_{AB} characterizing the binding strength of A to the promoter of B. The Hill coeffcent for a ligand to a binding site is assumed to be $n = 1$.

There are two ways to show that this system reports relative fold changes under certain conditions. The first is an argument based on dimensional scaling. Consider the case when X binds weakly and Y binds strongly to their respective binding sites. This means that $1 >> X/K_{XY}$ and $Y/K_{YZ} >> 1 + X/K_{XZ}$. This allows us to simplify the system as follows:

$$\dot{Y} = -\alpha_1 Y + \beta_1 X/K_{XY}$$
$$\dot{Z} = -\alpha_2 Z + \beta_2 \frac{K_{YZ}}{K_{XZ}} \frac{X}{Y}. \tag{13.11}$$

We then make the following substitutions to nondimensionalize the system:

$$x = X/X_0$$
$$y = \frac{\alpha_1 K_{XY}}{\beta_1 X_0} Y$$
$$z = \frac{\alpha_2 \beta_1}{\alpha_1 \beta_2} \frac{K_{XZ}}{K_{XY} K_{YZ}} Z \tag{13.12}$$
$$\tau = \alpha_1 t,$$

where X_0 is the basal level of the input X and x is the dimensionless variable denoting the fold change in X. This leads to

$$\dot{y} = -y + x$$
$$r\dot{z} = -z + x/y. \tag{13.13}$$

The only parameter of this system is $r = \alpha_1/\alpha_2$, which is a ratio of the half-lives of Y and Z. Note that the output of the system, z, depends on the fold change (x) rather than the absolute change of the input (X), and therefore, it will act like a fold change detector (figure 13.5). Furthermore, we can see that $\dot{y} = 0$ when $y = x$. So the concentration of the repressor Y will track the fold change and serve as something like an internal representation of the input at steady state.

13.2.3 Numerical Simulation

The second way to see this is to directly simulate equations (13.10) and explore when the system behaves like a fold change detector (figure 13.6). We assume that X goes from 1 to 2 at $t = 20$ and from 2 to 4 at $t = 40$. See the code in the online resources for details.

13.2.4 What Is the Utility of Fold Change Detection in Biological Systems?

Fold change detection cannot be achieved with linear circuits—those only detect absolute changes. So what use does this have in biology? For one thing, it can provide robustness against noise. In many cases, the noise that corrupts an input signal scales with the mean, so a higher basal level means higher fluctuations. Detecting fold change can adjust for this increase in noise. Another benefit is that it increases the dynamic range of the system. For example, our visual system must perform throughout the 10 log units of change in ambient light intensity that we experience from dawn to dusk. At the same time, the useful visual information lies in the *relative* intensities at different

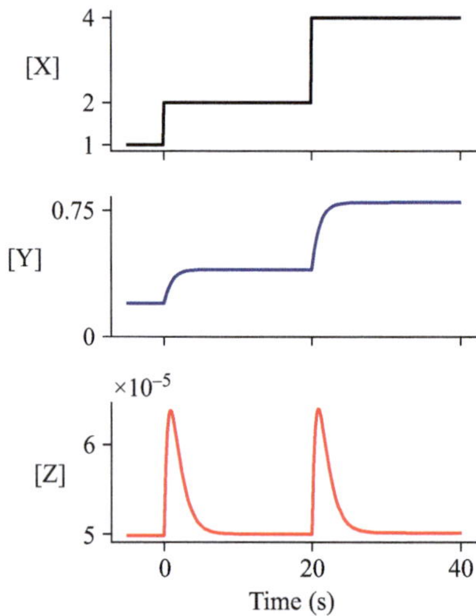

Figure 13.6
Numerical solution. Note that the output [Z] is the same for both when the input [X] goes from 1 to 2 and 2 to 4, demonstrating fold change detection.

points in the scene, as those indicate the reflectance of different objects. Here, a sensory system that computes relative changes serves to encode the relevant information, while rejecting the useless signal about absolute light intensity.

13.3 Bistability

In this section, we will examine the dynamics of a system with two components that inhibit each other (figure 13.7). Examples of such systems are common in the biological sciences; here, we have picked a few from systems neuroscience. This will lead us back to the concept of bifurcations (section 11.5.2.1).

Before delving in, we might ask ourselves: What kind of dynamics should we expect from this system? A range of outcomes are possible, it seems. For example, the two components might both go silent because they inhibit each other; or maybe they will oscillate, if the conditions are just right; or maybe one will inhibit the other more strongly by chance and emerge as the "winner." Keep these possibilities in mind as you read on.

13.3.1 Binocular Rivalry

Our first example involves binocular rivalry (exercise 8.1.14 from Strogatz, 2015). When two images are presented to our eyes at the same time (i.e. one image to the left eye, the other to the right eye, using two separate displays), we often see one or the other image rather than their superposition. If we continue to look at the images, our perception might switch back and forth between the two; but remarkably, we always see only one image (note that this is not true of every sensory system; if I play two sounds to your ears, for example, you will hear them together). A simple model of this phenomenon posits that there are two populations of neurons in our brain (x_1 and x_2), each receiving excitatory input (I) from one eye. These two populations also send inhibitory connections to each other. As a result, the activity of the two populations evolves based on

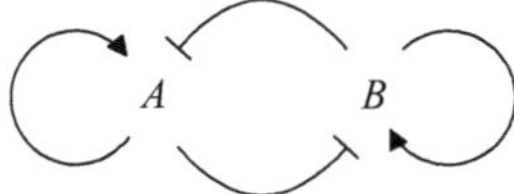

Figure 13.7
A two-component system with self-excitation and mutual inhibition.

the sensory excitation and the mutual inhibition, plus "leakage," or the tendency to bring the system back to baseline. The following set of equations summarizes these ideas:

$$\dot{x}_1 = -x_1 + F(I - bx_2)$$
$$\dot{x}_2 = -x_2 + F(I - bx_1). \tag{13.14}$$

Here, x_i represents the activity of the ith neural population; b is a parameter that governs how strongly one neural population inhibits the other; and

$$F(x) = \frac{1}{1 + e^{-x}} \tag{13.15}$$

is a sigmoid function that transforms the sum of the excitation from the eye and the inhibition from the other neural population.

How do we proceed? We could try to solve for the fixed point (x_1^*, x_2^*); given the symmetry, it looks like it will satisfy the condition $x_1^* = x_2^*$. We could also simulate the time evolution of the system numerically, as we have done before. But let us try something different this time and simply plot the vector field generated by the system.

In figure 13.8, we plotted the phase plane for when I is fixed at 1 and b increases from 0 (no inhibition) to 7 (strong inhibition). When there is no inhibition, the two populations do not interact, and they settle to a stable fixed point where the excitatory input is balanced by the leakage. As b gets larger, we see that the fixed point remains stable, but most of the dynamics is now confined to a single dimension from the top-left to the bottom-right of the plot. The dynamics in the orthogonal direction simply serves to push the flow onto that line (what does this say about the relative magnitudes of the eigenvalues of the Jacobian?). At the fixed point, $x_1^* = x_2^*$ as expected; this means that both neural populations are active at about the same level. Finally, when b is large, the stable fixed point becomes unstable (technically, a saddle point) and gives rise to two new stable fixed points. In other words, the system undergoes a **(supercritical) pitchfork bifurcation**. Note that one new stable fixed point is on the horizontal axis and the other is on the vertical axis. Since the trajectory in the phase plane will end up in only one of these two, this implies that only one population would be active. This is why we see only one percept. Which one wins? That depends on the initial condition; if the noise in the brain makes one population slighly more active than the other (i.e., slightly to one side of the **separatrix** defined by $x_1 = x_2$), then that will be the winner.

It is interesting that such a simple model of mutual inhibition can explain the competition and dominance of one percept over the other. What about the oscillation between the two percepts over time? This can be achieved by introducing another dynamic variable for adaptation; we will not go into details here (see exercise 8.2.17

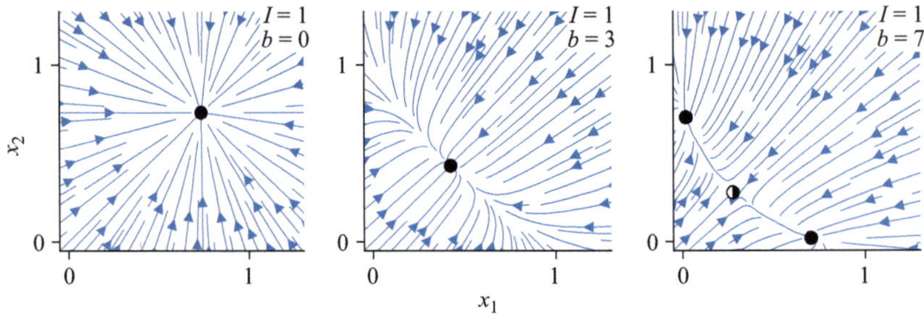

Figure 13.8
Dynamics of binocular rivalry. The parameter b that controls the strength of inhibitory coupling is increased from 0 to 7, giving rise to new stable fixed points.

in Strogatz, 2015, for a full treatment), but the idea is that the dominant neuron gets fatigued after a while, allowing the other neuron to take over, and this cycle repeats.

13.3.2　Flexible Control of Mutual Inhibition

The next example extends the basic model for binocular rivalry to the domain of cognition: the maintenance of a recent past experience in the working memory for decision making (Machens et al., 2005). First, here is some background on the behavior ("two-interval discrimination") that the animal is performing, which is a little more complicated than passively viewing two images presented to your eyes:

1. A mechanical vibration of frequency f1 is applied to the tip of the finger of a macaque monkey.
2. After 3 s, another vibration of frequency f2 is presented.
3. The monkey reports (after much training) whether f1 is greater than f2 (yes or no); if it does so correctly, it receives a reward (figure 13.9A–B).

While the monkey is engaged in the task, neuroscientists perform electrophysiological recordings in the monkey's brain, specifically the somatosensory (S2) and prefrontal cortices (PFC). The S2 neurons are of two types: the (+)-neuron, whose firing rate increases as a function of the stimulus frequency; and the (−)-neuron, whose firing rate decreases. These neurons are purely sensory in that they respond transiently to the stimulus onset and remain silent during the interstimulus interval (figure 13.9E–F). In the PFC, the researchers also find two types of neurons, but they are more interesting. The PFC (+)-neuron responds like the S2 (+)-neuron to f1, but during the interstimulus interval (when there is no stimulus), its firing rate is maintained. Then once f2 is presented, its response either shoots up or plummets to baseline, depending on whether the monkey reports yes or no. The PFC (−)-neuron is the opposite; it is driven by low f1, and the firing rate is maintained during interstimulus interval and shoots up if the monkey reports no (figure 13.9C–D). To summarize, the PFC neurons are different from the S2 neurons in that they maintain information about f1 throughout the interstimulus interval and signal the monkey's decision once f2 arrives.

Given the two populations with opposing response properties ("tuning"), Machens et al. (2005) propose a model of mutual inhibition, much like the one from the binocular rivalry example. Although the most sophisticated version includes a network of

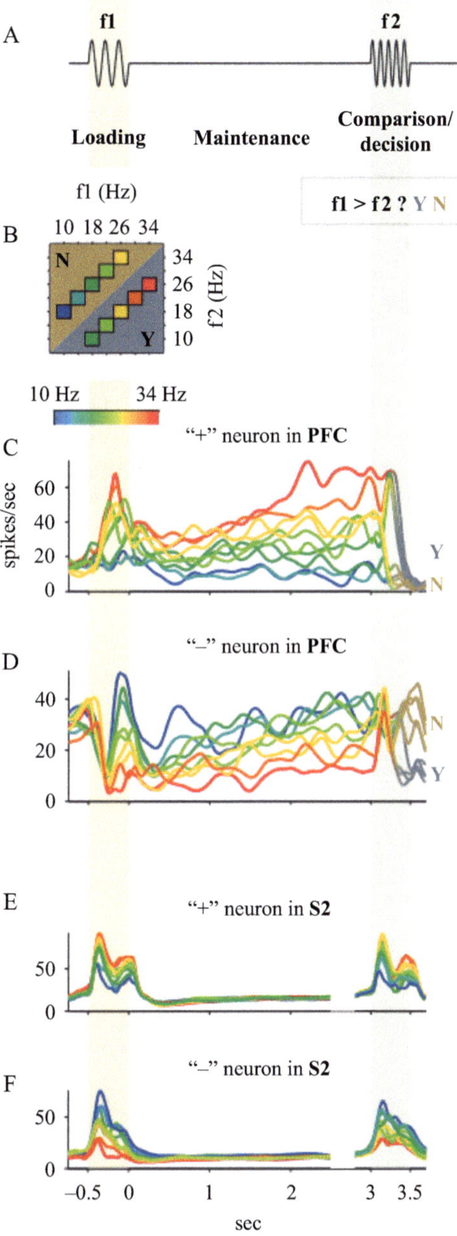

Figure 13.9
Response profiles of two types of neurons in the PFC and S2 during two-interval discrimination. A and B depict the task structure and the stimulus values for f1 and f2. C and D display a representative response of a (+) and a (−) neuron in PFC, respectively. E and F display a representative response of a (+) and (−) neuron in S2, respectively (adapted from Machens et al. (2005)).

spiking neurons, a two-node network suffices to capture the dynamics:

$$\tau\dot{x} = -x + f(-\omega y + E_x)$$
$$\tau\dot{y} = -y + f(-\omega x + E_y). \tag{13.16}$$

Here, x and y are the dynamic variables that correspond to the two PFC neuron types. E_i is the excitatory input to neuron type i and ω is the strength of the inhibitory synaptic connection between x and y, assumed to be symmetric. As you can see, equation (13.16) is very similar to equation (13.14) and gives rise to much the same dynamics. Again, f is a sigmoid-like curve that monotonically increases and saturates at the top.

The authors characterize the system with a **nullcline analysis**—if you set equation (13.16) to zero, you get two curves whose intersection gives you the fixed points (i.e., when $\dot{x} = \dot{y} = 0$). What makes this system different from the binocular rivalry example is that the input to the system at different phases of the task changes, thereby changing the nullclines. For example, in the "loading" phase (when f1 is presented), the two neuron types receive input from the corresponding neuron types in S2 (figure 13.10, left). Depending on the strength of stimulus f1, the S2 inputs to the two populations are different, and the dynamics settles to a different stable fixed point. During the "maintenance" phase (the interstimulus interval), the curves shift because the excitatory input from S2 neurons are now absent, and this generates nullclines that overlap at infinitely many fixed poins: a *line attractor* (figure 13.10, middle). Finally, in the "decision" phase (when f2 arrives), the nullclines shift yet again, giving rise to two new stable fixed points that correspond to the yes / no decision (figure 13.10, right).

The model relies on many assumptions. For example, it assumes that with no S2 input, the dynamics of PFC neurons generates a line attractor. This is not easy to get,

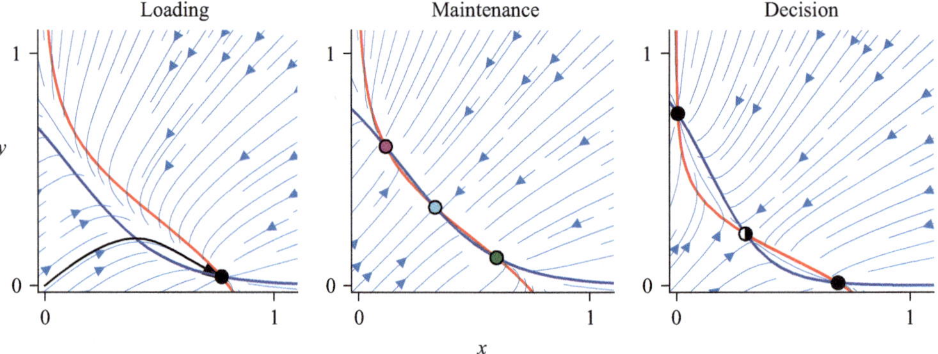

Figure 13.10
Changing input to the system during the three phases of the behavior causes shifts in the nullclines (red and blue). When the first stimulus arrives (loading phase, left), the two neural populations receive different inputs (one strong, the other weak). As a result, the system settles at a stable fixed point, as shown by the example trajectory in black. During the interstimulus interval (maintenance phase, middle) there is no S2 input and the overlapping nullclines generate a line attractor (three fixed points are shown, but technically, there is an infinite number of them along the line). Finally, when the second stimulus arrives (decision phase, right), the nullclines shift again, and the system ends up at one of the two new stable fixed points.

as it requires the nullclines to line up just right to generate an infinite number of fixed points (figure 13.10). Furthermore, the nullclines must "bend" in the decision phase to generate a bifurcation. Figure 13.10 achieves this by increasing the mutual inhibition, but Machens et al. (2005) hypothesize that this takes place via the crossing of S2 input, such that S2 neurons of one type (e.g., (+)) now excite PFC neurons of the other type (e.g. (−)). This is probably arranged by an unspecified control system that is outside the parts being modeled here. The empirical evidence for these details is yet to emerge, but the model still serves as a parsimonious explanation of how biological neurons may conspire to implement memory-guided decisions.

13.3.3 Other Possible Behaviors

So far, we have examined two examples where mutual inhibition gives rise to switch like dynamics. But other kind of behaviors are also possible—for example, such a network of mutually inhibiting neurons can display synchronized oscillation (chapter 5.9 from Dayan and Abbott, 2005) as follows.

$$\tau_m \dot{V}_m = (E_L - V_m) + k\alpha(E_s - V_m) + E_{ext}. \tag{13.17}$$

Equation (13.17) describes the dynamics of the membrane potential V_m of a neuron. We will delve deeper into this topic in sections 13.6.3 and 13.6.4, so we will be rather brief here. E_L is the baseline membrane potential; if the neuron receives no input, its membrane potential V settles at E_L. The third term E_{ext} is the external input injected into the neuron by the experimenter. The second term represents the synaptic input from the presynaptic neuron. E_s controls the "sign" of the synapse: the synapse is excitatory if $E_s > E_L$, and inhibitory if $E_s < E_L$. An important parameter here is α, which modulates the synaptic input and is itself a dynamic variable with the following form:

$$\alpha(t) = P_{max} \frac{t}{\tau_s} \exp(1 - t/\tau_s), \tag{13.18}$$

where P_{max} is the maximum synaptic conductance, τ_s is the synaptic time constant, and $t = 0$ marks the timing of the action potential from the presynaptic neuron. In other words, it converts an action potential from the presynaptic neuron to a postsynaptic conductance as shown in figure 13.11.

Note that this function reaches a peak when $t = \tau_s$ and then decays with time constant τ_s. Finally, k is a constant that scales the amplitude of the postsynaptic potential and τ_m is the membrane time constant.

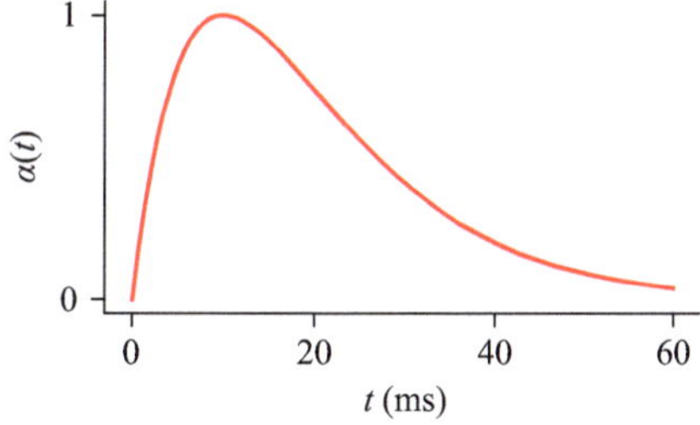

Figure 13.11
An α function for modeling synaptic conductance, with $P_{max} = 1$ and $\tau_s = 10$ ms.

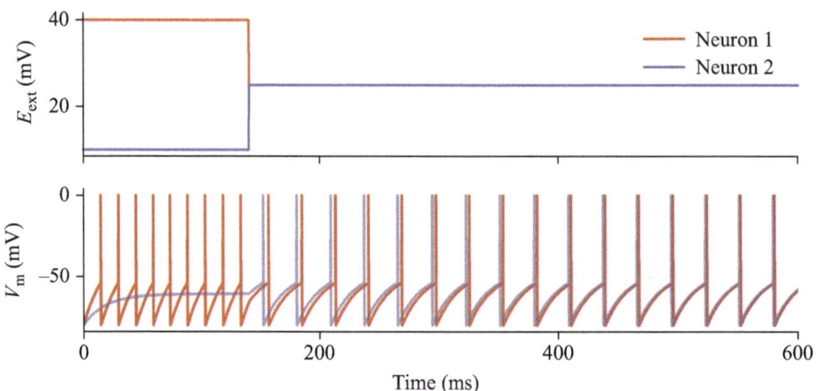

Figure 13.12
Synchronized oscillation from mutual inhibition. The two neurons are initially injected with different levels of current, and as a result, neuron 1 is depolarized and neuron 2 is hyperpolarized. At time 140 ms, the input current becomes equal and the two neurons start firing synchronously.

In addition, we require a peculiar kind of nonlinearity on V: whenever $V > V_{th}$, it is reset to V_{reset} in the next time step. This qualitatively captures the firing of an action potential by the neuron when the membrane potential reaches a certain threshold. We will encounter this again in section 13.6.4.

We now simulate the membrane potential for the two neurons in figure 13.12 (implementation is available in the online resources). We start by making E_{ext} different: neuron 1 will receive high input, while neuron 2 will receive low input. This will make the two neurons asynchronous. Then, at $t = 140$ ms, the external input will be equalized.

Interestingly, the two neurons do not compete in this case; rather, they become perfectly synchronized once the external input is matched. This is counterintuitive, as one might expect that when one neuron is active, it will silence the other neuron via the inhibitory synapse, and as a result, they will fire out of phase. Feel free to play with the code and discover for yourself which parameters are important; for example, what happens when the synaptic time constant changes? What if the synapse is excitatory?

13.4   Turing Patterns

We learned in section 12.3 about the reaction-diffusion equations, which specify a dynamical system of two **morphogens** (activator and inhibitor) that react and diffuse to generate a spatiotemporal pattern from an unstructured initial state. This was proposed as a possible mechanism of biological pattern formation by Turing (1952), and some empirical evidence exists to support this claim (e.g., Müller et al., 2012). But is it really capable of producing the diverse and elaborate patterns that we see in biology? After all, the analysis in section 12.3.2 was mostly on a linearized system in one dimension, whereas real biological patterns typically take place in two or three dimensions. Furthermore, the pattern on an animal may change over time during development, and it is unclear how Turing's model can explain this. In this example, we will explore

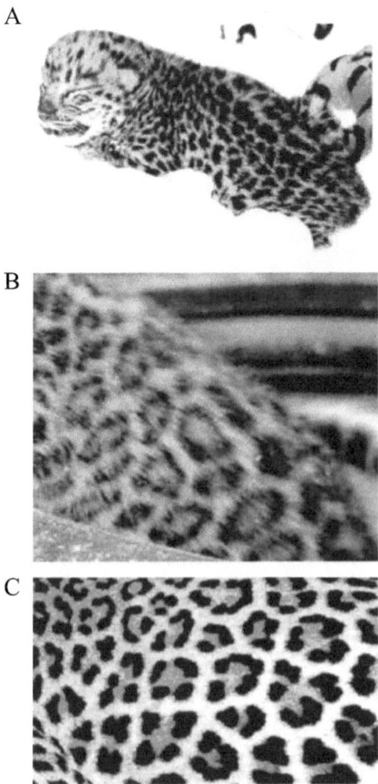

Figure 13.13
Pigmentation pattern on the leopard's coat: A. spots, B. rings, and C. rosettes. Adapted from Liu et al. (2006).

these issues by using the reaction-diffusion mechanism to generate a realistic biological pattern: the leopard's coat (Liu et al., 2006).

The pigmentation pattern on the leopard's coat is often called spots, but it actually varies throughout the animal's life. When it is young, it is indeed spotty (figure 13.13A). As it grows larger, however, the inside of the spots become lighter, making them more ringlike (figure 13.13B). Finally, the adult leopard sports "rosettes," in which the rings break into segments and arrange themselves like flower petals (figure 13.13C).

Liu et al. (2006) use a reaction-diffusion system with evolving parameters to capture this progression. Note that ∇^2 refers to the Laplacian operator (equation 8.25):

$$\dot{u} = D\delta\nabla^2 u + \alpha u + v - r_2 uv - \alpha r_3 uv^2 \tag{13.19}$$

$$\dot{v} = \delta\nabla^2 v + \gamma u + \beta v + r_2 uv + \alpha r_3 uv^2. \tag{13.20}$$

The reaction part here differs from the standard model discussed previously. In particular, it has quadratic and cubic terms, which represent nonlinear interactions between the two morphogens. We will simulate this dynamical system numerically to produce the three types of pigmentation patterns. For the code, please see the online resources.

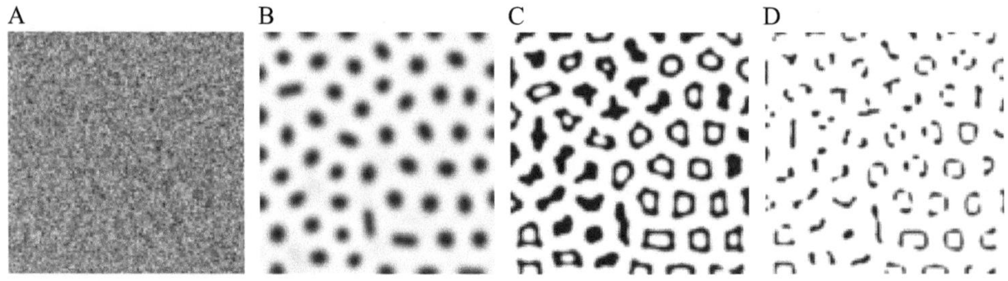

Figure 13.14
Generation of distinct coat patterns that a leopard sports during development. The four
panels depict random initialization (A), spots (B), rings (C), and rosettes (D).

First, the model is run with a random initial condition (figure 13.14A) and parame-
ters $D = 0.45$, $\delta = 6.0$, $\alpha = 0.899$, $\beta = -0.91$, $\gamma = -\alpha = -0.899$, $r_2 = 2.0$, and $r_3 = 3.5$. This
produces spots (figure 13.14B). Next, we run the model again, this time using the spots
as the initial condition and varying the parameter that corresponds to the strength of
the nonlinear interaction, specifically the quadratic term (r_2). This leads to a ringlike
pattern (figure 13.14C). Finally, we run the model again, with the rings as the initial
condition and lowering the parameter that controls the strength of the diffusion term
(δ). This gives us the rosettes (figure 13.14D).

We have seen that the reaction-diffusion mechanism is capable of generating a
biologically realistic sequence of patterns, given a more complex reaction and allowing
for changes in the parameters during development. The change in parameters may
correspond to physiological changes the cells experience as the organism grows larger,
or the different transcription programs that are turned on or off. So in principle, Turing's
model remains a viable hypothesis of pattern formation. It also has the virtue of being
simple: the system consists of just the two morphogens that react and diffuse. Is this
really the mechanism behind the leopard's coat pigmentation? Here, we quote Phillips
et al. (2012, p. 919):

Mechanistically, however, it is unlikely that animal skin patterns actually arise from the
mechanism that Turing envisioned. In general, the sizes of spots and stripes in animal skin
are much too large to be consistent with molecular diffusion. Furthermore, as the developmental biology of animal patterning has been explored, it has become clear that the
color patterns are due to migration and proliferation of dedicated pigment-carrying cells
called melanophores that arise from specific cellular precursors and are not subject to a
chemical "reaction" that can change their colors. Because these Turing-like patterns are
most likely not generated by reaction–diffusion mechanisms, the Turing model is widely
considered to be "wrong." But…the underlying concept is tremendously useful, and slight
modifications of the idea do clearly operate in biological pattern formation. Furthermore,
the elegant mathematics developed by Turing may well still apply to animal skin patterns,
even if the underlying physical processes are distinct. For example, random cell migration
may effectively take the place of diffusion.

13.5 Circadian Rhythms

Perhaps the most universal instance of a biological oscillator is the circadian clock.
Almost every living thing has a circadian rhythm, adapted for life on this planet,

where the environmental conditions, such as light and temperature, cycle with a 24 hour period. The rhythm manifests itself in many ways, such as the sleep schedule of animals. Importantly, these rhythms are not just driven passively by the cycling environmental conditions. Instead, each organism has an endogenous clock that cycles with a period of approximately ("circa") one day ("dian"). This allows the organism's internal functions to prepare for the next cycle of light and dark, rather than having to wait for confirmation that yet again the Sun has risen in the morning. This clock continues to run even when all external cues are taken away.

For example, figure 13.15 shows a sleep log of a typical human observed in a laboratory. For the first 20 days, sleep (dark bars) occurs on a circadian schedule with period of 24 h. Then all external time cues were removed. He continued sleeping on a periodic schedule, but now with a period of 25.3 h. This is a common occurrence—most people's free-running period is a bit longer than a day. So we have an internal clock that is tuned to approximately a one-day period. And that clock can be entrained by physical cues from the environment, notably light and temperature, so it accelerates a bit and follows a daily cycle under normal conditions.

For some people, called "Non-24" subjects, that entrainment doesn't seem to happen. Their sleep schedule slips every day, much like that of a normal person in the isolation lab. These people spend most of their lives out of synch with their surroundings. Frequently, this happens to blind people, reflecting the fact that entrainment is

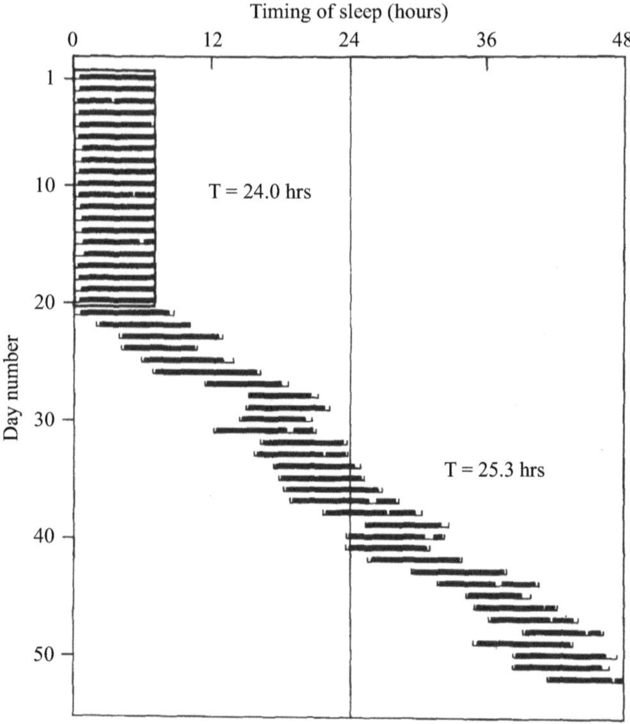

Figure 13.15
Sleep-wake pattern of a normal 22-year-old male subject, with cues about circadian time present (top) and absent (bottom). Adapted from Czeisler et al. (1981).

largely driven by light entering the eyes. However, some of the "Non-24" subjects have no overt sensory deficit. One obvious hypothesis is that they are somehow lacking the signaling pathway that connects the sensory cues to the clock mechanism. However, there is a more subtle possibility, which traces back to a bifurcation. Here, we explore this idea further.

13.5.1 A Toy Model of the Circadian Oscillator

Actually, a good amount is known about the biochemical and cellular mechanisms of the circadian clock (e.g. see the review by Foster, 2020). Fundamentally, it boils down to a negative feedback whereby certain proteins suppress the transcription of their own genes. Qualitatively, this resembles the repressilator system described in section 13.1. So let us consider the case of a two-component system (e.g. a protein and its repressor) and suppose that the dynamics of the system produce a stable limit cycle. Perhaps the simplest case of such a limit cycle is the Poincaré oscillator (figure 13.16), with variables r and ϕ:

$$\dot{r} = \lambda r (1 - r)$$
$$\dot{\phi} = 2\pi. \tag{13.21}$$

The system's phase point is given by radial coordinates (r, ϕ). The phase point revolves around the origin with a constant angular velocity and a period of 1. Meanwhile, radius r undergoes logistic dynamics as in section (11.3.1) with a fixed point at $r = 1$. Thus all the trajectories converge on the limit cycle given by the unit circle.

Next, we specify how the circadian external cues act on this oscillator. Let us suppose that these cues consist of discrete light flashes that happen with a period of T (figure 13.17). When a flash happens, that perturbs the phase point by displacing it to the right by amount A:

$$x \xrightarrow{\text{flash}} x + A. \tag{13.22}$$

For example, the light flash may produce more of one of the two reactants. This moves the phase point transiently off the limit cycle. Then it continues to follow the usual dynamics and eventually regains the limit cycle. Note that the oscillator's phase ϕ gets delayed somewhat if the flash happens early in the cycle ($\phi < \pi$), whereas it gets advanced if the flash happens late in the cycle ($\phi > \pi$). There are three parameters: the rate constant λ determines how fast any radial perturbations relax back to

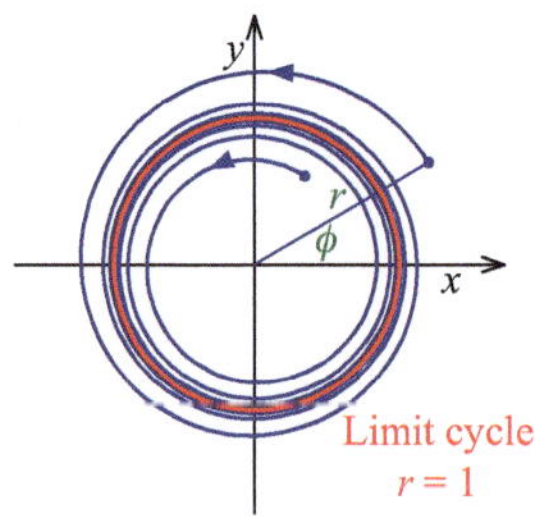

Figure 13.16
A simple limit cycle oscillator.

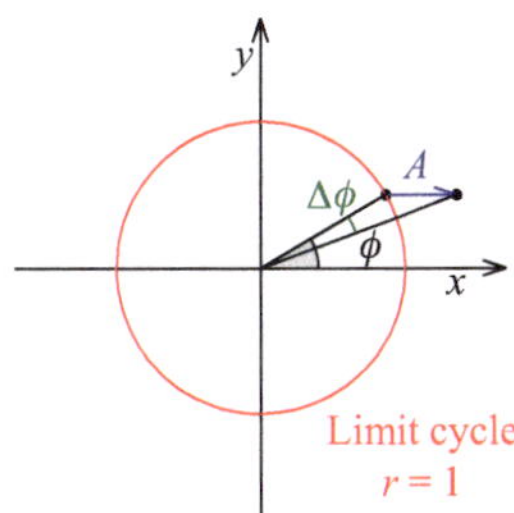

Figure 13.17
A light flash shifts the oscillator's phase point along the x-axis.

the limit cycle; and A and T spell out the strength and the period of the perturbation, respectively. All the dimensional parameters have been distilled away already.

13.5.2 An Even Simpler Model

Next, we suppose that the rate constant λ is very large, so any radial perturbations of the phase point relax almost immediately back to $r = 1$. So we only need to consider how the flashes perturb the phase ϕ. For convenience, let us imagine that the periodic flashes are produced by another oscillator, traditionally called "Zeitgeber" (time giver). The Zeitgeber also has a circular phase $\theta(t)$ and a constant period:

$$\dot{\theta} = \frac{2\pi}{T}. \tag{13.23}$$

The flash happens whenever θ crosses 0. At that point, the oscillator's phase ϕ gets shifted by the following amount:

$$\Delta\phi \approx -A\sin(\phi(t) - \theta(t)), \tag{13.24}$$

where the approximation holds if $A \ll 1$ (figure 13.17). Suppose that the oscillator is running ahead of the Zeitgeber, $\phi > \theta$. Then the next flash will set ϕ back a little, bringing it closer to θ. On the other hand, if the oscillator is behind, with $\phi < \theta$, the next flash will advance it a bit, again bringing it closer to θ.

Next, we take a coarse-grained approach to time, and ask how the phase difference $\Phi(t) = \phi(t) - \theta(t)$ develops from one period of the flash to the next. In one period T, the Zeitgeber phase $\theta(t)$ advances by 2π. Meanwhile, the oscillator phase $\phi(t)$ advances by $2\pi T$, and then the flash resets it by $\Delta\phi$. Setting $\tau = \frac{t}{T}$, one gets

$$\frac{d\Phi}{d\tau} = 2\pi(T - 1) - A\sin(\Phi). \tag{13.25}$$

So we are down to a 1D flow. The dynamical system of figure 13.18 looks familiar. The phase portrait is governed by a single parameter:

$$c = \frac{2\pi(T - 1)}{A}. \tag{13.26}$$

If $|c| < 1$, then there are two fixed points: one stable, the other unstable. The system ends up at the stable fixed point, where the relative phase is

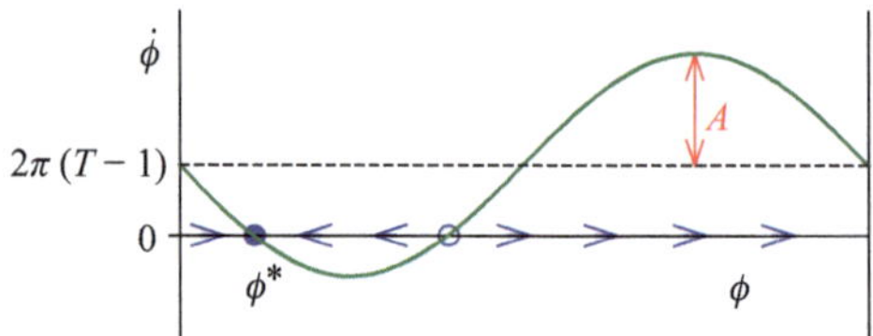

Figure 13.18
Flow diagram for the relative phase between the oscillator and the Zeitgeber.

$$\Phi^* = -\arcsin 2\pi\,\frac{T-1}{A}.\tag{13.27}$$

This corresponds to successful entrainment of the oscillator to the Zeitgeber's period. If the oscillator's intrinsic period is longer than the Zeitgeber's, then the oscillator will lag ($\Phi^* < 0$); otherwise, it will run ahead ($\Phi^* > 0$).

On the other hand, if $|c| > 1$, then there are no fixed points. The relative phase is unbounded, and the oscillator runs away on its own period. It may slow down and speed up a little according to the Zeitgeber's pulses, but on average, it follows its own period. No entrainment is possible.

We recognize the loss of entrainment as the outcome of a saddle-node bifurcation, governed by the parameter c with critical values at ± 1. Mechanistically, this may happen because the coupling A is too weak or the frequency difference $T-1$ is too large.

Returning to the unfortunate N on 24 individuals: if their lack of entrainment is due to the subtle effects considered here, rather than an all-or-nothing deficit in coupling, one would predict that they should be able to entrain to a Zeitgeber frequency that is closer, but not identical, to their own.

13.5.3   Further Reading
Czeisler, C.A., Richardson, G.S., Coleman, R.M., Zimmerman, J.C., Moore-Ede, M.C., Dement, W.C., and Weitzman, E.D. (1981). Chronotherapy: Resetting the circadian clocks of patients with delayed sleep phase insomnia. *Sleep* 4, 1–21.

Ermentrout, G.B., and Rinzel, J. (1984). Beyond a pacemaker's entrainment limit: Phase walk-through. *American Journal of Physiology* 246, R102–106.

13.6   How Do Neurons Communicate?

13.6.1   Why Do Neurons Use Electrical Signals?
The nervous system needs to control the body in response to external events, and some of these events happen quite suddenly. When you are driving down the road and a brake lights up on the car in front of you, the delay until your foot moves to the brake is only ~200 ms. And that requires processing through many neurons, and over a total eye-foot distance of more than 1 m. What are the biophysical options for such rapid signal transmission?

Over short distances, much of biological communication is handled by the passive transport of chemicals: a molecule diffuses from one place to another to elicit an effect. But, as discussed in section 8.1, diffusion occurs slowly. Furthermore, it gets slower the greater the distance that needs to be covered. The highest diffusion coefficients for small molecules in water are $\sim 10^{-5}$ cm^2/s. So in 200 ms, such a signal will spread about 10 μm—too slow by at least 5 log units.

Another option is mechanical force, and that option gets used during the brake pedal action. Muscles in your upper leg contract using active transport of myosin proteins on actin filaments. Those filaments are connected to tendons, which send the force down the leg to the joint where it is needed. In principle, one could imagine a nervous system based on tugging cables.

Instead, nature has employed a different option—namely, electrical signaling. Electrical pulses travel down the nerve fibers of the spinal cord and ultimately cause muscle contraction. Electricity can spread rapidly, up to the speed of light, because it doesn't require mass transport: the ions coming out of the bottom end of a spinal nerve fiber are not the same ions that entered the nerve fiber at the top.

13.6.2 Why Do Neurons Produce Electrical Pulses?

A long nerve fiber acts like a leaky cable. The cell membrane is an electrically insulating tube, but not perfectly so. The cell cytoplasm is electrically conductive, but again not perfectly. So if an electric voltage is applied between the inside and outside of the fiber at one end, current will flow through the fiber, and some of that voltage appears at the other end. But along the way, some of the current leaks through the cell membrane, so the output signal is much weaker than the input signal. Typically, the signal decays exponentially with distance, and the decay length is about 1 mm:

$$V(x) = V(0)\, e^{-x/\lambda}, \quad \lambda \approx 1 \text{ mm}. \tag{13.28}$$

You get to explore the origins and implications of this in exercise 10.25. So after that 1 m of spinal cord, the signal would be diminished by a factor $e^{-100} \approx 10^{-43}$. This is clearly not an option! The nervous system does use passive conduction over short distances of order λ, but for everything over 1 mm or so, it switches to an active mode.

The **action potential** is a short pulse of voltage across the cell membrane, only ~ 1 ms long. The pulse travels rapidly down the nerve fiber, at speeds around 1–10 m/s. It keeps the same shape along the way without attenuation because the voltage gets regenerated continuously, using free energy stored in the cell's transmembrane gradients. The cost of such reliable transmission is that the pulse amplitude is now standardized, and it no longer conveys any information about the input signal at the origin of the fiber. Instead, all the usable information lies in the timing of the pulses.

13.6.3 What Is the Mechanism of the Action Potential?

The explanation of the action potential was the first high-profile contribution of mathematical methods to neurobiology. In a startling series of papers, Alan Hodgkin and Andrew Huxley laid out a mathematical model that beautifully accounted for a slew of experimental observations that they had begun before the World War II broke out (Hodgkin and Huxley, 1952).

A good starting point is the so-called **equivalent circuit** of a neuron (figure 13.19). It represents the electrical state of the cell membrane based on three batteries (E_K, E_{Na} and E_L), the associated conductances (g_K, g_{Na} and g_L), and a capacitor (C). The batteries represent the chemical gradients of the relevant ions that are set up by ionic pumps using ATP (adenosine triphosphate). For example, sodium ions (Na$^+$) are pumped out of the cell, which creates an opposing **reversal potential** of $E_{Na} \approx 50$ mV (inside minus outside). By contrast, potassium ions are pumped into the cell, so $E_K \approx -77$ mV. E_L refers

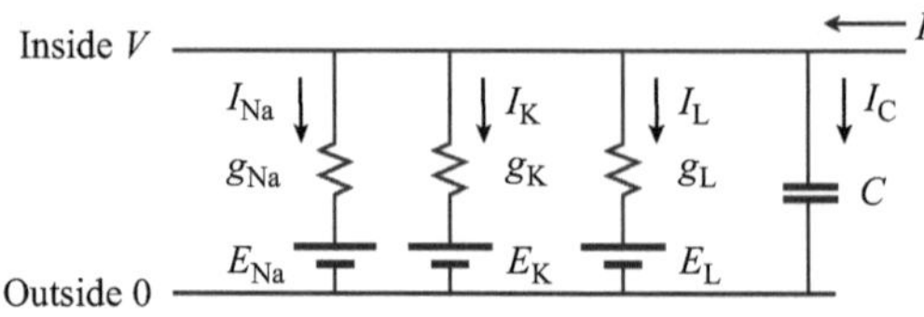

Figure 13.19
The equivalent circuit model of a neuron.

to a mix of ionic currents, commonly described as a "leak." The resistors represent the **conductance of the membrane** specific to each of the ions, namely, the change in current produced by a given change in membrane voltage. The capacitor represents the electrical **capacitance of the membrane**, namely, the amount of charge needed for a given change in the membrane voltage. Finally the neuron is driven by an **injected current** I, which typically comes from the synaptic inputs or from transduction of a sensory stimulus.

Using Kirchhoff's rules for currents and voltages in a circuit, one can derive the dynamic equation for the membrane voltage V as a function of the various reversal potentials, their conductances, and the injected current:

$$C \cdot \frac{dV}{dt} + g_{Na} \cdot (V - E_{Na}) + g_K \cdot (V - E_K) + g_L \cdot (V - E_L) = I. \tag{13.29}$$

If the conductances and reversal potentials remain constant, then this is a rather boring system. One can see immediately that $V(t)$ will be a linear transform of $I(t)$. In fact, it will take the form of a simple low-pass filter (section 4.5.1):

$$\tau \frac{dV}{dt} + V = R(I - I_0). \tag{13.30}$$

The magic of neural processing arises because the conductances g_i are **not constant**. In the Hodgkin-Huxley model, g_{Na} and g_K both vary with voltage V, and in a time-dependent way. For example, suppose that one suddenly steps the membrane voltage up from -65 mV to -25 mV. Figure 13.20 shows the resulting time course of the conductances. Note that the sodium conductance increases when the voltage is more positive, but only transiently. Later, the potassium conductance increases, and it stays high.

Now consider the neuron in its normal state, where the voltage is not clamped to a constant value. Suppose that some external current—such as a synaptic input— increases the membrane voltage a bit. Then the sodium conductance increases, Na$^+$ ions enter the cell, charging it to even more positive potential, increasing the sodium conductance further, which leads to more Na$^+$ rushing into the cell. This positive feedback loop leads to a sudden upswing in the membrane voltage. How does this runaway amplification terminate? After a short delay, the sodium conductance shuts down again. Also, the potassium conductance increases and K$^+$ flows *out* of the cell, which brings the voltage back down. Eventually, all the conductances return back to their resting values and the cell is ready to fire another action potential. Figure 13.21 illustrates these dynamics for different amounts of injected current.

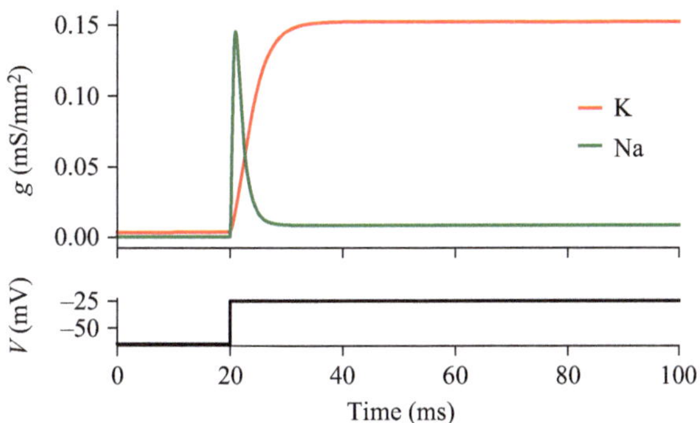

Figure 13.20
Response of the Hodgkin-Huxley model to a voltage step (bottom trace). Time course of the conductances g_{Na} and g_K for sodium and potassium.

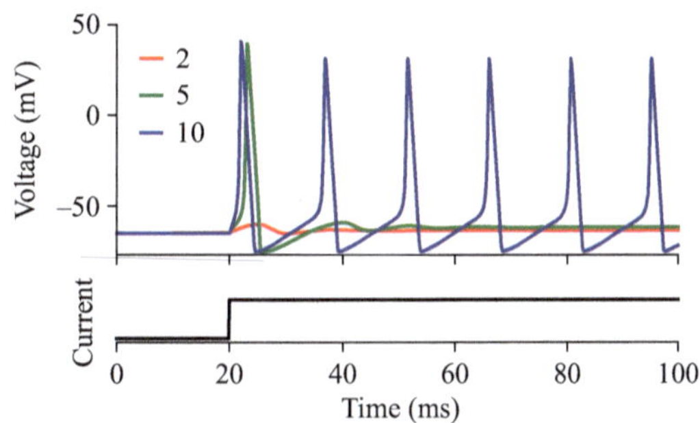

Figure 13.21
Simulation of the Hodgkin-Huxley model of the action potential, based on parameters for the squid giant axon. Bottom: A constant current is injected into the cell starting at $t = 20$ ms. Top: Depending on the size of the current (legend, in nA), the membrane potential produces a small blip, a single spike, or a periodic train of spikes.

A small current causes only a brief blip in the membrane voltage that is not sufficient to cross the threshold that launches the avalanche of Na^+ influx. A slightly larger current does cross that threshold and produces a single spike, after which the membrane settles down again.[2] A larger current still produces a periodic string of action potentials. Between two spikes, the membrane charges up in a linear ramp until the voltage crosses the threshold again.

The full Hodgkin-Huxley model is a complex interaction of four dynamic variables. At first glance, it seems to contain an overabundance of parameters to explain something as simple as a pulse. However, Hodgkin and Huxley designed experiments to isolate many of their postulated components and then measure the corresponding parameters independently. When they all got linked together through numerical simulation, the waveform of the action potential was predicted correctly, along with

---

2. Why does it not fire again? The reason is that the Na-conductance is in a desensitized state.

other experimental observables, like the threshold at which a spike gets produced or the minimum time that needs to elapse before the next spike.

At the time, Hodgkin and Huxley integrated the differential equations by hand using a mechanical calculating machine (Hodgkin, 1994). Simulating 4 ms in the life of a neuron required about 8 hours of setting levers and turning the crank of the calculator. Today, the computing time required for this simulation is negligible. By replicating these simulations, you get a sense of how the various parameters of the neuronal membrane interact. You may wish to start with a textbook, such as chapter 5 of Dayan and Abbott (2005) or chapter 2 of Gerstner et al. (2014), and build your simulation from first principles. Or explore the online resources for this chapter.

The full genius of Hodgkin and Huxley was appreciated only many years later, when selective ion channels were discovered (Hille, 2001; MacKinnon, 2004). These are protein pores that span the cell membrane and can be highly selective for specific ions, like Na^+ or K^+. At any time, the pore can be open or closed, and its open probability may depend on the voltage across the membrane, as well as on the time since the last opening. Nature has built hundreds of such specialized ion channels that shape the nonlinear behavior of neuronal membranes. They respond not only to voltage, but to temperature, mechanical strain, and many chemical ligands on the internal or external faces. Many cellular functions of the neuron, such as sensory transduction and synaptic transmission, can be understood based on the equivalent circuit model of figure 13.19, combined with specialized functional dependencies of the ionic conductances.

13.6.4 The Integrate-and-Fire Model of a Neuron

As discussed in section 13.6.3, an excitable cell exists in two regimes. With the membrane voltage below threshold, the dynamics are orderly, and typically, the voltage is a low-pass filtered version of the input current. But when the membrane voltage reaches a certain threshold, the voltage-activated conductances take over for a millisecond and produce a fast action potential, after which the voltage returns to somewhere below threshold again.

The **integrate-and-fire model** of the neuron implements this dichotomy explicitly without referring to the underlying biophysical details, such as ionic conductances: Below a threshold voltage V_{th} the dynamics are linear, and the neuron essentially integrates its input current. When the threshold is reached, the neuron emits a spike and the voltage is reset to V_{re}. Specifically, the dynamic equations are

$$\tau \frac{dV}{dt} = -V(t) + RI(t) + N(t),$$

where

$$V(t) = \text{membrane potential}$$

$$\tau = \text{membrane time constant}$$

$$R = \text{membrane resistance}$$

$$I(t) = \text{injected current}$$

$$N(t) = \text{noise voltage.}$$

When $V(t) = V_{th}$, emit a spike at time t and reset V to V_{re}.

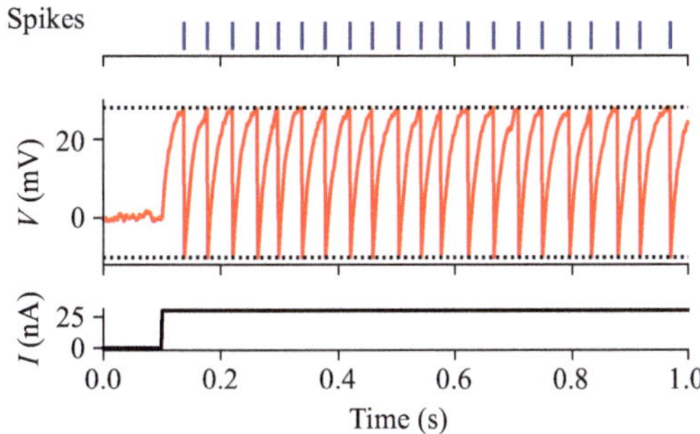

Figure 13.22
Simulation of the integrate-and-fire model of a neuron. Bottom: Input current. Middle: Membrane voltage, with V_{th} and V_{re} (dotted lines). Top: Emitted spikes.

Figure 13.22 shows a simulation of this process. Once the current injection starts, the membrane charges up with exponential dynamics. When the voltage crosses the threshold, a spike gets emitted, and integration resumes from the reset voltage. Because of the noise source, the intervals between spikes are somewhat variable.

Effectively, this model converts a continuous input current $I(t)$ into a point process of spike times $\{t_k\}$, and thus serves as one formulation of the neuronal encoding process. In exercise 14.9, we will see that this model can explain some intricate aspects of neuronal spike trains.

Exercise 14.1 (Analytical solution to bimolecular reaction) Find an analytical solution for the concentrations in a bimolecular reaction:

$$A + B \xrightarrow{k} AB. \qquad (14.1)$$

Exercise 14.2 (Dynamical systems with higher time derivatives) Consider a one-dimensional (1D) harmonic oscillator:

$$\frac{d^2 x}{dt^2} = -r \cdot x.$$

Show how this can be turned into a two-dimensional (2D) system with dynamical equations in standard form.

Exercise 14.3 (Dynamical systems with explicit time-dependence) Consider the driven system

$$\dot{x} = \sin \omega t.$$

Show how this can be turned into standard-form equations (Hint: Define time as a dynamic variable).

Exercise 14.4 (Degenerate node) Explain the shape of trajectories around a degenerate node in two dimensions. Hint: Note that the Jacobian matrix of this system can be written as

$$A = \begin{bmatrix} \lambda & 0 \\ a & \lambda \end{bmatrix}, a \neq 0.$$

Exercise 14.5 (Equal species) Explain what happens in the "rabbits vs sheep" system (section 11.5.1) when both species are equally voracious, such that $\beta = \delta = 1$.

Exercise 14.6 (Analytical solution of the predator-prey system) Find an analytical solution to the Lotka-Volterra equations for the predator-prey system: equations (11.44) and (11.45).

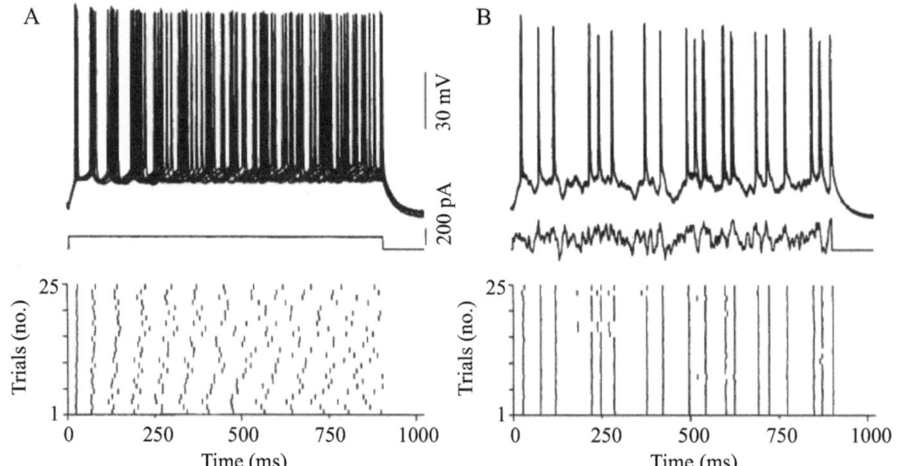

Figure 14.1
Figure 1 of Mainen and Sejnowski (1995). A constant input gives rise to spikes whose timing varies significantly across trials (A), whereas a noisy input gives rise to a much more reproducible spike trains (B).

Exercise 14.7 (Numerical solution of the predator-prey system) Although we have emphasized an analysis of dynamical systems that can be pursued with paper and pencil, there are powerful numerical tools available, for both finding actual trajectories and plotting phase portraits. Find these tools in your favorite programming language and use them to plot a phase portrait for the predator-prey system, including nullclines, flow vectors, and sample trajectories.

Exercise 14.8 (Circadian oscillator)

1. Return to the 2D model of figure 13.16, which includes the rate constant λ. Explore what role λ plays in the entrainment of the oscillator.
2. Reconsider the approximate expression in equation (13.24) for the phase shift $\Delta\phi$. What happens to the conditions for entrainment if A is not small? What if $A > 1$?
3. Consider what happens if the Zeitgeber has a period near an integer multiple of the oscillator period (i.e. $T \approx n$).

Exercise 14.9 (The integrate-and-fire neuron) This exercise explores the response of a neuron to a rapidly varying input current. Specifically, the goal is to replicate the experimental results of Mainen and Sejnowski (1995) on a simple model neuron, which is reproduced in figure 14.1.

The "startling" result of this paper was that stimulating a neuron with a constant current (A) produces spike trains that are different from trial to trial, whereas stimulating with a fluctuating current (B) leads to very reproducible spike trains. Your task is to explain this result based on a simple model.

(a) Write code to simulate an integrate-and-fire model (see section 13.6.4). The dynamic equation is reproduced here:

$$\tau \frac{dV}{dt} = -V(t) + RI(t) + N(t),$$

where

$$V(t) = \text{membrane potential}$$

$$\tau = \text{membrane time constant}$$

$$R = \text{membrane resistance}$$

$$I(t) = \text{injected current}$$

$$N(t) = \text{noise voltage.}$$

When $V(t) = V_{\text{th}}$, emit a spike at time t and reset V to V_{re}.

(b) Simulate the two sides of figure 14.1. The constant input in A was applied at 150 nA for 900 ms. The fluctuating input in B was a Gaussian white noise (mean 150 pA, standard deviation 100 pA) passed through a low-pass filter with exponential impulse response $\exp(-t/\tau_s)$, $\tau_s = 3$ ms.

(c) Play with the parameters of the model and see if you can find conditions where the spike trains look as they do in the figure.

(d) Explain in a few words why the spike train in B is more predictable.

14.1 Further Exercises

We highly recommend Strogatz (2015) for further study, which also contains an excellent collection of exercises.

References

Abramowitz, M., and Stegun, I. A. (1964). *Handbook of Mathematical Functions with Formulas, Graphs, and Mathematical Tables*. Dover Publications, 9th Dover printing, 10th GPO printing edition.

Alon, U. (2019). *An Introduction to Systems Biology: Design Principles of Biological Circuits*. 2nd ed. Chapman and Hall/CRC.

American Institute of Physics. (1973). Interview of Richard Feynman by Charles Weiner on 1973 February 4, Niels Bohr Library & Archives, American Institute of Physics, College Park, MD USA. www.aip.org/history-programs/niels-bohr-library/oral-histories/5020-5

Bartol, T. M., Bromer, C., Kinney, J. P., et al. (2015). Nanoconnectomic upper bound on the variability of synaptic plasticity. *Elife*, 4.

Bastian, J., Schniederjan, S., and Nguyenkim, J. (2001). Arginine vasotocin modulates a sexually dimorphic communication behavior in the weakly electric fish Apteronotus leptorhynchus. *Journal of Experimental Biology*, 204(11): 1909–1923.

Baylor, D. A., Lamb, T. D., and Yau, K. W. (1979). Responses of retinal rods to single photons. *The Journal of Physiology*, 288: 613–634.

Berry, M. J., Warland, D. K., and Meister, M. (1997). The structure and precision of retinal spike trains. *Proc Natl Acad Sci U S A*, 94: 5411–6.

Block, S. M., Segall, J. E., and Berg, H. C. (1982). Impulse responses in bacterial chemotaxis. *Cell*, 31: 215–26.

Borst, A., and Theunissen, F. E. (1999). Information theory and neural coding. *Nature Neuroscience*, 2(11): 947–957.

Brown, E. N., Kass, R. E., and Mitra, P. P. (2004). Multiple neural spike train data analysis: State-of-the-art and future challenges. *Nature Neuroscience*, 7(5): 456–461.

Brown, R. (1828). XXVII. A brief account of microscopical observations made in the months of June, July and August 1827, on the particles contained in the pollen of plants; and on the general existence of active molecules in organic and inorganic bodies. *The Philosophical Magazine*, 4(21): 161–173.

Brush, S. G. (1968). A history of random processes: I. Brownian movement from Brown to Perrin. *Archive for History of Exact Sciences*, 5(1): 1–36.

Carslaw, H. S., and Jaeger, J. C. (1986). *Conduction of Heat in Solids*. 2nd ed. Oxford University Press.

Chichilnisky, E. J. (2001). A simple white noise analysis of neuronal light responses. *Network*, 12: 199–213.

Chichilnisky, E. J., and Kalmar, R. S. (2002). Functional asymmetries in ON and OFF ganglion cells of primate retina. *Journal of Neuroscience: The Official Journal of the Society for Neuroscience*, 22(7): 2737–2747.

Clark, D. A., Benichou, R., Meister, M., and Azeredo da Silveira, R. (2013). Dynamical adaptation in photoreceptors. *PLoS Computational Biology*, 9: e1003289.

Council, N. R. (2003). *BIO2010: Transforming Undergraduate Education for Future Research Biologists*. National Academies Press.

Cover, T. M., and Thomas, J. A. (2012). *Elements of Information Theory*. John Wiley & Sons.

Crevier, D. W., and Meister, M. (1998). Synchronous period-doubling in flicker vision of salamander and man. *Journal of Neurophysiology*, 79(4): 1869–1878. PMID: 9535954.

Czeisler, C. A., Richardson, G. S., Coleman, R. M., et al. (1981). Chronotherapy: Resetting the circadian clocks of patients with delayed sleep phase insomnia. *Sleep*, 4(1): 1–21.

Dahlquist, F. W., Lovely, P., and Koshland, D.E. (1972). Quantitative Analysis of Bacterial Migration in Chemotaxis. *Nature New Biology*, 236: 120–123.

Daley, D. J., and Vere-Jones, D. (2013). *An Introduction to the Theory of Point Processes: Volume I: Elementary Theory and Methods*. 2nd ed. Springer (2003).

Dayan, P., and Abbott, L. F. (2005). *Theoretical Neuroscience: Computational and Mathematical Modeling of Neural Systems*. MIT Press.

de Ruyter van Steveninck, R., Bialek, W., and Barlow, H. B. (1988). Real-time performance of a movement-sensitive neuron in the blowfly visual system: Coding and information transfer in short spike sequences. *Proceedings of the Royal Society of London. Series B. Biological Sciences*, 234(1277): 379–414.

Digman, M. A., and Gratton, E. (2011). Lessons in fluctuation correlation spectroscopy. *Annual Review of Physical Chemistry*, 62(1): 645–668.

Drenth, J. (2007). *Principles of Protein X-ray Crystallography*. Springer.

Durbin, R., Eddy, S. R., Krogh, A., and Mitchison, G. (1998). *Biological Sequence Analysis: Probabilistic Models of Proteins and Nucleic Acids*. Cambridge University Press.

Dymarski, P. (2011). *Hidden Markov Models, Theory and Applications*. IntechOpen.

Eddy, S. R. (2004). What is a hidden Markov model? *Nature Biotechnology*, 22(10): 1315–1316.

Edwards, A. W. (1986). Are Mendel's results really too close? *Biological Reviews of the Cambridge Philosophical Society*, 61(4): 295–312.

Efron, B., and Tibshirani, R. J. (1993). *An Introduction to the Bootstrap*. Chapman & Hall/CRC.

Eigen, M. (2019). *From Strange Simplicity to Complex Familiarity: A Treatise on Matter, Information, Life and Thought*. Oxford University Press.

Einstein, A. (1905). Über die von der molekularkinetischen Theorie der Wärme geforderte Bewegung von in ruhenden Flüssigkeiten suspendierten Teilchen. *Annalen der Physik*, 322(8): 549–560.

Elowitz, M. B., and Leibler, S. (2000). A synthetic oscillatory network of transcriptional regulators. *Nature*, 403(6767): 335–338.

Entrenas Castillo, M., Entrenas Costa, L. M., Vaquero Barrios, J. M., et al. (2020). Effect of calcifediol treatment and best available therapy versus best available therapy on intensive care unit admission and mortality among patients hospitalized for COVID-19: A pilot randomized clinical study. *Journal of Steroid Biochemistry and Molecular Biology*, 203: 105751.

Fain, G. L., Matthews, H. R., Cornwall, M. C., and Koutalos, Y. (2001). Adaptation in vertebrate photoreceptors. *Physiological Reviews*, 81: 117–151.

Fisher, R. (1936). Has Mendel's work been rediscovered? *Annals of Science*, 1(2): 115–137.

Foster, R. G. (2020). Sleep, circadian rhythms and health. *Interface Focus*, 10(3): 20190098.

Franklin, A., Edwards, A. W. F., Fairbanks, D. J., and Hartl, D. L. (2008). *Ending the Mendel-Fisher Controversy*. University of Pittsburgh Press.

Frey, T. G., Chan, S. H., and Schatz, G. (1978). Structure and orientation of cytochrome c oxidase in crystalline membranes. Studies by electron microscopy and by labeling with subunit-specific antibodies. *Journal of Biological Chemistry*, 253(12): 4389–4395.

Geffen, M. N., Broome, B. M., Laurent, G., and Meister, M. (2009). Neural encoding of rapidly fluctuating odors. *Neuron*, 61: 570–586.

Gerstner, W., Kistler, W. M., Naud, R., and Paninski, L. (2014). *Neuronal Dynamics: From Single Neurons to Networks and Models of Cognition*. Cambridge University Press.

Goentoro, L., Shoval, O., Kirschner, M. W., and Alon, U. (2009). The incoherent feedforward loop can provide fold-change detection in gene regulation. *Molecular Cell*, 36(5): 894–899.

Goodman, J. (2005). *Introduction to Fourier Optics*. 3rd ed. Roberts & Co.

Hamada, H., Watanabe, M., Lau, H. E., et al. (2014). Involvement of delta/notch signaling in zebrafish adult pigment stripe patterning. *Development*, 141(2): 318–324.

Hecht, S., Shlaer, S., and Pirenne, M. H. (1942). Energy, quanta, and vision. *Journal of General Physiology*, 25: 819–840.

Henderson, J., Salzberg, S., and Fasman, K. H. (1997). Finding genes in DNA with a hidden Markov model. *Journal of Computational Biology*, 4(2): 127–141.

Hille, B. (2001). *Ion Channels of Excitable Membranes*. 3rd ed. Sinauer Associates.

Hiscock, T. W., and Megason, S. G. (2015). Mathematically guided approaches to distinguish models of periodic patterning. *Development*, 142(3): 409–419.

Hodgkin, A. (1994). *Chance and Design: Reminiscences of Science in Peace and War*. Illustrated ed. Cambridge University Press.

Hodgkin, A., and Huxley, A. (1952). A quantitative description of membrane current and its application to conduction and excitation in nerve. *Journal of Physiology*, 117: 500–544.

Josef, K., Saranak, J., and Foster, K. W. (2005). Ciliary behavior of a negatively phototactic Chlamydomonas reinhardtii. *Cell Motility and the Cytoskeleton*, 61(2): 97–111.

Kanji, G. K. (2006). *100 Statistical Tests*. 3rd ed. SAGE Publications.

Kawasaki, M., Rose, G., and Heiligenberg, W. (1988). Temporal hyperacuity in single neurons of electric fish. *Nature*, 336: 173–176.

Konopka, R. J., and Benzer, S. (1971). Clock mutants of Drosophila melanogaster. *Proceedings of the National Academy of Sciences*, 68(9): 2112–2116.

Lipson, E. D. (1975). White noise analysis of Phycomyces light growth response system. I. Normal intensity range. *Biophysical Journal*, 15(10): 989–1011.

Liu, R. T., Liaw, S. S., and Maini, P. K. (2006). Two-stage Turing model for generating pigment patterns on the leopard and the jaguar. *Physical Review E*, 74: 011914.

Lorenz, E. N. (1963). Deterministic nonperiodic flow. *Journal of Atmospheric Sciences*, 20(2): 130–141.

Ludwig, D., Jones, D. D., and Holling, C. S. (1978). Qualitative analysis of insect outbreak systems: The spruce budworm and forest. *Journal of Animal Ecology*, 47(1): 315–332.

Luria, S. E., and Delbrück, M. (1943). Mutations of bacteria from virus sensitivity to virus resistance. *Genetics*, 28(6): 491–511.

Machens, C. K., Romo, R., and Brody, C. D. (2005). Flexible control of mutual inhibition: A neural model of two-interval discrimination. *Science*, 307(5712): 1121–1124.

MacKinnon, R. (2004). Potassium channels and the atomic basis of selective ion conduction. *Bioscience Reports*, 24(2): 75–100.

Mainen, Z. F., and Sejnowski, T. J. (1995). Reliability of spike timing in neocortical neurons. *Science (New York, N.Y.)*, 268(5216): 1503–1506.

Meister, M., and Tessier-Lavigne, M. (2021). Low-level visual processing: The retina. In Kandel, E., Koester, J. D., Mack, S. H., and Siegelbaum, S., editors, *Principles of Neural Science* 6th ed. (pp. 521–544). McGraw Hill / Medical.

Müller, P., Rogers, K. W., Jordan, B. M., et al. (2012). Differential diffusivity of Nodal and Lefty underlies a reaction-diffusion patterning system. *Science (New York, N.Y.)*, 336(6082): 721–724.

Murray, J. D. (2002). *Mathematical Biology: I. An Introduction*. 3rd ed. Interdisciplinary Applied Mathematics, Mathematical Biology. Springer-Verlag, New York.

Murray, J. (2003). *Mathematical Biology II: Spatial Models and Biomedical Applications*. Springer.

Neher, E., and Stevens, C. F. (1977). Conductance fluctuations and ionic pores in membranes. *Annual Review of Biophysics and Bioengineering*, 6(1): 345–381.

Nelson, P. (2022). *Physical Models of Living Systems: Probability, Simulation, Dynamics*. 2nd ed. Philadelphia: Chiliagon Science.

Paninski, L. (2004). Maximum likelihood estimation of cascade point-process neural encoding models. *Network: Computation in Neural Systems*, 15(4): 243–262.

Petrucco, L., Lavian, H., Wu, Y., Svara, F., Stih, V., and Portugues, R. (2023). "Neural dynamics and architecture of the heading direction circuit in zebrafish". *Nature Neuroscience*, 26: 765–773.

Phillips, R., Kondev, J., Theriot, J., and Garcia, H. G. (2012). *Physical Biology of the Cell*. 2nd ed. CRC Press.

Pitkow, X. and Meister, M. (2014). Neural computation in sensory systems. In Gazzaniga, M. S. and Mangun, G. R., editors, *The Cognitive Neurosciences* 5th ed. (pp. 305–318). MIT Press.

Pugh, E. N. Jr. (2018). The discovery of the ability of rod photoreceptors to signal single photons. *Journal of General Physiology*, 150(3): 383–388.

Rice, J. A. (2006). *Mathematical Statistics and Data Analysis*. 3rd ed. Cengage Learning.

Rieke, F., Bodnar, D. A., and Bialek, W. (1995). Naturalistic stimuli increase the rate and efficiency of information transmission by primary auditory afferents. *Proceedings: Biological Sciences*, 262(1365): 259–265.

Rieke, F., Warland, D., de Ruyter van Steveninck, R., and Bialek, W. (1997). *Spikes: Exploring the Neural Code*. MIT Press.

Roubinet, B., Bischoff, M., Nizamov, S., et al. (2018). Photoactivatable rhodamine spiroamides and diazoketones decorated with "universal hydrophilizer" or hydroxyl groups. *Journal of Organic Chemistry*, 83(12): 6466–6476.

Sakitt, B. (1972). Counting every quantum. *Journal of Physiology*, 223: 131–150.

Sakmann, B. and Neher, E., eds. (1995). *Single-Channel Recording*. 2nd ed. Springer.

Schnapf, J. L., Nunn, B. J., Meister, M., and Baylor, D. A. (1990). Visual transduction in cones of the monkey Macaca fascicularis. *Journal of Physiology*, 427: 681–713.

Schneeweis, D. M., and Schnapf, J. L. (2000). Noise and light adaptation in rods of the macaque monkey. *Visual Neuroscience*, 17(5): 659–666.

Shannon, C. E. (1948). A mathematical theory of communication. *Bell System Technical Journal*, 27(3): 379–423.

Shannon, C. E. (1951). Prediction and entropy of printed english. *Bell System Technical Journal*, 30(1): 50–64.

Shannon, C. E., and Weaver, W. (1964). *The Mathematical Theory of Communication*. University of Illinois Press.

Smith, A., Frank, L., Wirth, S., et al. (2004). Dynamic analysis of learning in behavioral experiments. *Journal of Neuroscience*, 24(2): 447–461.

Spearman, C. (1904). "General intelligence," objectively determined and measured. *American Journal of Psychology*, 15: 201–293.

Spellman, P., Sherlock, G., Zhang, M., et al. (1998). Comprehensive identification of cell cycle-regulated genes of the yeast saccharomyces cerevisiae by microarray hybridization. *Molecular Biology of the Cell*, 9: 3273–3297.

Strogatz, S. H. (2015). *Nonlinear Dynamics and Chaos: With Applications to Physics, Biology, and Engineering*. 2nd ed. CRC Press.

Svoboda, K., Schmidt, C. F., Schnapp, B. J., and Block, S. M. (1993). Direct observation of kinesin stepping by optical trapping interferometry. *Nature*, 365: 721–727.

Tkacik, G., and Bialek, W. (2016). Information processing in living systems. *Annual Review of Condensed Matter Physics*, 7(1): 89–117.

Turing, A. M. (1952). The chemical basis of morphogenesis. *Philosophical Transactions of the Royal Society of London: Series B, Biological Sciences*, 237(641): 37–72.

Weinstein, M. S. (1977). Hares, lynx, and trappers. *The American Naturalist*, 111(980): 806–808.

Werner, G., and Mountcastle, V. B. (1965). Neural activity in mechanoreceptive cutaneous afferents: Stimulus-response relations, Weber functions, and information transmission. *Journal of Neurophysiology*, 28: 359–397.

Wiener, N. (1949). *Extrapolation, Interpolation, and Smoothing of Stationary Time Series, with Engineering Applications*. Wiley.